T0229544

A Concise Introduction to Machine Learning

Chapman & Hall/CRC Machine Learning & Pattern Recognition

For more information on this series please visit: https://www.crcpress.com/Chapman--HallCRC-Machine-Learning--Pattern-Recognition/book-series/

A Concise Introduction to Machine Learning

A. C. Faul

CRC Press
Taylor & Francis Group
Boca Raton London New York

CRC Press is an imprint of the
Taylor & Francis Group, an **informa** business

A CHAPMAN & HALL BOOK

MATLAB® is a trademark of The MathWorks, Inc. and is used with permission. The MathWorks does not warrant the accuracy of the text or exercises in this book. This book's use or discussion of MATLAB® software or related products does not constitute endorsement or sponsorship by The MathWorks of a particular pedagogical approach or particular use of the MATLAB® software.

CRC Press
Taylor & Francis Group
6000 Broken Sound Parkway NW, Suite 300
Boca Raton, FL 33487-2742

© 2020 by Taylor & Francis Group, LLC
CRC Press is an imprint of Taylor & Francis Group, an Informa business

No claim to original U.S. Government works

Printed on acid-free paper

International Standard Book Number-13: 978-0-8153-8410-6 (Paperback)
978-0-8153-8420-5 (Hardback)

Library of Congress Cataloging-in-Publication Data

Names: Faul, A. C. (Anita C.), author.
Title: A concise introduction to machine learning / Anita Faul.
Description: Boca Raton, Florida : CRC Press, [2019] | Series: Chapman & Hall/CRC machine learning & pattern recognition | Includes bibliographical references and index.
Identifiers: LCCN 2019015915| ISBN 9780815384205 (hbk : alk. paper) | ISBN 9780815384106 (pbk : alk. paper) | ISBN 9781351204750 (ebk)
Subjects: LCSH: Machine learning--Textbooks.
Classification: LCC Q325.5 .F38 2020 | DDC 006.3/1--dc23
LC record available at https://lccn.loc.gov/2019015915

Visit the Taylor & Francis Web site at
http://www.taylorandfrancis.com

and the CRC Press Web site at
http://www.crcpress.com

To Helmut and Marieluise, and their caring love, especially his loving care.

Contents

List of Figures

Preface

Machine Learning is known under many names such as Machine Learning, Artificial Intelligence, Pattern Recognition, Data Mining, Data Assimilation, and Big Data, to list but a few. It developed in many areas of science, such as in physics, engineering, computer science, and mathematics in parallel and independently. For example it is used for Spam Filtering, Optical Character Recognition (OCR), Search Engines, Computer Vision, Natural Language Processing (NLP), Advertising, Fraud Detection, Robotics, Data Prediction, Material Discovery, Astronomy. This makes it sometimes difficult to find a solution for a particular problem in the literature, simply because different words and phrases are used for the same concept.

This book aims to alleviate this. A common concept, but known in several disciplines under different names, is described using mathematics as the common language. Readers will find the index useful to find a particular topic as it is known to them. The index is comprehensive, making it easy to find the required information. Hopefully, the book will prove useful as a reference and make it an essential on the bookshelves of anybody employing machine learning techniques.

Often, in teaching Machine Learning, the emphasis is on the questions "What?" and "How?", and the question "Why?" is neglected. In my opinion, however, this is the most important question. Only if why an algorithm is successful is understood can it be properly applied. On the other hand, if why it arrived at its results is not understood, these results cannot be trusted.

Algorithms are often taught side by side without showing the similarities and differences between them. This book addresses this by introducing the commonalities. Most techniques try to find an approximate model generating the data in a finite, low-dimensional space. They differ in how a solution to this approximate model is found and how the model space is chosen. This approach helps to keep the book concise while still giving a thorough and in depth treatment. In some places, where further detail was felt to be beyond the scope of this book, the reader is referred to further reading.

Techniques are illustrated by MATLAB® implementations. The main purpose is to show the inner workings of the method in order to develop an intuition. In most cases, the listings are printed in the book, but all are available online at https://www.crcpress.com/A-Concise-Introduction-to-Machine-Learning/Faul/9780815384106 as part of the package K339637_Downloads.zip.

MATLAB and Simulink are registered trademarks of The Mathworks, Inc. For product information please contact:

The Mathworks, Inc.
3 Apple Hill Drive
Natick, MA 01760-2098 USA
Tel: 508-647-7000
Fax: 508-647-7001
E–mail: info@mathworks.com
Web: https://www.mathworks.com

How to buy: https://www.mathworks.com/store Find your local office: https://www.mathworks.com/company/worldwide

Acknowledgments

First, I have to give special thanks Dr Nikos Nikiforakis and Professor James Elliot, whose decision to exclude me from most activities freed up time to develop additional teaching material. This book is based on the lecture course taught within the MPhil in Scientific Computing at the University of Cambridge, UK, and additional material. Lectured at the Centre for Mathematical Sciences in Cambridge, since lecturing this was not approved by the director of the MPhil in Scientific Computing. Thanks go to Dr James Fergusson who made the additional lectures possible.

Next, thanks are also due to my PhD supervisor Professor Mike Powell who sadly passed away in April 2014. From him I learned that one should strive for understanding and simplicity. I more often saw him working through an algorithm with pen and paper than sitting at a computer. He wanted to know why a particular algorithm was successful. Especially, when machine learning systems reach greater and greater complexity, it is important to continue to strive for understanding.

Of course, I also need to thank all the graduate students who attended the lectures and whose questions and quest to further their understanding improved this book.

I would also like to express my gratitude to Cambridge University and the staff and fellows at Selwyn College for creating such a wonderful atmosphere in which to learn and teach.

This book would not have been written without the support of many people in my private life, foremost my parents Helmut and Marieluise Faul, who instilled a love for knowledge in me, and taught me to appreciate the value of integrity, especially the integrity of science.

A.C. Faul

CHAPTER 1

Introduction

When thinking about machine learning, it seems prudent to start thinking about how we learn. King Frederick II (26 December 1194 - 13 December 1250) was Holy Roman Emperor and King of Sicily in the 13th century. King Frederick was a passionate patron of the sciences and arts. He spoke six languages which were Latin, Sicilian, German, French, Greek and Arabic. He desired to determine the "god given" language. The Italian Franciscan friar Salimbene de Adam writes in his Cronica [25] that Frederick bade "foster-mothers and nurses to suckle and bathe and wash the children, but in no ways to prattle or speak with them; for he would have learned whether they would speak the Hebrew language (which had been the first), or Greek, or Latin, or Arabic, or perchance the tongue of their parents of whom they had been born. But he laboured in vain, for the children could not live without clappings of the hands, and gestures, and gladness of countenance, and blandishments." In other words the physical needs of the children were satisfied, but they were raised without any human interaction. It is doubtful whether this is a true account, since this is the only account and Salimbene was a political opponent of Frederick II. Nevertheless, nobody doubts that sensory stimulation and experiences are essential for learning in any respect.

Blakemore and Cooper [4] experimented with kittens. The kittens were brought up in a dark room and only brought out at certain times and then placed in an environment with either only horizontal or only vertical lines. Kittens brought up in the horizontal environment showed no reaction to vertical lines. There was no brain activity. Indeed, when the inclination of a horizontal line was changed gradually towards a vertical line the brain activity became less and less.

The experiment showed that only what the brain is presented with by the environment is learned. This is an efficient preparation for the future. This is also true for human vision. Australian Aborigines have the sharpest vision ever measured, about four times better than the vision of those of white ethnicity. This means that they can see objects sharply at six meters distance which the average white person can see clearly at only 1.5 meters, a quarter

Figure 1.1: Vertical environment. Reprinted by permission from Macmillan
Publishers Ltd: Nature [4], copyright (1970).

of the distance. Often the eyesight deteriorates with old age. Ophthalmologist Professor Fred Hollows [22] corrected the vision of an elderly Aboriginal man back to the average white person's vision with glasses. The reaction was "Thank you for trying, but this is hopeless. I used to be able to see much better." The Australian outback is a wide open landscape and good vision in the distance is vital for survival.

We can conclude that there is a need for experiences. However, how does a machine "experience"? We can view our senses as taking measurements and our brain interprets these and draws conclusions. A machine can take various measurements and then perform calculations, but can it emulate the power of a human brain?

Perhaps to answer this question, we need to take a step back and not look at how we learn, but how we teach. Traditional *teaching* is from the front of the classroom. The pupils are given a set of instructions and are expected to reproduce these. This is very similar to *procedural programming* where the flow of instructions is encoded. *Object oriented programming* was a further development where instructions depend on the nature of the data.

However, our brain retains information much better when it has the positive experience of discovery. So often there is teacher-led learning. Galileo illustrates this beautifully in his book "Discourses and Mathematical Demonstrations Relating to Two New Sciences" [15] (Figure 1.2). The two sciences are the science of motion and the science of materials and construction. The ideas are developed as a dialogue between three characters, Salviati, Sagredo and Simplicio. The latter is portrayed as a simpleton, the pupil to be instructed. In the science of motion, the starting point is the observation that even though objects have different masses, they reach the ground at the same time. However, this is very difficult to quantify, since it is over all too fast. Galileo therefore developed the inclined plane experiment which he describes as such:

> "A piece of wooden moulding or scantling, about 12 cubits long, half a cubit wide, and three finger-breadths thick, was taken; on its edge was cut a channel a little more than one finger in breadth; having made this groove very straight, smooth, and polished, and having lined it with parchment, also as smooth and polished as possible, we rolled along it a hard, smooth, and very round bronze ball. Having placed this board in a sloping position, by raising one end some one or two cubits above the other, we rolled the ball, as I was just saying, along the channel, noting, in a manner presently to be described, the time required to make the descent. We repeated this experiment more than once in order to measure the time with an accuracy such that the deviation between two observations never exceeded one-tenth of a pulse-beat. Having performed this operation and having assured ourselves of its reliability, we now rolled the ball only one-quarter the length of the channel; and having measured the time of its descent, we found it precisely one-half of the former. Next we tried other distances, compared the time for the whole length with that for the half, or with that for two-thirds, or three-fourths, or indeed for any fraction; in such experiments, repeated a full hundred times, we always found that the spaces traversed were to each other as the squares of the times, and this was true for all inclinations of the plane, i.e., of the channel, along which we rolled the ball. We also observed that the times of descent, for various inclinations of the plane, bore to one another precisely that ratio which, as we shall see later, the Author had predicted and demonstrated for them.
>
> For the measurement of time, we employed a large vessel of water placed in an elevated position; to the bottom of this vessel was soldered a pipe of small diameter giving a thin jet of water which we collected in a small glass during the time of each descent, whether for the whole length of the channel or for part of its length; the water thus collected was weighed, after each descent, on a very accurate balance; the differences and ratios of these weights gave us the differences and ratios of the times, and this with such accuracy that although the operation was repeated many, many times, there was no appreciable discrepancy in the results."

In the 19th century an apparatus for this experiment was built and can now be seen in the Museo Galileo in Florence, Italy. This experiment is repeated

Figure 1.2: Discorsi e Dimostrazioni Matematiche Intorno a Due Nuove Scienze Image in the public domain.

by school children all over the world again and again, stacking their books to create an inclined plane. Only stop watches have replaced the water clock. This is *supervised learning* and a *regression* problem. In regression we try to find a relationship between two or more parameters. In this case it is distance, d, and time, t.

Note that the relationship between the two parameters, distance and time, is not *linear*. In fact,

$$\frac{d_1}{d_2} = \frac{t_1^2}{t_2^2}.$$

We can rephrase this so that distance and time are primary parameters and the square of the time is a secondary parameter. Then we have found a linear relationship between a primary parameter, the distance, and a secondary parameter, the square of time. We will encounter this again when discussing the *kernel trick*. Another way of viewing this is as an instance of *deep learning* since another layer of abstraction is added by the square. Deep learning is trying to uncover hidden relationships.

However, how did Galileo arrive at this experiment? Remember that the starting point was the observation that objects of different masses reach the ground at the same time. He realized that air resistance is a factor, becoming dominant for extremely light objects with a lot of air resistance such as feathers. Thus he needed to make his experiment as independent from air resistance

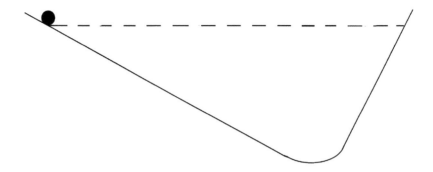

Figure 1.3: Two smoothly connected ramps.

as possible and arrived at a bronze ball. More importantly, however, he needed to slow the experiment down in order to make accurate measurements. Galileo noticed that a ball rolling down a ramp which is smoothly connected to an upward ramp will reach essentially the same level it started from. It will roll backwards and forwards until it comes to a rest because of friction and air resistance. If both ramps have the same inclination, this is not that surprising, but the ball will reach the same level even if the second ramp is much steeper. This is illustrated in Figure 1.3. Making this second ramp steeper and steeper, we approach a vertical ramp. This means that the experiment of a falling ball and a rolling ball are equivalent. In the latter the motion is however slower and thus measurements can be obtained.

This is an instance of *unsupervised learning*. Galileo realized by himself that all the experiments belong to the same class, independent of the angles of the two ramps.

Returning to the experiment with one ramp. Galileo showed that balls of different weights would travel the same distance in the same amount of time. Hence he proved that weight was not a parameter influencing the experiment. However the times of descent for the same distance depend on the inclinations of the plane. The steeper the plane, the sooner the experiment is over. Thus the inclination angle is a parameter influencing the outcome. This is an example of *feature detection* where we try to find which parameters, or in other words features, are relevant.

Human learning is a combination of all these:

- Feature detection: Our senses experience the world around us. The cat in the Blakemore and Cooper experiments never experienced the feature of a vertical line.

- Unsupervised learning: Here we try to make sense of our experiences.

- Deep learning: We put our experiences in a wider context and draw conclusions.

- Supervised learning: Here we have some external input, but are also allowed to discover things by ourselves.

- Teaching: This is completely governed by external input.

Once we have learned something, we can recognize anomalies. A child below a certain age will not be impressed by a levitation act. However, once it has discovered gravity (though in a more rudimentary way than Galileo), it wants to experiment on it again and again and again as any parent picking up the toy the child has dropped from its highchair for the thousandth time will testify.

Machine learning is all about transferring the various modes of learning we have identified here to machines. We are already very good at writing procedural or object oriented programs, teaching machines this way. But what about the other modes?

CHAPTER 2

Probability Theory

The concepts in probability theory needed for machine learning are introduced. These include independence, rules of probability, Simpson's paradox, probability mass and density functions, cumulative distribution functions, and the definitions of expectation, variance and moments. The probability mass and density functions used throughout the book are introduced and their connections explained. Further functions of random variables are explored as well as conjugacy of probability distributions and their application to prior and posterior probability distributions. The chapter concludes with graphical representations of random variables and their dependencies and parameters.

2.1 Independence, Probability Rules and Simpson's Paradox

In the introduction we have seen that learning is not possible without experiences which can be obtained from repeating experiments. To formalize this we need to quantify the different outcomes of the experiments, and probability theory lends itself to that. Ross gives a very good introduction in [36]. An example is given of a fair die; each number is equally probable with the probability being 1/6. The probabilities for getting heads or tails when flipping a coin are equal, 1/2. Here the coin and die are *random variables*. They are *discrete variables*, since there are only finitely many different outcomes. That is when repeating the experiment, different outcomes are possible. To describe more complex situations, probabilities are combined. For example, the probability of getting heads and also rolling the number 3, is $1/2 \times 1/6 = 1/12$. This is called the *joint probability* of the coin and die and can be expressed in a grid.

	⚀	⚁	⚂	⚃	⚄	⚅
heads	1/12	1/12	1/12	1/12	1/12	1/12
tails	1/12	1/12	1/12	1/12	1/12	1/12

Note that the rows sum to $1/2$, which are the *marginal* probabilities of the coin, while the columns sum to $1/6$, the marginal probabilities of the die. The term marginal comes from *marginalizing*, that is rendering the other random variable unimportant.

When learning, we can repeat an experiment with random variables N times and record the number of experiments where the random variables take a certain value. For example let X and Y be two random variables where X can take the values $x_1, \ldots x_M$ and Y can take the values $y_1, \ldots y_L$. The probabilities are denoted

$$p(X = x) \text{ and } p(Y = y),$$

while the joint probability is denoted by

$$p(X = x, Y = y).$$

However, this notation is cumbersome, and we write instead $p(x), p(y)$ and $p(x, y)$. In the above example X is the coin and Y is the die with $p(\text{heads}) = 1/2$, $p(3) = 1/6$ and $p(\text{heads}, 3) = 1/12$ for example.

The coin and the die cannot influence each other, but other random variables can, for example the height and weight of a person. We expect a very tall person to not be feather light. Of course, there are also examples proving this expectation wrong. When two random variables do not influence each other, they are called *independent*, otherwise *dependent*. Two random variables X and Y are independent, if and only if

$$p(X, Y) = p(X)p(Y) \tag{2.1}$$

for all possible outcomes x_i, $i = 1, \ldots, M$ and y_j, $j = 1, \ldots, L$.

When learning from experiments, instead of recording probabilities in the grid we record how often a particular outcome occurs. Let n_{ij} be the number the outcome is $X = x_i$ and $Y = y_j$. We denote m_i the number of occurrences of $X = x_i$ disregarding Y, and l_j the number of occurrences where $Y = y_j$ whatever X is. The grid then looks like

		Y					
		y_1	$\cdots$	y_j	$\cdots$	y_L	
	x_1	n_{11}		$\cdots$		n_{1L}	m_1
	$\vdots$		$\ddots$		$\cdots$		$\vdots$
X	x_i	$\vdots$		n_{ij}		$\vdots$	m_i
	$\vdots$		$\cdots$		$\ddots$		$\vdots$
	x_M	n_{M1}		$\cdots$		n_{ML}	m_M
		l_1	$\cdots$	l_j	$\cdots$	l_L	

We see that

$$m_i = \sum_{j=1}^{L} n_{ij} \text{ and } l_j = \sum_{i=1}^{M} n_{ij}, \tag{2.2}$$

since marginalizing over one random variable means summing over all possible outcomes of the other.

We can now estimate the probabilities as fractions of the total number of experiments N:

$$p(x_i) = \frac{m_i}{N}, p(y_j) = \frac{l_j}{N} \text{ and } p(x_i, y_j) = \frac{n_{ij}}{N}.$$

Combining this with (2.2), we see that

$$p(x_i) = \sum_{j=1}^{L} p(x_i, y_j) \tag{2.3}$$

Equation (2.3) is known as the *sum rule* of probability. The random variable Y is marginalized.

If we only consider experiments, where the outcome for X is x_i, then the fraction of those, where the outcome for Y is y_j, is written as $p(y_j|x_i)$ and is called the *conditional probability* of y_j given x_i. Looking at the ith row of the grid, it can be calculated as

$$p(y_j|x_i) = \frac{n_{ij}}{m_i}.$$

On the other hand, we have

$$p(x_i, y_j) = \frac{n_{ij}}{N} = \frac{n_{ij}}{m_i} \frac{m_i}{N} = p(y_j|x_i)p(x_i).$$

This is known as the *product rule* of probability. Written in terms of the random variables, the *sum* and *product rule of probability* are

$$p(X) = \sum_{Y} p(X, Y) \text{ and } p(X, Y) = p(Y|X)p(X),$$

where the sum is over all possible instances of Y.

From the definition of independence in (2.1), we see that Y is independent of X, if the conditional probability equals the marginal probability for all outcomes. That is, the outcome of X does not influence the outcome of Y at all.

The joint probability of two random variables is symmetric. That is

$$p(X, Y) = p(Y, X).$$

Combining this with the product rule gives $p(Y|X)p(X) = p(X|Y)p(Y)$. Dividing by $p(X)$ leads to *Bayes' rule*, also known as *Bayes' law* or *theorem* which relates the conditional probabilities of random variables to each other:

$$p(Y|X) = \frac{p(X|Y)p(Y)}{p(X)}.$$

Bayes' rule plays an important role within machine learning, sparking many *Bayesian* subbranches. A very good account of its historic influence is given by McGrayne in [28].

Within machine learning, the random variable X is often the available data, which either has been gathered or is generated by experiments. X can be multidimensional, since different kinds of information can be given for one data point. The random variable Y on the other hand encompasses the variables which we believe influence the data. Again this can be multidimensional. The probability $p(Y)$ gives our *prior* belief. This can be experience or the opinion of experts, or derived from other experiments. The prior has a controversial role, since it quantifies a subjective opinion. The conditional probability $p(X|Y)$ is in some way a measure of how well different choices for Y explain X. It is sometimes referred to as the *likelihood* and is a function of Y. The overall probability of the data X is given by $p(X)$. It can be calculated using a combination of the sum and product rule

$$p(X) = \sum_Y p(X|Y)p(Y).$$

We see that it is the sum of the numerator in Bayes' rule over all instances of Y. Therefore it is also sometimes referred to as the *normalizing factor*, since it ensures that

$$\sum_Y p(Y|X) = 1.$$

The sum of probabilities over all instances always has to add to one. Lastly, the probability $p(Y|X)$ quantifies our *posterior* belief once we have observed and taken into account the data. The human equivalent is revising one's opinion once some experiences have been made. While many other factors influence whether a human revises their opinion or not (e.g., their stubbornness), Bayes' rule is a mathematical process to do so. If there are different hypotheses with associated variables Y of explaining the data, $p(Y|X)$ can be used to select a hypothesis. It is then also called the *evidence*. More on the Bayesian methodology can be found in [35], while [28] gives many examples of how the repeated application of Bayes' rule solved real world problems.

However, care needs to be taken when gathering data as the example by Appleton, French and Vanderpump in [1] of *Simpson's paradox* shows. They gathered data on smoking behaviour in the seventies and followed this up twenty years later. They restricted their view to women who either had never smoked or were current smokers at the time of the original study. There were

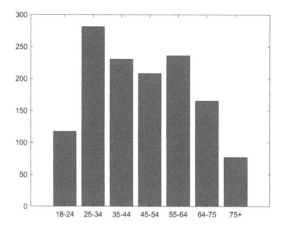

Figure 2.1: Distribution of participants across age groups in [1].

582 smokers of which 139 had died resulting in a mortality rate of $139/582 \approx 23.9\%$. In the non-smoking group 230 died out of 732 giving a mortality rate of $230/732 \approx 31.4\%$. This contradicts what health professionals tell us.

Let Y be the random variable taking the values {smoker, non-smoker}, and let X be the random variable with values {died, alive}. Our grid then looks like

	smoker	non-smoker	
died	139	230	369
alive	443	502	945
	582	732	1314

The explanation reveals itself when looking at the data more closely by taking into account the women's age at the time of the original study. They were grouped into seven age groups: $18 - 24, 25 - 34, 35 - 44, 45 - 54, 55 - 64, 65 - 74$ and older than 75. We now let the variable Y have 14 possible values, the age group together with whether they were smokers (S) or not (NS). The table now becomes

	$18 - 24$		$25 - 34$		$35 - 44$		$45 - 54$		$55 - 64$		$65 - 74$		$75+$	
	S	NS	S	NS	S	NS	S	NS	S	NS	S	NS	S	NS
died	2	1	3	5	14	7	27	12	51	40	29	101	13	64
alive	53	61	121	152	95	114	103	66	64	81	7	28	0	0
	55	62	124	157	109	121	130	78	114	121	36	129	13	64

The first thing to note is that the participants are not equally distributed across the age ranges. This can be visualized by plotting the *histogram* of the participants in the study with regards to their age groups as seen in Figure 2.1.

Ideally, one would like the data to be *uniformly distributed*. That is each category is equally likely. If enough samples are taken, then each category should contain approximately the same number of samples. This is a big "if". In many practical situations, particularly when participants are self-selecting, this is not possible. This is why white students are over-represented in psychological experiments, since the participants are recruited among the student population of the university where the experiment is conducted. Another oddity about this data is that all the age ranges encompass ten years, apart from the youngest and oldest. There are of course practical reasons for this, since for the youngest age range only legal adults could take part, and at the other end of the range, it needed to be ensured there are enough participants in the oldest. These are just some examples of the care which needs to be taken when gathering data.

As well as the above, in this example the percentage of smokers varies across the age ranges:

	$18-24$	$25-34$	$35-44$	$45-54$	$55-64$	$65-74$	$75+$
smokers	47%	44%	47%	63%	49%	22%	17%

The non-smokers are disproportionally overrepresented in the two oldest age groups which are naturally more likely to die within the next twenty years. This skews the overall mortality in favour of the smokers, when the data is aggregated into just two groups of smokers and non-smokers.

We can now estimate the mortality for smokers and non-smokers for each of the age groups. Or in other words the estimated conditional probability of death given the age group and whether they were smokers or not, $p(\text{died}|Y)$:

		mortality
18–24	smokers	3.6%
	non-smokers	1.6%
25–34	smokers	2.4%
	non-smokers	3.2%
35–44	smokers	12.8%
	non-smokers	5.8%
45–54	smokers	20.8%
	non-smokers	15.4%
55–64	smokers	44.3%
	non-smokers	33.1%
65–74	smokers	80.6%
	non-smokers	78.3%
75+	smokers	100%
	non-smokers	100%

Now the mortality reflects the health advice given. In all age groups but two, the mortality of smokers is higher. Smokers die younger on average which might in part explain their under-representation in the two oldest age groups.

Simpson's paradox is due to another variable influencing the data, but hiding. This is called the *confounding variable*. In this case, it is the ages of the women in the original study and the distribution of smokers and non-smokers within the different age groups. Whenever the distributions within a population (the women in the original study) varies for different parts of the population (the different age groups), data analysis has to be done carefully, taking these variations into account.

Another example of Simpson's paradox is the correlation between price and demand. Here the confounding variable is time. If time is not taken into account, price and demand can appear positively correlated, while in fact they are negatively correlated which becomes apparent when plotting both demand and price against time.

2.2 Probability Densities, Expectation, Variance and Moments

In the previous sections, the random variables were discrete. However, one of them, the age, is fundamentally a continuous variable, if it is considered as time passed since birth. It was *discretized* by considering age groups. The choice of how to discretize can influence the interpretation as well. We have already seen that no discretization at all leads to the wrong interpretation of mortality with smokers being less likely to die than non-smokers. On the other hand, the discretization shall not be too fine grained, because then the number of samples in each category becomes too small for interpretation. This is also known as *over-fitting*, where the model fits the data too exactly and does not generalize to unseen data. For example, assume that we discretized the ages so finely that in each age group there is only one woman, or possibly none. We can then either make no prediction at all for a new participant of the study, if that age group is empty, or we will make the prediction according to the mortality of that one woman in that age group, which could be completely wrong.

Discretization is one way of dealing with probabilities of a continuous real-valued variable x. Another way is the *probability density function (pdf)* $f(x)$ over x which describes the distribution. Formally, it is defined by $f(x)\Delta x$ being the probability of a sample falling inside the interval $(x, x + \Delta x)$ as the interval size Δx approaches zero. The probability density function is always non-negative. It can be viewed as taking discretization to the limit. There is a valid probability value for x falling into any interval (a, b) and it is given by

$$p(x \in (a, b)) = \int_a^b f(x)dx.$$

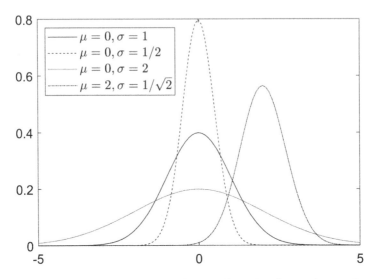

Figure 2.2: The normal probability density function for various values of μ and σ.

If $a = -\infty$ and $b = \infty$, that is we integrate over all possible values of x, the probability has to be one.

As continuous variables can be discretized, discrete variables can be viewed as continuous. In this case, $f(x)$ is called the *probability mass function (pmf)*, because the probabilities for each discrete possible value of x are concentrated as point masses at these values. For all other values of x the probability is zero.

A well-known example of a probability density function is the *bell curve* of the *normal distribution* also known as the *Gaussian distribution* $x \sim \mathcal{N}(\mu, \sigma^2)$, where $\sim$ means "is distributed as". Its probability density function is given by

$$\mathcal{N}(x|\mu, \sigma^2) = \frac{1}{\sqrt{2\pi}\sigma} \exp\left(-\frac{1}{2\sigma^2}(x - \mu)^2\right). \tag{2.4}$$

Its position and shape are determined by the parameters μ, called the *mean*, and σ^2, known as the *variance*. The square root of the variance, σ, is called the *standard deviation*. Figure 2.2 illustrates the shapes of the bell curve for various values of the mean and variance. When $\mu = 0$ and $\sigma = \sigma^2 = 1$, it is known as the *standard normal distribution*.

There is a common misconception that "normal" means common. Gauss introduced the naming with "normal" referring to a norm as in the method of Ordinary Least Squares discussed in Section 8.4. One technique to minimize the norm is to solve the "normal" equations and these were what Gauss in his choice of words referred to. Over the years, "normal" took on the meaning of "usual".

Normal distributions are indeed quite common. Lyon elaborates on the explanations of this in [27]. The *central limit theorem* is often quoted. It states that, if $x_1, \ldots, x_n$ is a sequence of *independent and identically distributed* (i.i.d.) random variables with mean 0 and variance σ^2, then the distribution of the normalized sum

$$S_n = \frac{x_1 + \ldots + x_n}{\sqrt{n}}$$

is close to a normal distribution for large enough n. As n tends to infinity and $S_n \to S$, it approaches the normal distribution $\mathcal{N}(S|0, \sigma^2)$. This holds however the distribution of $x_1, \ldots, x_n$ appeared in the first place, as long as they are independent and identically distributed. This, however, is the crux of this explanation.

While some phenomena we want to describe using probability density functions might be the result of independent and identically distributed summands, many are influenced by different factors, all of which have their own distinct distributions. Take for example a person's height. There is their potential for growth determined by their genetic make-up, where also different genes play a role. This is the nature part. Next is the nurture part. Nutrition, pollution, diseases, etc. all are factors. All these are not independent and identically distributed, but height is still normally distributed.

Another explanation involves the *entropy* of the probability density $f(x)$ defined as

$$-\int_{-\infty}^{\infty} f(x) \log f(x) dx,$$

where $f(x) \log f(x)$ is zero, wherever $f(x) = 0$. A change of base for the logarithm results in scaling of the entropy. As long as the same base is used throughout, entropies of different distributions can be compared. Information theorists prefer to use base 2 in line with bits. Since $0 \leq f(x) \leq 1$, the entropy is non-negative. Entropy is often interpreted as a measure of disorder or randomness or lack of information. The larger the entropy, the more disorder, the less information. For example, if there is only one possible value for x with probability 1, then $\log f(x) = 0$ there and the integrand is zero everywhere. Hence the entropy is zero. The information value is perfect, since the outcome is determined. Many physical systems strive to maximize the entropy over time. For a given mean μ and variance σ^2, $f(x) = \mathcal{N}(x|\mu, \sigma^2)$ maximizes the entropy. So this choice mirrors a natural process. Also, it is wise to choose a distribution with the least information, if nothing else is known.

To show that the normal distribution integrates to one, we consider

$$\int_{-\infty}^{\infty} \frac{1}{\sqrt{2\pi}\sigma} \exp\left(-\frac{1}{2\sigma^2}(x - \mu)^2\right) dx.$$

Letting $x - \mu = \sigma y$, we have $dx = \sigma dy$. The range of integration does not change with this change of variables. Thus we arrive at

$$\frac{1}{\sqrt{2\pi}} \int_{-\infty}^{\infty} \exp\left(-\frac{y^2}{2}\right) dy.$$

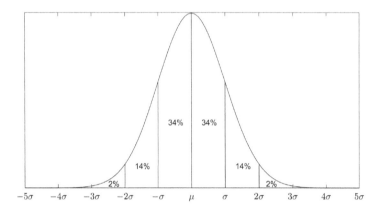

Figure 2.3: Approximate percentages of data falling within ranges determined by σ.

Next we consider the square of the integral:

$$
\left(\frac{1}{\sqrt{2\pi}} \int_{-\infty}^{\infty} \exp\left(-\frac{y^2}{2}\right) dy \right)^2 = \frac{1}{2\pi} \left(\int_{-\infty}^{\infty} \exp\left(-\frac{y^2}{2}\right) dy \right) \times
$$

$$
\left(\int_{-\infty}^{\infty} \exp\left(-\frac{z^2}{2}\right) dz \right)
$$

$$
= \frac{1}{2\pi} \int_{-\infty}^{\infty} \int_{-\infty}^{\infty} \exp\left(-\frac{y^2 + z^2}{2}\right) dy dz.
$$

$$(2.5)$$

Changing to polar coordinates, $y = r\cos\theta$, $z = r\sin\theta$ and $dy dz = r dr d\theta$, gives

$$
\frac{1}{2\pi} \int_0^{2\pi} \int_0^{\infty} \exp\left(\frac{-r^2(\cos^2\theta + \sin^2\theta)}{2}\right) r dr d\theta = \frac{1}{2\pi} \int_0^{2\pi} \int_0^{\infty} r \exp\left(\frac{-r^2}{2}\right) dr d\theta,
$$

$$(2.6)$$

where we used $\cos^2\theta + \sin^2\theta = 1$. The integrand is independent of θ. Thus the integral with respect to θ evaluates to 2π, the length of the interval, which cancels with the denominator. The integrand is the derivative of $-\exp\left(-r^2/2\right)$ with respect to r which evaluates to 0 at the upper limit ∞ and -1 at the lower limit 0 which we subtract. Hence the integral of the normal distribution is one as required.

The standard deviation entered mainstream news in 2012, when the discovery of the Higgs boson was announced at a significance of five sigma. This means that the data observed in the particle colliders did not fall within five standard deviations to the left or right of the mean of the distribution describing the *null hypothesis*. The null hypothesis assumes that the particle does not exist. In other words, the data falls in the extremes at the end of the bell curve. The probability of data that extreme or even more is very low. Figure 2.3 shows the bell curve and gives the rough percentages of how

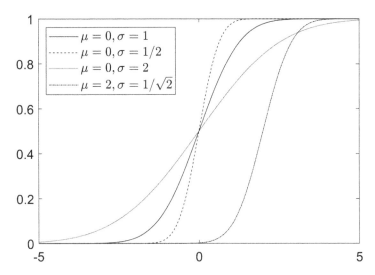

Figure 2.4: The normal cumulative distribution function for various values of μ and σ.

much data falls into each region. The following table gives more accurate percentages:

interval	expected percentage of data inside interval
$(\mu - \sigma, \mu + \sigma)$	68.26895%
$(\mu - 2\sigma, \mu + 2\sigma)$	95.44997%
$(\mu - 3\sigma, \mu + 3\sigma)$	99.73002%
$(\mu - 4\sigma, \mu + 4\sigma)$	99.99367%
$(\mu - 5\sigma, \mu + 5\sigma)$	99.99994%

The probability of the observed data (or even more extreme data) occurring by chance under the null hypothesis is known as the *p-value*. In the case of five sigma significance, the probability of data outside this region is $1 - 0.9999994 = 6 \times 10^{-7}$. There is a subtlety here, because it is equally likely for data to be in either the high or low tail of the bell curve. So the probability is halved to give a *p*-value of 3×10^{-7} or approximately 1 in 3.5 million. This is *not* the probability that the Higgs boson does not exist. It is the probability that the data would be observed, if it does not exist. The scientific story of the discovery of the Higgs boson is described in [38].

Setting the *level of significance* in terms of the standard deviation sigma is one way. Another way is to set it to an arbitrarily pre-defined threshold value α, for example $\alpha = 0.05, 0.01, 0.005$, or 0.001. The chosen value depends on the field of research. High-energy physics requires a threshold of 3×10^{-3} to announce there is evidence of a particle, and a threshold of 3×10^{-7} to claim a discovery. This stringent threshold is due to the *"look elsewhere effect"*. The probability that I win the lottery is quite small; the probability that somebody wins the lottery is, however, quite large. So when a lot of experiments are conducted and data gathered, it is not unusual to see a few statistical anomalies.

Associated with any probability density function is its *cumulative distribution function (cdf)* given by

$$F(x) = \int_{-\infty}^{x} f(t)dt.$$

It gives the probability of t falling in the interval $(-\infty, x)$. Hence the probability of t falling into the interval (a, b) can be calculated as $F(b) - F(a)$. The derivative of the cumulative distribution function satisfies $\frac{d}{dx}F(x) = F'(x) = f(x)$ by the *Fundamental Theorem of Calculus*. The cumulative distribution functions of the normal distribution for the same values of μ and σ as in Figure 2.2 are shown in Figure 2.4. All approach one as the argument increases, since the probability for t falling into the interval $(-\infty, \infty)$ is one.

The *rules of probability* also hold for continuous variables and probability density functions:

$$
\begin{aligned}
\textit{sum rule} \quad f(x) &= \int_{-\infty}^{\infty} f(x, y)dy, \\
\textit{product rule} \quad f(x, y) &= f(x|y)f(y), \\
\textit{Bayes' rule} \quad f(y|x) &= \frac{f(x|y)f(y)}{f(x)}.
\end{aligned}
$$

For a formal derivation of these rules see [13].

When dealing with a random variable, it is useful to know the average value it takes. This is known as the *expectation*, *mean* or *first moment* of the probability distribution, and is defined as

$$\mathbb{E}[x] = \int_{-\infty}^{\infty} f(x)x\,dx.$$

When x is discrete, integrating over the point masses at the possible values x can take reduces the expression to a sum,

$$\mathbb{E}[x] = \sum_{x} f(x)x.$$

Equally, it is useful to know how much we can *expect* the values to differ from the average. This is the *variance* or *second central moment*,

$$\text{var}[x] = \mathbb{E}[(x - \mathbb{E}[x])^2].$$

Using the definition for the expectation this becomes

$$
\begin{aligned}
\text{var}[x] &= \int_{-\infty}^{\infty} (x^2 - 2x\mathbb{E}[x] + \mathbb{E}[x]^2)f(x)dx \\
&= \int_{-\infty}^{\infty} x^2 f(x)dx - 2\mathbb{E}[x]\int_{-\infty}^{\infty} x\,f(x)dx + \mathbb{E}[x]^2\int_{-\infty}^{\infty} f(x)dx \\
&= \mathbb{E}[x^2] - \mathbb{E}[x]^2,
\end{aligned}
$$

since $f(x)$ integrates to one and $\int_{-\infty}^{\infty} x \, f(x)dx = \mathbb{E}[x]$. This formula also holds for discrete probability distributions with point masses.

We see that this formula involves the expectation of x^2. This is known as the *second raw* or *crude moment*. The ith raw (crude) moment is defined as

$$\int_{-\infty}^{\infty} f(x)x^i dx,$$

while the ith central moment is given by

$$\int_{-\infty}^{\infty} f(x)(x - \mathbb{E}[x])^i dx.$$

The *normalized* i^{th} central moment is the ith central moment divided by $\text{var}[x]^i$. The normalized central moments are invariant to any linear change of the random variable x. This means if $y = ax + b$, the normalized central moments of x and y are the same. The normalized central moments are used to quantify the shape of a probability density.

The third normalized central moment is known as *skewness* and measures the asymmetry around the mean of a probability distribution. If the distribution is symmetric, the skewness is zero. If more probability mass lies to the left of the mean, the skewness is negative, and positive, if more probability mass is to the right of the mean.

The forth normalized central moment is the *kurtosis* and measures how heavy tailed the probability distribution is. It is derived from the Greek word for "arching". A large value of the kurtosis means that extreme values of the random variable are more likely, while a small value of kurtosis means outliers are rare.

The moments can be used to inform the choice of which probability density function should be chosen for the distribution of a random variable.

More generally, often the random variable itself is not of interest, although a derivation $g(x)$ of it, which in itself is a random variable, is. In this case, the expectation and variance are written as $\mathbb{E}[g]$ and $\text{var}[g]$.

If there are more than one random variable, a subscript is used to indicate with respect to which variable the expectation is taken. For example, in

$$\mathbb{E}_y[g(x,y)]$$

the expectation is taken with respect to y, and the result is a function of x. The expectation with respect to x can then be taken. Since the order does not matter when integrating, the result is the same if the expectation with respect to x is taken first and then with respect to y, if x and y are independent, that is $f(x,y) = f(x)f(y)$.

Returning to the normal distribution, we calculate its expectation (first moment) as

$$\mathbb{E}[x] = \int_{-\infty}^{\infty} \frac{x}{\sqrt{2\pi}\sigma} \exp\left(-\frac{(x-\mu)^2}{2\sigma^2}\right) dx$$

Letting $y = x - \mu$ with $dy = dx$, the range of integration with this change of variable remains the same,

$$
\begin{aligned}
\mathbb{E}[x] &= \int_{-\infty}^{\infty} \frac{y + \mu}{\sqrt{2\pi}\sigma} \exp\left(-\frac{y^2}{2\sigma^2}\right) dy \\
&= \int_{-\infty}^{\infty} \frac{y}{\sqrt{2\pi}\sigma} \exp\left(-\frac{y^2}{2\sigma^2}\right) dy + \mu \int_{-\infty}^{\infty} \frac{1}{\sqrt{2\pi}\sigma} \exp\left(-\frac{y^2}{2\sigma^2}\right) dy.
\end{aligned}
$$

The integrand of the first integral is an odd function, that is $f(-y) = -f(y)$. Integrals of odd functions over an interval symmetric about the origin evaluate to zero. As we have seen above, the second integral evaluates to one. Hence for the normal distribution $\mathbb{E}[x] = \mu$.

For the variance we calculate

$$
\mathrm{var}[x] = \int_{-\infty}^{\infty} \frac{(x-\mu)^2}{\sqrt{2\pi}\sigma} \exp\left(-\frac{(x-\mu)^2}{2\sigma^2}\right) dx.
$$

Again letting $y = x - \mu$ as before, we have

$$
\mathrm{var}[x] = \int_{-\infty}^{\infty} \frac{y^2}{\sqrt{2\pi}\sigma} \exp\left(-\frac{y^2}{2\sigma^2}\right) dy.
$$

This integral can be calculated by *integration by parts* which can be thought of as the inverse of the *product rule of differentiation* which is

$$
\frac{d(uv)}{dy} = \frac{du}{dy}v + u\frac{dv}{dy}.
$$

Integration by parts is then

$$
\int_a^b u\,dv = [uv]_a^b - \int_a^b v\,du.
$$

Let $u = y$ and $dv = y\exp\left(-\frac{y^2}{2\sigma^2}\right) dy$, then $du = dy$ and $v = -\sigma^2 \exp\left(-\frac{y^2}{2\sigma^2}\right)$. With this we get

$$
\mathrm{var}[x] = \left[\frac{-\sigma^2 y}{\sqrt{2\pi}\sigma} \exp\left(-\frac{y^2}{2\sigma^2}\right)\right]_{-\infty}^{\infty} - \int_{-\infty}^{\infty} \frac{-\sigma^2}{\sqrt{2\pi}\sigma} \exp\left(-\frac{y^2}{2\sigma^2}\right) dy.
$$

The first expression is dominated by the exponential as the argument tends to $\pm\infty$ and thus is zero. The integrand is $-\sigma^2$ times the probability density function of a normal distribution with mean 0 and variance σ^2. The integral of a probability density function from $-\infty$ to ∞ is one. Thus for the normal distribution $\mathrm{var}[x] = \sigma^2$.

As for the third central moment, this is zero for the normal distribution, since the integrand again becomes an odd function. This means the skewness is also zero.

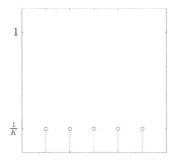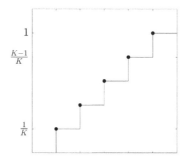

Figure 2.5: The discrete uniform probability mass function and its cumulative distribution function.

The forth central moment of the normal distribution can be calculated by *integration by parts* similar to the calculation of the variance. It is given by $3\sigma^4$. Hence the kurtosis of the normal distribution is 3. Often the kurtosis of other distributions is compared to this value. If it is larger the distribution is known as *leptokurtic*. If it is smaller, it is *platykurtic*. Both present their own difficulties, if an underlying normal distribution is assumed, but the data does not support this. In the case of leptokurtic distributions, the risk of extreme events is underestimated and is not sufficiently prepared for. Examples for this can be found in the history of financial theory. Extreme events are rare in platykurtic distributions. However, if these events are catastrophic, dismissing their possibility can be literally fatal.

2.3 Examples of Discrete Probability Mass Functions

In the following we give a short overview of commonly used probability distributions starting with discrete variables. We have already encountered the uniform distribution in the experiments of flipping a coin or rolling a die. All outcomes are equally likely as long as the die and coin are fair. If there are K possible outcomes, the probability for each is $1/K$. Figure 2.5 illustrates the discrete uniform probability density as point masses. The cumulative probability density function is a stair function where the left point of each step is included, but the end point is not.

Next we consider *binary* variables used when experiments can only have one of two outcomes, e.g yes or no, true or false, head or tail, 0 or 1. These are also known as *Boolean* variables. For simplicity we use the outcomes 0 and 1. The probability of $x = 1$ is denoted by $0 \leq \mu \leq 1$. The probability of $x = 0$ is then $1 - \mu$. This distribution is known as the *Bernoulli distribution* $x \sim \text{Bern}(\mu)$. Its probability density is given by

$$\text{Bern}(x|\mu) = \mu^x (1 - \mu)^{1-x}.$$

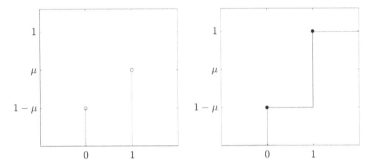

Figure 2.6: The Bernoulli probability mass function and its cumulative distribution function.

This formula might look daunting, but remember that it is only evaluated for $x = 0$ or 1, since no other outcomes are possible, and thus

$$\text{Bern}(0|\mu) = \mu^0(1-\mu)^1 = 1-\mu \text{ and } \text{Bern}(1|\mu) = \mu^1(1-\mu)^0 = \mu.$$

In Figure 2.6 the probability mass function is two point masses, while the cumulative distribution function is a simple step function.

The expectation and the second crude moment of the Bernoulli distribution are given by

$$\mathbb{E}[x] = \mu \times 1 + (1-\mu) \times 0 = \mu,$$

$$\mathbb{E}[x^2] = \mu \times 1^2 + (1-\mu) \times 0^2 = \mu.$$

From this we can calculate the variance

$$\text{var}[x] = \mathbb{E}[x^2] - \mathbb{E}[x]^2 = \mu - \mu^2 = \mu(1-\mu).$$

The Bernoulli distribution describes an experiment where the outcome is binary. We can now consider the question of how many successes ($x = 1$) are in a set of N experiments. When considering m successes, there are N possibilities for the first success, $N-1$ for the second, and so on until there are $N - m + 1$ possibilities for the last success. The resulting number $N(N-1)\cdots(N-m+1)$ is the number of m-permutations. It gives the number of possibilities of choosing an ordered set of size m from a set of size N. However, the ordering in our case is unimportant, and we therefore divide by the number of possible permutations of size m which is given by the *factorial* $m! = m(m-1)\cdots 1$. The resulting number of possibilities to achieve m times $x = 1$ in N experiments is given by the *binomial coefficient*,

$$\binom{N}{m} = \frac{N(N-1)\cdots(N-m+1)}{m(m-1)\cdots 1}.$$

It is read as "N choose m" and is the coefficient of the a^m term in the expansion of $(1+a)^N$.

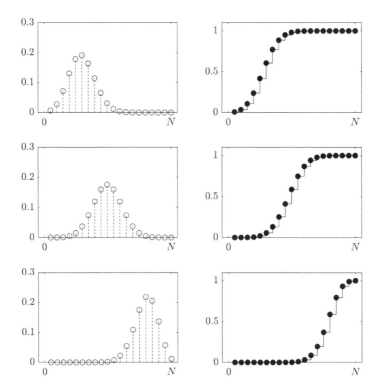

Figure 2.7: The binomial probability mass function and its cumulative distribution function for $N = 20$ and $\mu = 0.3$ (top row), $\mu = 0.5$ (middle row), and $\mu = 0.8$ (bottom row).

This leads to the *binomial distribution* $m \sim \text{Bin}(N, \mu)$ with probability mass function

$$\text{Bin}(m|N, \mu) = \binom{N}{m} \mu^m (1 - \mu)^{N-m}, \tag{2.7}$$

which depends on the number of experiments N and the probability of success μ. Figure 2.7 shows the binomial probability mass function and its cumulative distribution function for $\mu = 0.3, 0.5$ and 0.8. If the probability of $x = 1$ is small, the probability of achieving a large number of successes is small. It increases as μ increases.

The expectation of the binomial distribution is

$$\mathbb{E}[m] = \mu N,$$

since we expect the portion of N given by μ to be successful. The variance on the other hand is

$$\text{var}[m] = \mu(1 - \mu)N.$$

It is N times the variance of a single experiment.

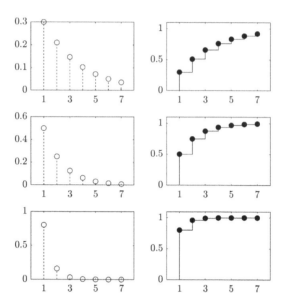

Figure 2.8: The geometric probability mass function and its cumulative distribution function for $\mu = 0.3$ (top row), $\mu = 0.5$ (middle row), and $\mu = 0.8$ (bottom row).

The other question is, how many attempts $m \geq 1$ are necessary to succeed. This is described by the *geometric distribution* $m \sim \text{Geo}(\mu)$. Its probability mass function is

$$\text{Geo}(m|\mu) = \mu(1 - \mu)^m - 1.$$

Its expectation is

$$\mathbb{E}[m] = \frac{1}{\mu}.$$

This means that the smaller μ, the more attempts are necessary to succeed. The variance on the other hand is given by

$$\text{var}[m] = \frac{1 - \mu}{\mu^2}.$$

Sometimes, e.g. in MATLAB, the geometric distribution is defined as the number of failures $\hat{m} = m - 1$ before the first success. In this case, the probability mass function is $\mu(1 - \mu)^{\hat{m}}$ and the expectation is $\mathbb{E}[\hat{m}] = (1 - \mu)/\mu$, while the variance remains the same.

Instead of using the probability of success, the rate λ of events occurring over a certain time interval can be used. It is often given as a positive number, but does not need to be an integer. It needs to be in the context of a time interval. Examples for λ are the number of phone calls to a call center per hour or the number of cars passing a traffic checkpoint per day.

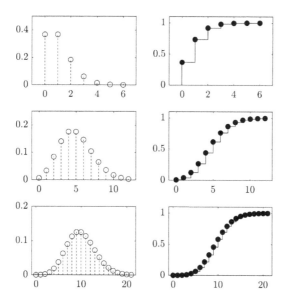

Figure 2.9: The Poisson probability mass function and its cumulative distribution function for $\lambda = 1$ (top row), $\lambda = 5$ (middle row), and $\lambda = 10$ (bottom row).

The *Poisson distribution* $m \sim \text{Po}(\lambda)$ gives the probability of m events occurring in the given time interval. The probability mass function is

$$\text{Po}(m|\lambda) = \frac{\lambda^m \exp(-\lambda)}{m!}. \tag{2.8}$$

Both its expectation and variance are equal to λ,

$$\mathbb{E}[m] = \text{var}[m] = \lambda.$$

Of course, we are not just interested in experiments where the outcome is binary, but where the outcome can be any of K possible, mutually distinct outcomes. For example, the roll of a die would be one such experiment. The outcomes could be denoted $\{1, 2, 3, 4, 5, 6\}$ or $\{\boxdot, \boxdot, \boxdot, \boxdot, \boxdot, \boxdot\}$. However, it is convention to standardize the representation by using a *1-of-K representation*. That is the random variable $\mathbf{x}$ describing the outcome is a K-dimensional vector, where exactly one element x_k equals one and all the others are zero. For example, if in the die experiment $x_4 = 1$, then $\mathbf{x} = (0, 0, 0, 1, 0, 0)^T$ and it means that a four was rolled. The transpose ^{T} is used, since all vector variables are column vectors throughout, but it is easier to write row vectors. Any experiments of a *categorical* nature can be described in this way.

The probability to get the kth possible result, which means $x_k = 1$, is denoted by μ_k. All probabilities are summarized in the vector $\boldsymbol{\mu} = (\mu_1, \ldots, \mu_K)^T$.

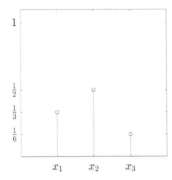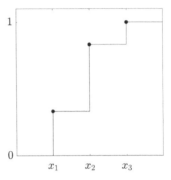

Figure 2.10: The categorical probability mass function and its cumulative distribution function for $\boldsymbol{\mu} = (1/3, 1/2, 1/6)^T$.

The probabilities of all outcomes must sum to one,

$$\sum_{k=1}^{K} \mu_k = 1.$$

This defines the *categorical distribution* denoted $\mathbf{x} \sim \text{Cat}(\boldsymbol{\mu})$ with probability density function

$$\text{Cat}(\mathbf{x}|\boldsymbol{\mu}) = p(x_k = 1|\boldsymbol{\mu}) = \prod_{k=1}^{K} \mu_k^{x_k}. \tag{2.9}$$

Note that in the product all factors are 1 apart from one, since in $\mathbf{x}$ all entries are 0 apart from one, which is 1. The categorical distribution can be viewed as a generalization of the Bernoulli distribution. The uniform distribution is a special case of it, as $\mu_1 = \cdots = \mu_k = 1/K$.

Figure 2.10 illustrates the probability mass function of a categorical distribution where $\boldsymbol{\mu} = (1/3, 1/2, 1/6)^T$. The cumulative distribution function is an irregular stair function. Both are drawn with regards to the indices in $\mathbf{x}$. If the possible outcomes are ordered differently in $\mathbf{x}$, the graphs will be different.

The expectation of the categorical distribution is calculated as

$$\mathbb{E}[\mathbf{x}] = \sum_{\mathbf{x}} \text{Cat}(\mathbf{x}|\boldsymbol{\mu})\mathbf{x} = \boldsymbol{\mu},$$

where the sum is over all possible values $\mathbf{x}$ can take. Note that here we have made the step into a *multivariate distribution*, since $\mathbf{x}$ is a vector, though one where all elements are zero apart from one, which is one.

Recall that the variance of a *univariate distribution* of the random variable x is given by $\text{var}[x] = \mathbb{E}[(x - \mathbb{E}[x])^2] = \mathbb{E}[x^2] - \mathbb{E}[x]^2$. The multivariate equivalent is

$$\text{var}[\mathbf{x}] = \mathbb{E}[(\mathbf{x} - \mathbb{E}[\mathbf{x}])(\mathbf{x} - \mathbb{E}[\mathbf{x}])^T] = \mathbb{E}[\mathbf{x}\mathbf{x}^T] - \mathbb{E}[\mathbf{x}]\mathbb{E}[\mathbf{x}]^T. \tag{2.10}$$

The result is a matrix, where the kth diagonal entry gives how much x_k varies around the mean μ_k. The off-diagonal entries measure how much different components of $\mathbf{x}$ influence each other.

Now, $\mathbf{x}$ is a 1-of-K representation. If $x_k = 1$ in $\mathbf{x}$, then $\mathbf{x}\mathbf{x}^T$ is a matrix with zero entries everywhere apart from the kth diagonal element which has the entry 1. From this

$$\text{var}[\mathbf{x}] = \sum_{\mathbf{x}} \text{Cat}(\mathbf{x}|\boldsymbol{\mu})\mathbf{x}\mathbf{x}^T - \boldsymbol{\mu}\boldsymbol{\mu}^T = \text{diag}(\boldsymbol{\mu}) - \boldsymbol{\mu}\boldsymbol{\mu}^T$$

follows, where $\text{diag}(\boldsymbol{\mu})$ denotes the matrix, where the off-diagonal elements are zero and the diagonal is given by the entries of $\boldsymbol{\mu}$.

Note that the kth diagonal entry of $\text{var}[\mathbf{x}]$ is $\mu_k - \mu_k^2 = \mu_k(1 - \mu_k)$. This is the variance of a Bernoulli distribution, where success is defined as the outcome being $x_k = 1$ and failure as $x_k = 0$. The (k, l) off-diagonal entry is known as the *covariance* $\text{cov}[x_k, x_l]$ between k^{th} and l^{th} element and is $-\mu_k\mu_l$. This shows that the kth and l^{th} component influence each other in a negative way, since if one is 1, the other has to be 0.

As with binary variables, we can also ask for the distribution of different combinations of outcomes, when the experiment is repeated N times. Each experiment gives a result vector $\mathbf{x}_n$, $n = 1, \ldots, N$. Let m_k be the number of times the outcome of the experiment is $x_{nk} = 1$. Since x_{nk} is zero otherwise, we have

$$m_k = \sum_{n=1}^{N} x_{nk}.$$

We are interested in the distribution of the *tuple* $\mathbf{m} = (m_1, \ldots, m_K)^T$ of length K. It is actually of length $K - 1$, since the numbers m_k are constrained by

$$\sum_{k=1}^{N} m_k = N, \tag{2.11}$$

and hence m_K is given by $m_K = N - m_1 - \ldots - m_{K-1}$. The number of possibilities to generate a particular tuple is given by the *multinomial coefficient*,

$$\binom{N}{m_1 \cdots m_K} = \frac{N!}{m_1! \cdots m_K!}.$$

Combinatorially, it is the number of ways one can put N balls into K bins where bin 1 contains m_1 balls, bin 2 contains m_2 objects and so on. Here $0!$ is defined to be one. The binomial coefficient is the multinomial coefficient with $m_1 = m$ and $m_2 = N - m$.

The probability density function of the *multinomial distribution* $\mathbf{m} \sim \text{Mult}(N, \boldsymbol{\mu})$ is then

$$\text{Mult}(\mathbf{m}|N, \boldsymbol{\mu}) = \binom{N}{m_1 \cdots m_K} \prod_{k=1}^{K} \mu_k^{m_k}. \tag{2.12}$$

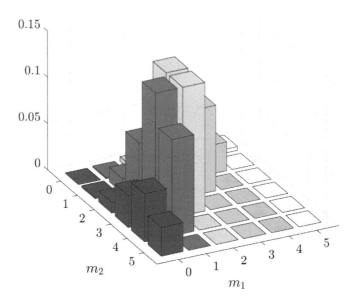

Figure 2.11: The multinomial probability mass function for $N = 5$ and $\boldsymbol{\mu} = (1/3, 1/2, 1/6)^T$.

Figure 2.11 shows the probability mass function of the multinomial distribution for $N = 5$ and $\boldsymbol{\mu} = (1/3, 1/2, 1/6)^T$ as a bar graph. The graph only plots against m_1 and m_2, since $m_3 = N - m_1 - m_2$ is given implicitly. On the right hand side we see that the probability mass is zero, whenever $m_1 + m_2 > 5$ as required by the constraint given in 2.11. After zero, the smallest probability is for the tuple $(0, 0, 5)^T$ and is $1/6^5$. The largest probability occurs three times for the tuples $(2, 2, 1)^T, (2, 3, 0)^T$ and $(1, 3, 1)^T$, since

$$\frac{5!}{2!\,2!\,1!}\left(\frac{1}{3}\right)^2\left(\frac{1}{2}\right)^2\left(\frac{1}{6}\right)^1 = \frac{5!}{2!\,3!\,0!}\left(\frac{1}{3}\right)^2\left(\frac{1}{2}\right)^3\left(\frac{1}{6}\right)^0$$

$$= \frac{5!}{1!\,3!\,1!}\left(\frac{1}{3}\right)^1\left(\frac{1}{2}\right)^3\left(\frac{1}{6}\right)^1 = \frac{5}{6^2}.$$

The expectation of the binomial distribution is

$$\mathbb{E}[\boldsymbol{m}] = N\boldsymbol{\mu},$$

since we expect the $\mu_k N$ experiments to result in the outcome $x_k = 1$. The variance on the other hand is

$$\text{var}[\mathbf{m}] = N\,\text{diag}(\boldsymbol{\mu}) - \boldsymbol{\mu}\boldsymbol{\mu}^T.$$

It is N times the variance of a single experiment, which is described by the categorical distribution.

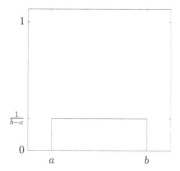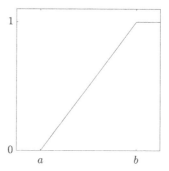

Figure 2.12: The uniform probability density function and its cumulative distribution function.

2.4 Examples of Continuous Probability Density Functions

As with the discrete distributions, the simplest continuous probability density function is the uniform distribution. The random variable x can take any value in the interval (a, b) with the same probability, $x \sim \mathrm{Unif}(a, b)$. Since the probability has to integrate to one over the interval, it has to be $1/(b-a)$. The uniform density function is given by

$$\mathrm{Unif}(x|a, b) = \begin{cases} \dfrac{1}{b-a} & \text{for } a < x < b \\ 0 & \text{otherwise} \end{cases},$$

while the cumulative distribution function is

$$\begin{cases} 0 & \text{for } x < a \\ \dfrac{x-a}{b-a} & \text{for } a \le x \le b \\ 1 & \text{for } x > b \end{cases}.$$

Figure 2.12 gives a graphical illustration. The continuous uniform density function can be viewed as the limit of its discrete equivalent, when the set of all possible results includes every point in the interval (a, b).

Its expectation is given by

$$\mathbb{E}[x] = \int_a^b \frac{1}{b-a} x \, dx = \left[\frac{x^2}{2(b-a)} \right]_a^b = \frac{b^2 - a^2}{2(b-a)} = \frac{a+b}{2}.$$

The variance on the other hand is

$$\mathrm{var}[x] = \int_a^b \frac{1}{b-a} \left(x - \frac{a+b}{2} \right)^2 dx = \left[\frac{1}{3(b-a)} \left(x - \frac{a+b}{2} \right)^3 \right]_a^b = \frac{1}{12}(b-a)^2.$$

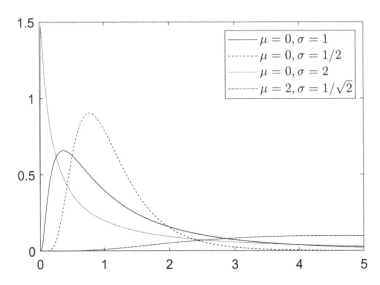

Figure 2.13: The log-normal probability density function for various values of μ and σ.

We already encountered the *univariate normal* distribution in Equation (2.4) and its probability density function and cumulative distribution function are depicted in Figures 2.2 and 2.4 respectively. Its multivariate counterpart has the probability density function

$$f(\mathbf{x}) = \frac{1}{\sqrt{|2\pi\boldsymbol{\Sigma}|}} \exp\left(-\frac{1}{2}(\mathbf{x} - \boldsymbol{\mu})^T \boldsymbol{\Sigma}^{-1}(\mathbf{x} - \boldsymbol{\mu})\right),$$

where $|\cdot|$ denotes the determinant and $\cdot^{-1}$ is the inverse of the matrix $\boldsymbol{\Sigma}$. The mean of the multivariate normal distribution is $\boldsymbol{\mu}$, while its variance is $\boldsymbol{\Sigma}$.

A random variable x, whose logarithm is normally distributed, that is $y = \log x$ has a normal distribution, has a *log-normal distribution*, also known as the *Galton distribution*. Likewise, since $x = \exp y$, the exponential of a normally distributed variable is log-normally distributed. Since the exponential is always positive, x will only take positive values.

Many growth processes are described by percentual changes. For example the annual growth rate of the global human population varies around 1.1%. Let z_1 be the actual growth rate in year 1, z_2 in year 2, and so on. If a is the population size at the beginning, then after N years, the population will have size

$$x = a \prod_{n=1}^{N} z_n.$$

Taking the logarithm results in

$$y = \log a + \sum_{n=1}^{N} \log z_n.$$

Hence the change in the logarithm of the population size is a sum of independent and identically distributed variables, the $\log z_n$. By the central limit theorem, such a sum tends to be normally distributed, irrespective of the distribution of $\log z_n$. The addition of $\log a$ only shifts the mean of this distribution. Hence the logarithm of the population size y is normally distributed. Thus the population size x is log-normally distributed. Instead of letting the numbers of years grow, the limit can also be achieved by considering smaller and smaller time intervals. It needs to be noted, though, that then the variance in the percentual growth increases, since longer time intervals have an averaging effect.

Growth plays a major role in areas such as biology, medicine and economics. There are many other examples, where a log-normal distribution is used. For example the dwell time on online content and the length of comments online follow a log-normal distribution, as does the length of chess games. The *Black-Scholes model* for option pricing assumed that the returns of a stock are log-normally distributed. However, this assumption is flawed, because extreme price changes such as in the event of a stock market crash need a distribution with a larger kurtosis. In practice, such extreme events are more likely than the log-normal distribution indicates.

If μ and σ^2 are the mean and variance of the associated normal distribution, the log-normal density function for $x \sim \log \mathcal{N}(\mu, \sigma^2)$ is given by

$$\log \mathcal{N}(x|\mu, \sigma^2) = \frac{1}{x\sqrt{2\pi}\sigma} \exp\left(-\frac{(\log x - \mu)^2}{2\sigma^2}\right).$$

It is shown in Figure 2.13, while Figure 2.14 depicts the cumulative distribution function for various values of μ and σ. For $\mu = 0$ and $\sigma = 2$, the density function is peaked very close to zero. It is not visible that it is zero at the origin.

The expectation is calculated as

$$\mathbb{E}[x] = \int_0^{\infty} \frac{x}{x\sqrt{2\pi}\sigma} \exp\left(-\frac{(\log x - \mu)^2}{2\sigma^2}\right) dx.$$

The factor of x cancels with the denominator in the density function. We let $y = \log x$ or equivalently $x = \exp y$. With this choice, $dx = \exp y \, dy$ and the range of integration changes to $(-\infty, \infty)$,

$$\begin{aligned}
\mathbb{E}[x] &= \int_{-\infty}^{\infty} \frac{1}{\sqrt{2\pi}\sigma} \exp\left(-\frac{(y - \mu)^2}{2\sigma^2}\right) \exp y \, dy \\
&= \int_{-\infty}^{\infty} \frac{1}{\sqrt{2\pi}\sigma} \exp\left(\frac{-(y - \mu)^2 + 2\sigma^2 y}{2\sigma^2}\right) dy.
\end{aligned}$$

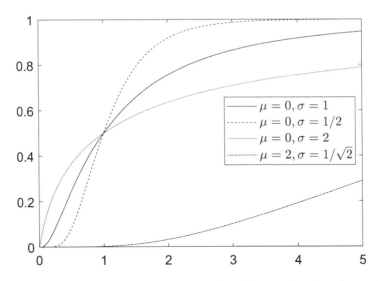

Figure 2.14: The log-normal cumulative distribution function for various values of μ and σ.

We now use a technique called *completion of squares*:

$$
\begin{aligned}
-(y-\mu)^2 + 2\sigma^2 y &= -y^2 + 2y\mu - \mu^2 + 2\sigma^2\mu = -y^2 + 2(\mu+\sigma^2)y - \mu^2 \\
&= -(y-(\mu+\sigma^2))^2 + (\mu+\sigma^2)^2 - \mu^2 \\
&= -(y-(\mu+\sigma^2))^2 + 2\mu\sigma^2 + \sigma^4.
\end{aligned}
$$

With this, the expectation becomes

$$
\begin{aligned}
\mathbb{E}[x] &= \int_{-\infty}^{\infty} \frac{1}{\sqrt{2\pi}\sigma} \exp\left(\frac{-(y-(\mu+\sigma^2))^2 + 2\mu\sigma^2 + \sigma^4}{2\sigma^2}\right) dy \\
&= \exp\left(\mu + \frac{\sigma^2}{2}\right) \int_{-\infty}^{\infty} \frac{1}{\sqrt{2\pi}\sigma} \exp\left(\frac{-(y-(\mu+\sigma^2))^2}{2\sigma^2}\right) dy.
\end{aligned}
$$

The integral evaluates to 1, since the integrand is the normal probability density function with mean $\mu + \sigma^2$ and variance σ^2. Thus the expectation is

$$
\mathbb{E}[x] = \exp\left(\mu + \frac{\sigma^2}{2}\right).
$$

The variance can be similarly calculated and is

$$
\mathrm{var}[x] = \left(\exp(\sigma^2) - 1\right)\exp\left(2\mu + \sigma^2\right).
$$

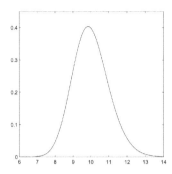 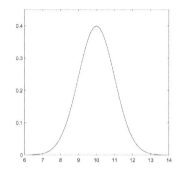

Figure 2.15: The log-normal density function for expectation $\mathbb{E}[x] = 100$ and variance $\mathrm{var}[x] = 1$ on the left and the normal density function with the same expectation and variance on the right.

If the variance $\mathrm{var}[x]$ and expectation $\mathbb{E}[x]$ of the log-normal distribution are known, the mean and variance of the associated normal distribution can be calculated as

$$
\begin{aligned}
\mu &= \log\left(\frac{\mathbb{E}[x]^2}{\sqrt{\mathrm{var}[x] + \mathbb{E}[x]^2}}\right) \\
\sigma^2 &= \log\left(1 + \frac{\mathrm{var}[x]}{\mathbb{E}[x]^2}\right).
\end{aligned}
\tag{2.13}
$$

We already gave one word of caution, when using the log-normal distribution. In general, assuming an underlying distribution, is a very strong assumption, which needs to be well justified and communicated, in case new data makes this assumption invalid. Figure 2.15 shows the log-normal density function for expectation $\mathbb{E}[x] = 100$ and variance $\mathrm{var}[x] = 1$ on the left and the normal density function with the same expectation and variance on the right. They are essentially indistinguishable.

The log-normal probability density function has its only maximum at $\exp(\mu - \sigma^2)$. This is also known as its *mode*. Both the log-normal and normal distribution are *unimodal*, since they only have one maximum. If a probability density function has several local maxima, which can take different values, it is called *multimodal*. For the normal distribution the mode coincides with the mean. The log-normal probability density function is looking more and more like the normal one, the closer the mode gets to the expectation, that is the mean. Using Equations (2.13), the mode can be calculated as

$$
\begin{aligned}
\exp(\mu - \sigma^2) &= \exp\left[\log\left(\frac{\mathbb{E}[x]}{\sqrt{1 + \frac{\mathrm{var}[x]}{\mathbb{E}[x]^2}}}\right) - \log\left(1 + \frac{\mathrm{var}[x]}{\mathbb{E}[x]^2}\right)\right] \\
&= \left(1 + \frac{\mathrm{var}[x]}{\mathbb{E}[x]^2}\right)^{-3/2} \mathbb{E}[x].
\end{aligned}
$$

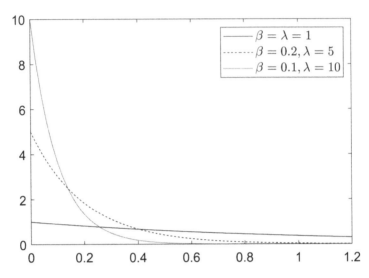

Figure 2.16: The exponential probability density function for various values of $\beta = 1/\lambda$ in accordance with Figure 2.9 of the Poisson distribution.

The factor by which the mode differs from the expectation is very close to 1, if the variance is much smaller than the expectation. In our case with $\mathbb{E}[x] = 100$ and $\text{var}[x] = 1$, the factor is approximately 0.99985.

It is important that underlying assumptions are clear and documented as the following example illustrates. In 2G mobile phone networks, each mobile is assigned its own frequency, and served by an antenna. It might hop to another frequency, if it travels away from the coverage area of this antenna and into the coverage area of another antenna. It might also get a new frequency from another antenna further away, if the antenna which it is connected to currently needs to serve more mobiles. To assess the coverage of the mobile network of a large area, mobile phone usage would be simulated in this area, including handovers between antennae. When the simulations were first coded, programmers did not include code paths for cases, which were possible, but improbable. With increased mobile phone usage, the probabilities changed. Inserting these now necessary code paths into large often undocumented pieces of code caused some problems. The moral of the story is that all cases, even the improbable ones, need to be coded and documented. It is even wise to include the impossible ones, because somewhere someone will change the code and they become possible.

The *exponential distribution*, $x \sim \text{Exp}(\beta)$, has probability density function

$$\text{Exp}(x|\beta) = \begin{cases} \dfrac{1}{\beta} \exp\left(-\dfrac{x}{\beta}\right) & \text{for} \quad x \geq 0, \\ 0 & \text{for} \quad x < 0, \end{cases} \tag{2.14}$$

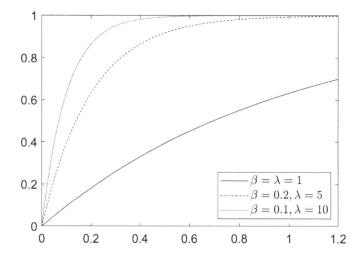

Figure 2.17: The exponential cumulative distribution function for various values of $\beta = 1/\lambda$ in accordance with Figure 2.9 of the Poisson distribution.

which is parametrized by the *scale parameter* β. Its inverse, $\lambda = 1/\beta$, is the *rate parameter*. It is related to the Poisson distribution defined in Equation (2.8) with rate parameter $\lambda = 1/\beta$ in that it describes the time between occurrences of events described by the Poisson distribution.

The cumulative distribution function is zero for $x < 0$ and for $x \geq 0$ is given by

$$F(x) = p(t \leq x) = \int_{-\infty}^{x} \text{Exp}(t|\beta)dt = \int_{0}^{x} \frac{1}{\beta} \exp\left(-\frac{t}{\beta}\right) dt = 1 - \exp\left(-\frac{x}{\beta}\right).$$

It gives the probability that the event has happened in this time frame. On the other hand, $1 - F(x) = \exp\left(-\frac{x}{\beta}\right) = p(t > x)$ is the probability that the event has not occurred yet.

Examples of the probability density function and the cumulative distribution function for various values of β are given in Figures 2.16 and 2.17 respectively.

The probability that the waiting time is $x + y$ given that already x amount of time has passed is by the product rule

$$p(t > x + y|t > x)p(t > x) = p(t > x + y \quad \text{and} \quad t > x) = p(t > x + y).$$

Solving for $p(t > x + y|t > x)$, we arrive at

$$p(t > x + y|t > x) = \frac{p(t > x + y)}{p(t > x)} = \frac{\exp(-\frac{x+y}{\beta})}{\exp(-\frac{x}{\beta})} = \exp\left(-\frac{y}{\beta}\right) = p(t > y).$$

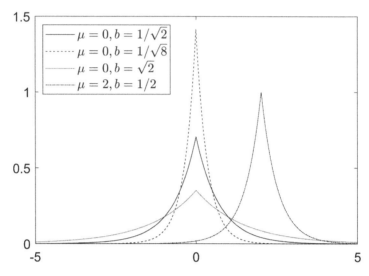

Figure 2.18: The Laplace probability density function for various values of μ and b, chosen such that the mean and variance are the same as in Figure 2.2.

In practical terms, this means that the time one has to wait for an event is not influenced at all by the time spent waiting so far. This means the exponential distribution is *memoryless*, a property it shares with the geometric distribution. The exponential distribution is the continuous equivalent of the geometric distribution which describes the number of attempts of a binary experiment to succeed. Instead of thinking about failures and successes, it is more convenient to think about a change of state. λ gives the rate of change over a unit time interval, and the exponential distribution describes the probability of how much time passes until the states changes. Constant rate of change is a crucial assumption not often satisfied by real-world processes. For example, the rate of emergency calls increases on Friday and Saturday nights and staffing levels need to be adjusted compared to other times. If the rate is constant, it is used in queuing theory. The exponential distribution is inappropriate to model the overall lifetime of organisms or technical devices, where the event of death or failure depends on the lifetime so far.

Using *integration by parts*, the expectation is given by

$$
\begin{aligned}
\mathbb{E}[x] &= \int_0^\infty \frac{x}{\beta} \exp\left(-\frac{x}{\beta}\right) dx \\
&= \left[\frac{x}{\beta}\left(-\beta \exp\left(-\frac{x}{\beta}\right)\right)\right]_0^\infty + \int_0^\infty \exp\left(-\frac{x}{\beta}\right) dx \\
&= \left[-\beta \exp\left(-\frac{x}{\beta}\right)\right]_0^\infty = \beta.
\end{aligned}
$$

The variance can be calculated similarly and is

$$
\text{var}[x] = \beta^2.
$$

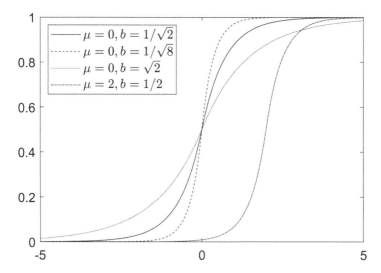

Figure 2.19: The Laplace cumulative distribution function for various values of μ and b, chosen such that the mean and variance are the same as in Figure 2.2.

If the positive part of exponential probability density function is mirrored across the vertical axis at zero and shifted and scaled such that it integrates to one, it becomes the Laplace probability density function as seen in Figure 2.18. The probability density function of the *Laplace distribution* $x \sim$ Laplace(μ, b) is given by

$$\text{Laplace}(x|\mu, b) = \frac{1}{2b} \exp\left(-\frac{|x - \mu|}{b}\right),$$

where μ is known as the *location parameter* and b is a *scale parameter* called the *diversity*.

Its expectation is $\mathbb{E}[x] = \mu$ and the variance is $\text{var}[x] = 2b^2$. Examples of the cumulative distribution function are shown in Figure 2.19. They can be compared to Figures 2.2 and 2.4 illustrating the normal distribution, since the mean is the same and b is chosen such that the variance is the same as well.

Next, we consider a probability density known as the *beta distribution*, $x \sim \text{Beta}(\alpha, \beta)$. It is only defined on the interval $(0, 1)$ and its probability density function is

$$\text{Beta}(x|\alpha, \beta) = \frac{\Gamma(\alpha + \beta)}{\Gamma(\alpha)\Gamma(\beta)} x^{\alpha - 1}(1 - x)^{\beta - 1},$$

where α and β are real numbers and positive. The function $\Gamma(z)$ is known as the *gamma function* and defined as

$$\Gamma(z) = \int_0^\infty x^{z-1} \exp(-x)dx.$$

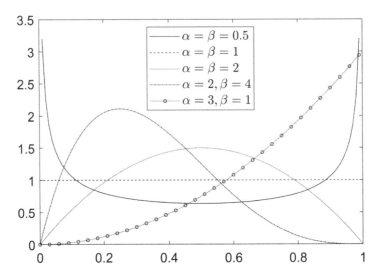

Figure 2.20: The beta probability density function for various values of β and α.

Using *integration by parts*, we can calculate

$$\Gamma(z+1) = \int_0^\infty x^z \exp(-x)dx = \left[- x^z \exp(-x) \right]_0^\infty + \int_0^\infty z\, x^{z-1} \exp(-x)dx.$$

The first term vanishes, since at zero we have a positive power of zero, while when x tends to infinity, $\exp(-x)$ tends to zero quicker than any power of x. Therefore,

$$\Gamma(z+1) = z \int_0^\infty x^{z-1} \exp(-x)dx = z\Gamma(z), \qquad (2.15)$$

and it is regarded as an extension of the factorial to real, positive numbers. Indeed, if $z = n$ is an integer, then using the above relation repeatedly leads to

$$\Gamma(n+1) = n\Gamma(n) = n(n-1)\Gamma(n-1) = n(n-1)\cdots 1\Gamma(1) = n!,$$

since $\Gamma(1) = 1$.

The fraction

$$B(\alpha, \beta) = \frac{\Gamma(\alpha)\Gamma(\beta)}{\Gamma(\alpha + \beta)}$$

is known as the *beta function*, giving the distribution its name. It can be regarded as an extension to the inverse of the binomial coefficient. Indeed, if $\alpha = m$ and $\beta = n$ are integers, then

$$\frac{1}{B(m,n)} = \frac{(m+n)!}{m!n!} = \binom{m+n}{m} = \binom{m+n}{n}.$$

With this in mind, we see that the beta density function has a functional form similar to the probability density function of the binomial distribution in (2.7). The main difference however is that in the binomial distribution the random variable is the number of successes m, while in the beta distribution the random variable is x. It takes the place of the probability of success. Because of this, the beta distribution is often seen as a distribution over probabilities or proportions. It gives the probability of seeing a certain probability x. We will see in Section 2.6 how we can use the functional similarity to our advantage.

Figure 2.20 shows the beta density function for different choices of α and β which encompasses many different shapes. For $\alpha = \beta = 1$, the beta distribution becomes the uniform distribution. When α or β is less than one, the density function will go to infinity at the corresponding edge of the interval.

The expectation is calculated as

$$
\begin{aligned}
\mathbb{E}[x] &= \int_0^1 x \, \text{Beta}(x|\alpha, \beta) dx = \frac{\Gamma(\alpha + \beta)}{\Gamma(\alpha)\Gamma(\beta)} \int_0^1 x^\alpha (1-x)^{\beta-1} dx \\
&= \frac{\Gamma(\alpha+\beta)}{\Gamma(\alpha)\Gamma(\beta)} \frac{\Gamma(\alpha+1)\Gamma(\beta)}{\Gamma(\alpha+1+\beta)} \int_0^1 \text{Beta}(x|\alpha+1, \beta) dx \\
&= \frac{\alpha}{\alpha+\beta},
\end{aligned}
$$

where we used the property given in 2.15 of the gamma function and the fact that the integral of any probability density function over its range is one. The variance is given by

$$
\text{var}[x] = \frac{\alpha\beta}{(\alpha+\beta)^2(\alpha+\beta+1)}.
$$

When $\alpha = \beta$, the expectation is always $1/2$, while the variance is $(4(2\alpha+1))^{-1}$. Therefore, the graphs of the probability density function for these values are symmetric about $1/2$, while the graphs of the cumulative distribution function in Figure 2.21 pass through $(1/2, 1/2)$.

The beta distribution can be viewed as the distribution over the probabilities of experiments with binary outcomes, where x is the probability of success and $1-x$ the probability of failure. If x has the distribution $\text{Beta}(x|\alpha, \beta)$, then $1-x$ has the distribution $\text{Beta}(1-x|\beta, \alpha)$.

Just as we moved from the Bernoulli distribution of binary variables to the categorical distribution for experiments with multiple, mutually distinct outcomes, we extend the beta distribution to a distribution of K probabilities $0 \le x_k \le 1$, $k = 1, \ldots, K$, which have to sum to one,

$$
\sum_{k=1}^{K} x_k = 1.
$$

The probabilities are gathered in the random variable vector $\mathbf{x} = (x_1, \ldots, x_K)^T$. Because of the constraints $\mathbf{x}$ can only lie in a subset of the

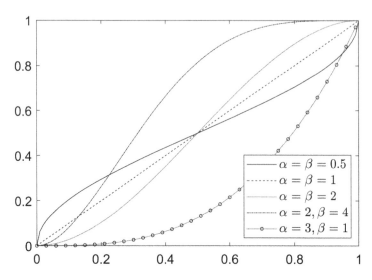

Figure 2.21: The beta cumulative distribution function for various values of β and α.

space of vectors of length K. For $K = 3$, this is illustrated in Figure 2.22. **x** on lies on the face of the shown triangle. The generalization to higher dimensions is known as a *simplex*.

The distribution is known as the *Dirichlet distribution* for $\mathbf{x} \sim \mathrm{Dir}(\boldsymbol{\alpha})$ and the probability density function is given by

$$\mathrm{Dir}(\mathbf{x}|\boldsymbol{\alpha}) = \frac{\Gamma(\alpha_1 + \cdots + \alpha_K)}{\Gamma(\alpha_1) \cdots \Gamma(\alpha_K)} \prod_{k=1}^{K} x_k^{\alpha_k - 1}. \tag{2.16}$$

The parameters $\boldsymbol{\alpha} = (\alpha_1, \ldots, \alpha_K)^T$ are known as *concentration parameters*. They are all positive. When $K = 2$, the Dirichlet distribution is the beta distribution.

Figure 2.23 shows the three dimensional Dirichlet probability density function for various values of $\boldsymbol{\alpha}$ on the simplex. The values the density function takes are indicated by the colour.

For $\alpha_1 = \cdots = \alpha_K$ the distribution is known as *symmetric Dirichlet distribution* or *flat Dirichlet distribution*. The latter naming is misleading, since it is only flat when all concentration parameters take the value 1. The Dirichlet distribution then becomes the uniform distribution over the simplex. In this case,

$$\mathrm{Dir}(\mathbf{x}|\boldsymbol{\alpha}) = \frac{\Gamma(K)}{\Gamma(1) \cdots \Gamma(1)} \prod_{k=1}^{K} x_k^0 = (K-1)!,$$

which is the inverse of the volume of the simplex. For $K = 3$, the triangle has area $1/2$.

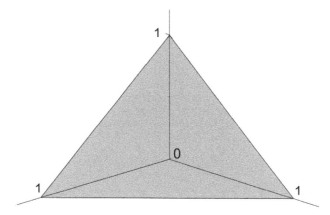

Figure 2.22: Simplex of the possible location of **x** due to the constraints on its elements.

We see in Figure 2.23 that for $\boldsymbol{\alpha} = (5, 5, 5)^T$, the probability mass is concentrated in the centre, while for $\boldsymbol{\alpha} = (5, 5, 2)^T$, it is concentrated towards the edge opposite of the corner with the smallest concentration parameter. For $\boldsymbol{\alpha} = (5, 2, 2)^T$, the probability mass is concentrated at the corner with the largest concentration parameter.

When one of the concentration parameters α_k is less than one, then the probability density function will tend to infinity at the corresponding edge $x_k = 0$. In Figure 2.23, this is visible for $\boldsymbol{\alpha} = (0.8, 2, 2)^T$, but hardly visible for $\boldsymbol{\alpha} = (0.8, 0.8, 2)^T$.

Listing 2.1 provides an implementation calculating the probability density function of the Dirichlet distribution.

We calculate the expectation of the components of **x** individually. Let

$$\alpha_0 = \sum_{k=1}^{K} \alpha_k.$$

Note that the integral is over the simplex in which **x** lies.

$$
\begin{aligned}
\mathbb{E}[x_j] &= \int x_j \, \mathrm{Dir}(\mathbf{x}|\boldsymbol{\alpha})d\mathbf{x} = \frac{\Gamma(\alpha_0)}{\Gamma(\alpha_1)\cdots\Gamma(\alpha_K)} \int x_j^{\alpha_j} \prod_{\substack{k=1 \\ k\neq i}}^{K} x_k^{\alpha_k-1} d\mathbf{x} \\
&= \frac{\Gamma(\alpha_0)}{\Gamma(\alpha_1)\cdots\Gamma(\alpha_K)} \frac{\Gamma(\alpha_1)\cdots\Gamma(\alpha_{j-1})\Gamma(\alpha_j+1)\Gamma(\alpha_{j+1})\cdots\Gamma(\alpha_K)}{\Gamma(\alpha_0+1)} \\
&\quad \int \mathrm{Dir}(\mathbf{x}|\alpha_1,\ldots,\alpha_{j-1},\alpha_j+1,\alpha_{j+1},\ldots,\alpha_K)dx \\
&= \frac{\alpha_j}{\alpha_0},
\end{aligned}
$$

where we again used the property given in Equation 2.15 of the gamma function and the fact that the integral of any probability density function over its

```
function y = dirpdf(x,a)
%   Dirichlet probability density function.
%   y = dirpdf(x,a) returns the Dirichlet probability density
%   function with concentration parameters in the row vector a
%   at each row of x. x must have the same number of columns
%   as a or one less. y is a column vector with the same number
%   of elements as rows in x.
%

if nargin ≠ 2
    error('Exactly two input arguments are required.');
end

[ra,ca] = size(a);
[rx,cx] = size(x);

if ra ≠1
    error('Second argument needs to be a row vector.');
end

if any(a<0)
    error('Second argument needs to be positive.');
end

if ca == cx+1
    % create last column of x
    x(:,cx+1) = 1 − sum(x,2);
    cx = cx+1;
end

if ca≠cx
    error('Number of columns mismatch');
end

% find rows of x which do not sum to 1 or have elements out of bounds
xOut = sum(x,2)>1 | any(x<0|x>1,2);
xIn = ¬xOut;

% Initialize y to zero.
y = zeros(rx,1);

% Since the gamma function increases quickly it is safer to use its
% logarithm and the implementation avoiding underflow or overflow.
y(xIn) = prod(x(xIn,:).^(a(1,:)−1),2).*exp(gammaln(sum(a(1,:),2))...
    −sum(gammaln(a(1,:)),2));
y(xOut) = NaN;
```

Listing 2.1: Dirichlet probability density function.

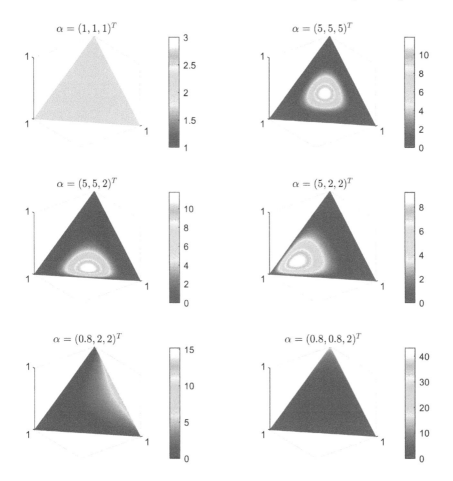

Figure 2.23: Dirichlet probability density function for various values of $\boldsymbol{\alpha}$.

range is one. The variance is given by

$$\text{var}[x_j] = \frac{\alpha_j(\alpha_0 - \alpha_j)}{\alpha_0^2(\alpha_0 + 1)},$$

while the covariance for $j \neq k$ is

$$\text{cov}[x_j, x_k] = \frac{-\alpha_j\alpha_k}{\alpha_0^2(\alpha_0 + 1)}.$$

As we have seen before with the categorical distribution, the influence is negative, since, if one component increases, another one has to decrease due to the constraints on $\mathbf{x}$.

Looking again at Figure 2.23, we see that the relative size of the concentration parameters α_k to their sum α_0 determines where most of the probability mass is concentrated within the simplex. Therefore, in some literature, the parameters of the Dirichlet distribution are given as $\alpha\boldsymbol{\mu}$, where

$$
\begin{aligned}
\alpha &= \alpha_0 = \sum_{k=1}^{K} \alpha_k, \\
\boldsymbol{\mu} &= \left(\frac{\alpha_1}{\alpha_0}, \ldots, \frac{\alpha_K}{\alpha_0} \right)^T.
\end{aligned}
\tag{2.17}
$$

In this case, α is known as the (single) *concentration parameter*, while $\boldsymbol{\mu}$ is the *base distribution*. It is also the expectation of the distribution and satisfies the constraint

$$
\sum_{k=1}^{K} \mu_k = 1.
$$

With this notation,

$$
\begin{aligned}
\mathrm{var}[x_j] &= \frac{1}{\alpha+1} \mu_j (1 - \mu_j), \\
\mathrm{cov}[x_j, x_k] &= \frac{1}{\alpha+1} \mu_j \mu_k.
\end{aligned}
$$

We have already seen that, if $\alpha = 1$, the Dirichlet distribution is the uniform distribution. This means all distributions $\mathbf{x}$ are equally likely. If α tends to infinity, then only distributions which are themselves nearly uniform distributions become likely. The Dirichlet probability density function becomes infinitely peaked around the uniform distribution given by $\mathbf{x} = (1/K, \ldots, 1/K)^T$. In Figure 2.23, $\alpha_0 = \alpha$ is largest for $\boldsymbol{\alpha} = (5, 5, 5)^T$ and the probability mass is concentrated around $(1/3, 1/3, 1/3)$, which describes the uniform distribution for three possible outcomes.

If α tends to zero, let μ_k be the component which is largest, i.e. $\mu_k > \mu_j$ for all $j \neq k$. Then the Dirichlet distribution tends to be peaked at this corner of the simplex. If there are several components with the same largest value, it tends to peaks at these corners, while it is zero everywhere else.

Both the beta distribution and the Dirichlet distribution can be illustrated by the *Pólya urn* model. Imagine an urn which contains balls of K different colours. For the beta distribution, these are two colours. Initially the urn contains α_1 balls of colour 1, α_2 balls of colour 2, and so on. As given in Equation 2.17, the total number of balls is the concentration α and the initial proportions are given by the base distribution $\boldsymbol{\mu}$.

Now, each time a ball is drawn from the urn and replaced, but also another ball of the same colour is added, until the urn contains N balls. Since the colour drawn is random, the final number of balls of each colour is a random variable. However, it is influenced by the initial proportion of balls. If initially there are only balls of one colour, then the outcome is determined, because then the urn will continue to contain only balls of one colour. If there are disproportionally more balls of one colour, we would expect there also to be more balls of that

```
function y = polya(alpha,N)
%    Polya urn simulation function
%    alpha is a vector contuining the initial number of balls of
%    each colour. The number of colours is the length of alpha.
%    After N −sum(alpha) draws, y contains the final number of
%    balls of each colour in the urn.
%
if nargin ≠ 2
    error('Exactly two input arguments are required.');
end

[ra,ca] = size(alpha);
if ra >1 && ca>1
    error('First argument needs to be a vector.');
end

y =alpha;

for n=1:N−sum(alpha)
    % calculate proportions
    p = y/sum(y);
    % randomly pick a ball
    idx = randsample(length(p),1,true,p);
    y(idx) = y(idx)+1;
end
```

Listing 2.2: Polya urn simulation.

colour in the final set, since it is more likely that this colour is drawn and then it becomes even more likely, since a ball of that colour is added. This is known as *rich-get-richer*. However, this is the expectation. Other proportions are also possible, just less likely. In fact, as N approaches infinity, the possibilities for the vector $\mathbf{x}$ of proportions of balls follow a Dirichlet distribution $\text{Dir}(\mathbf{x}|\boldsymbol{\alpha})$. This might seem counter-intuitive. Imagine there is initially only one black and one white ball, $K = 2$ and $\alpha_1 = \alpha_2 = 1$. Then it is equally likely that the urn contains exactly the same number of black and white balls in the end, as containing only white balls apart from the initial black ball.

The function in Listing 2.2 simulates the Pólya urn experiment. It was called 10000 times to generate the histograms in Figure 2.24, where the final number of balls in the urn was $N = 100$. The horizontal line is the number of balls of colour 1, while each bar of the histogram shows how often this occurred in the 10000 experiments. If the bars are scaled by $100/10000$, since 10000 experiments are distributed over 100 bins, the histograms agree well with the curves in Figure 2.20.

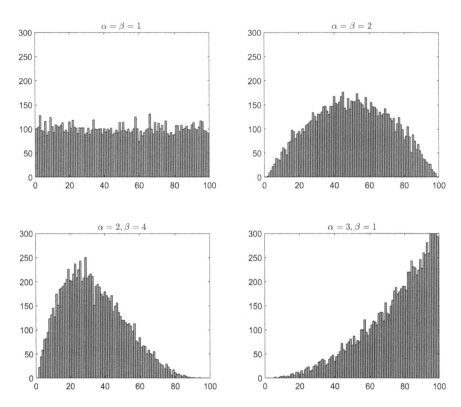

Figure 2.24: Histograms of Polya urn experiments for various values of α and β.

2.5 Functions of Continuous Random Variables

If x is a continuous random variable with probability density function $f_x(x)$, then, for a continuous function g, $y = g(x)$ is also a continuous random variable, but what is its probability density function $f_y(y)$? We include subscripts to be clear which random variable the probability density function and the cumulative distribution function refer to.

We first restrict our attention to the case, when g is a *one-to-one* transformation. That is each value of x generates exactly one value of y, and each value of y is generated by exactly one value of x. Because of this, an inverse function g^{-1} exists such that $x = g^{-1}(y)$. Since g is continuous, it has to be either strictly decreasing or increasing.

Taking the strictly increasing case first, we can derive the probability density function by using the cumulative distribution functions, F_x and F_y. Given a value y, then

$$F_y(y) = p(g(x) \le y).$$

Now, since g is strictly increasing, if $g(x) \leq y$, then $x \leq g^{-1}(y)$ at the same time. Hence,

$$F_y(y) = p(x \leq f^{-1}(y)) = F_x(f^{-1}(y)) = \int_{-\infty}^{g^{-1}(y)} f_x(x)dx.$$

Using integration by substitution, we get

$$F_y(y) = \int_{-\infty}^{y} f_x(g^{-1}(t))(g^{-1})'(t)dt.$$

To find the probability density function, we differentiate with respect to y. By the Fundamental Theorem of Calculus

$$f_y(y) = f_x(g^{-1}(y))(g^{-1})'(y).$$

The derivation for the strictly decreasing case starts off in the same way. However, $g(x)$ is less than or equal to y, if and only if $x \geq g^{-1}(y)$, due to g being decreasing. Thus,

$$F_y(y) = p(x \geq f^{-1}(y)) = 1 - F_x(g^{-1}(y)) = 1 - \int_{-\infty}^{g^{-1}(y)} f_x(x)dx.$$

Again, integrating by substitution and differentiating with respect to y gives

$$f_y(y) = -f_x(g^{-1}(y))(g^{-1})'(y).$$

Since g is decreasing, so is g^{-1} and thus its derivative is negative. This makes the expression overall positive.

Since in the strictly increasing case, the derivative of $g^{-1}(y)$ is always positive, both cases can be summarized as

$$f_y(y) = f_x(g^{-1}(y)) \left| (g^{-1})'(y) \right| = f_x(g^{-1}(y)) \left| \frac{dx}{dy} \right|, \qquad (2.18)$$

where the last expression uses a common short-hand notation. This is known as *change-of-variable technique*, since in its derivation we change the variable in the integral. Another name is *method of direct transformation*.

We illustrate the case, when g is not a one-to-one transformation, by an example first. Let x follow the standard normal distribution $\mathcal{N}(0,1)$ and let $y = x^2$. While $-\infty < x < \infty$, we have $0 \leq y < \infty$. Each value of y is generated by two values $x_1 = -\sqrt{y}$ and $x_2 = \sqrt{y}$. Taking the same approach as before

$$F_y(y) = p(x^2 \leq y) = p(-\sqrt{y} \leq x \leq \sqrt{y}) = F_x(\sqrt{y}) - F_x(-\sqrt{y})$$

$$= \int_{-\infty}^{\sqrt{y}} f_x(x)dx - \int_{-\infty}^{-\sqrt{y}} f_x(x)dx$$

$$= \int_{-\infty}^{y} f_x(\sqrt{t}) \frac{1}{2\sqrt{t}} dt - \int_{-\infty}^{y} f_x(-\sqrt{t}) \frac{-1}{2\sqrt{t}} dt,$$

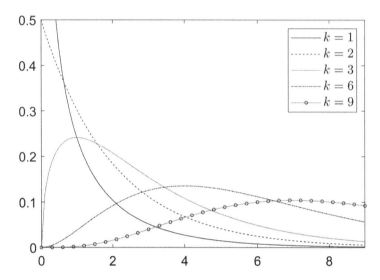

Figure 2.25: The χ^2 probability density function for various degrees of freedom k.

where the last line is the integration by substitution, but different substitutions are used for each integral. After differentiation with respect to y, we have:

$$f_y(y) = \frac{1}{2\sqrt{y}} \left[f_x(\sqrt{y}) + f_x(-\sqrt{y}) \right].$$

Using $f_x(x) = \exp(-x^2/2)/\sqrt{2\pi} = f_x(-x)$, we get

$$f_y(y) = \frac{1}{\sqrt{2\pi y}} \exp(-y/2).$$

This is the *chi-square(d) distribution*, also denoted as χ^2-*distribution*, with one degree of freedom. More generally, the probability density function of the χ^2-distribution with k degrees of freedom is given by

$$\chi_k^2(x) = \frac{1}{2^{k/2}\Gamma(k/2)} x^{k/2-1} \exp(-x/2).$$

The mean lies at k and the variance is $2k$. The previous result above for $k = 1$ follows from $\Gamma(1/2) = \sqrt{\pi}$. Figures 2.25 and 2.26 give examples of the probability density function and the cumulative distribution function for various degrees of freedom.

If g is a *many-to-one* mapping, that is one value of y might be generated by several values of x, the general approach to find $f_y(y)$ is:

- find all intervals $(a_1(y), b_1(y)), \ldots, (a_I(y), b_I(y))$ of x which cause $g(x)$ to be less than or equal to y (note that a_i could be $-\infty$ and b_i could be $+\infty$),

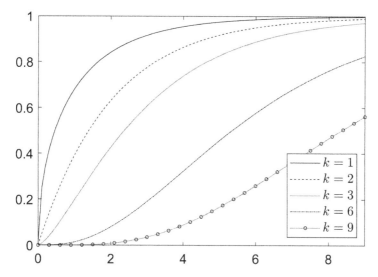

Figure 2.26: The χ^2 cumulative distribution function for various degrees of freedom k.

- calculate $F_y(y)$ as a sum over these intervals,

$$F_y(y) = \sum_{i=1}^{I} \left[\int_{-\infty}^{b_i(y)} f_x(x)dx - \int_{-\infty}^{a_i(y)} f_x(x)dx \right],$$

- use integration by substitution,

- differentiate $F_y(y)$ with respect to y.

In the above example, there was only one interval with $a_1(y) = -\sqrt{y}$ and $b_1(y) = \sqrt{y}$.

This can be extended to y being a function of multiple random variables $x_1, \ldots, x_K$ which are summarized as $\mathbf{x} = (x_1, \ldots, x_K)^T$ and follow a joint probability distribution $f_{\mathbf{x}}(\mathbf{x})$. The first step is to find the region $\mathcal{R}(y)$ where $\mathbf{x} \in \mathcal{R}(y)$ results in $g(\mathbf{x}) \leq y$. Then calculate

$$F_y(y) = \int_{\mathcal{R}(y)} f_{\mathbf{x}}(\mathbf{x})d\mathbf{x},$$

and differentiate with respect to y. However, this task might not be straight forward, since no closed solution for the integral involved, which is in several dimensions, might exist. We illustrate with some examples.

Let $y = g(x_1, x_2) = x_1 + x_2$, where x_1 and x_2 are independent variables. Because of independence, things simplify as $f_{\mathbf{x}}(\mathbf{x}) = f_{x_1}(x_1)f_{x_2}(x_2)$. The condition $g(x_1, x_2) = x_1 + x_2 \leq y$ can written as $x_1 \leq y$ and $x_2 \leq y - x_1$, where

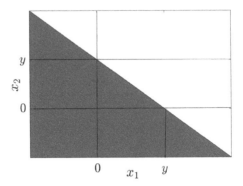

Figure 2.27: The shaded area is where $x_1 + x_2 \leq y$ for the given y.

all inequalities have to hold at the same time. While the interval of x_1 only depends on y, the interval of x_2 depends on both y and x_1. The shaded area in Figure 2.27 shows where x_1 and x_2 lie to have $x_1 + x_2 \leq y$.

We can now calculate

$$F_y(y) = \int_0^y f_{x_1}(x_1) \int_0^{y-x_1} f_{x_2}(x_2) dx_2 dx_1,$$

where everything independent of x_2 has been taken out of the inner integral.

Firstly, let x_1 and x_2 lie between 0 and ∞ and both follow the *exponential distribution*, $x_1, x_2 \sim \text{Exp}(\beta)$ with probability density function as defined in 2.14. For simplicity we let $\beta = 1$. Then

$$
\begin{aligned}
F_y(y) &= \int_0^y \exp(-x_1) \int_0^{y-x_1} \exp(-x_2) dx_2 dx_1 \\
&= \int_0^y \exp(-x_1) \left[-\exp(-x_2) \right]_0^{y-x_1} dx_1 \\
&= \int_0^y \exp(-x_1) \left[1 - \exp(-y + x_1) \right] dx_1 \\
&= \int_0^y \exp(-x_1) - \exp(-y) dx_1 \\
&= \left[-\exp(-x_1) - x_1 \exp(-y) \right]_0^y \\
&= \left[1 - \exp(-y) - y \exp(-y) \right].
\end{aligned}
$$

Differentiating with respect to y gives

$$f_y(y) = \exp(-y) - (\exp(-y) - y \exp(-y)) = y \exp(-y).$$

This probability density function describes the *gamma distribution* for parameters $\alpha = 2$ and $\beta = 1$, $y \sim \text{Gamma}(\alpha, \beta)$. More generally, the probability

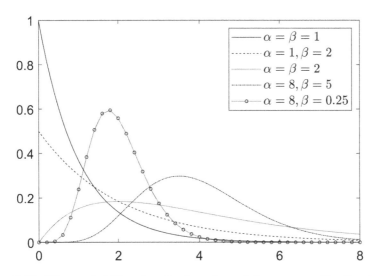

Figure 2.28: The probability density function of the gamma distribution for various choices of α and β. For $\alpha = 1$, it is also the exponential probability distribution.

density function of the gamma distribution is defined as

$$\text{Gamma}(x|\alpha, \beta) = \begin{cases} \dfrac{1}{\Gamma(\alpha)\beta^\alpha} x^{\alpha-1} \exp\left(-\dfrac{x}{\beta}\right) & \text{for} \quad x \geq 0, \\ 0 & \text{for} \quad x < 0. \end{cases}$$

The parameter α is known as the *shape parameter*, while β and $1/\beta$ are the *scale parameter* and *rate parameter* as for the exponential distribution. The mean is given by $\alpha\beta$ while the variance is $\alpha\beta^2$. The exponential distribution is a special case of the gamma distribution where $\alpha = 1$. The χ^2-distribution with k degrees of freedom is also a special case with $\alpha = k/2$ and $\beta = 2$.

Secondly, let x_1 and x_2 lie in all of $\mathbb{R}$ and both follow the standard normal distribution. Then

$$F_y(y) = \frac{1}{2\pi} \int_{-\infty}^{y} \exp(-x_1^2/2) \int_{-\infty}^{y-x_1} \exp(-x_2^2/2) dx_2 dx_1. \tag{2.19}$$

The inner integral is the cumulative distribution function $F_{x_2}(y - x_1)$, while the outer integral is the expectation of the function $F_{x_2}(y - x_1)$ with respect to the probability density function $f_{x_1}(x_1)$. There is no straight-forward way to calculate this. Approximations need to be used. We will look at some general techniques in Chapter 3 on sampling.

At the start of this section, we used the inverse of the function g. However, there are infinitely many possibilities to get y by adding two real numbers. Not even with splitting into regions will it be possible to generate an inverse.

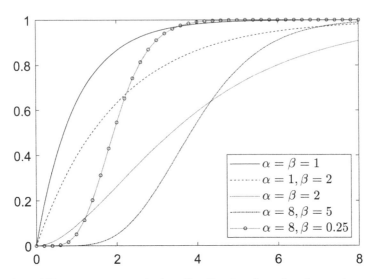

Figure 2.29: The gamma cumulative distribution function for various choices of α and β. For $\alpha = 1$, it is also the exponential probability distribution.

All is not lost however, if we introduce a further random variable. Let $y_1 = g_1(x_1, x_2) = x_1 + x_2$ as before, and let $y_2 = g_2(x_1, x_2) = x_1 - x_2$. In vector form this is

$$\mathbf{y} = (y_1, y_2)^T = \mathbf{g}(\mathbf{x}) = \mathbf{g}(x_1, x_2) = (g_1(x_1, x_2), g_2(x_1, x_2))^T .$$

We can then express x_1 and x_2 in terms of y_1 and y_2,

$$
\begin{aligned}
x_1 &= \frac{y_1 + y_2}{2} \\
x_2 &= \frac{y_1 - y_2}{2}.
\end{aligned}
\tag{2.20}
$$

We might not have an inverse of the individual components of $\mathbf{g}$, but we have an inverse of $\mathbf{g}$, which we denote by $\mathbf{g}^{-1}$.

Suppose we want to calculate the probability that $\mathbf{y}$ falls into a region defined by $\mathcal{Y}$. The set of $\mathbf{x}$, which are mapped into that region by $\mathbf{g}$, is given by $\mathbf{g}^{-1}(\mathcal{Y})$. Thus, the probability is given by

$$\int_{\mathbf{g}^{-1}(\mathcal{Y})} f_{\mathbf{x}}(\mathbf{x})d\mathbf{x}.$$

Integration by substitution for multiple variables yields

$$\int_{\mathcal{Y}} f_{\mathbf{x}}(\mathbf{g}^{-1}(\mathbf{y}))|J_{\mathbf{g}^{-1}}(\mathbf{y})|d\mathbf{y},$$

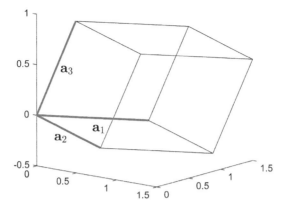

Figure 2.30: Parallelepiped spanned by the vectors $\mathbf{a}_1, \mathbf{a}_2$ and $\mathbf{a}_3$, which are the columns of a 3×3 matrix $\mathbf{A}$.

where $|\cdot|$ in this case denotes the modulus of the determinant of the *Jacobian matrix* $J_{\mathbf{g}^{-1}}(\mathbf{y})$ of $\mathbf{g}^{-1}$ evaluated at $\mathbf{y}$. In general, if $\mathbf{f}(\mathbf{x})$ is a function from $\mathbb{R}^b$ to $\mathbb{R}^a$, then the Jacobian matrix $J_{\mathbf{f}}(\mathbf{x})$ is the $a \times b$ matrix containing the partial derivatives of all components,

$$J_{\mathbf{f}}(\mathbf{x}) = \begin{pmatrix} \frac{\partial f_1}{\partial x_1}(\mathbf{x}) & \cdots & \frac{\partial f_1}{\partial x_b}(\mathbf{x}) \\ \vdots & \ddots & \vdots \\ \frac{\partial f_a}{\partial x_1}(\mathbf{x}) & \cdots & \frac{\partial f_a}{\partial x_b}(\mathbf{x}) \end{pmatrix}.$$

If $a = b$, the determinant of this matrix is called the *Jacobian determinant*. Confusingly, both, matrix and determinant, are often just referred to as the *Jacobian*. The matrix can be calculated for any a and b, while the determinant is only defined for $a = b$.

The modulus of the Jacobian determinant gives the factor by which volumes under the transformation by $\mathbf{f}$ shrink or expand. To visualize, the image of the unit cube in three dimensions under a 3×3 matrix $\mathbf{A}$ is the parallelepiped spanned by the columns of $\mathbf{A}$. For example let

$$\mathbf{a}_1 = \begin{pmatrix} 1 \\ 1/2 \\ 0 \end{pmatrix}, \mathbf{a}_2 = \begin{pmatrix} 0 \\ 1 \\ -1/2 \end{pmatrix}, \mathbf{a}_3 = \begin{pmatrix} 1/2 \\ 0 \\ 1 \end{pmatrix}$$

be the columns of $\mathbf{A}$ as seen in Figure 2.30. The volume of this parallelepiped is the modulus of the determinant of $\mathbf{A}$, which is also the *scalar triple product* $(\mathbf{a}_1 \times \mathbf{a}_2)^T \mathbf{a}_3 = 7/8$.

We can now deduce the multivariate *change-of-variable technique*. If $\mathbf{g} : \mathbb{R}^K \to \mathbb{R}^K$ is an invertible and differentiable function, and $\mathbf{x} \in \mathbb{R}^K$ is a random variable with probability density function $f_{\mathbf{x}}(\mathbf{x})$, then $\mathbf{y} = \mathbf{g}(\mathbf{x}) \in \mathbb{R}^K$ is also a random variable and its probability density function is

$$f_{\mathbf{y}}(\mathbf{y}) = f_{\mathbf{x}}(\mathbf{g}^{-1}(\mathbf{y}))|J_{\mathbf{g}^{-1}}(\mathbf{y})|.$$

Returning to our example, the inverse $\mathbf{g}^{-1}$ is given by Equation (2.20). The matrix of partial derivatives is

$$\begin{pmatrix} \frac{1}{2} & \frac{1}{2} \\ \frac{1}{2} & -\frac{1}{2} \end{pmatrix}$$

and the modulus of its determinant is $1/2$. Both x_1 and x_2 were assumed to be independent and follow the standard normal distribution. Hence,

$$
\begin{aligned}
f_{\mathbf{y}}(\mathbf{y}) &= \frac{1}{\sqrt{2\pi}} \exp\left(-\left(\frac{y_1+y_2}{2}\right)^2/2\right) \frac{1}{\sqrt{2\pi}} \exp\left(-\left(\frac{y_1-y_2}{2}\right)^2/2\right) \frac{1}{2} \\
&= \frac{1}{4\pi} \exp\left(\frac{y_1^2+y_2^2}{4}\right) \\
&= \frac{1}{\sqrt{2}\sqrt{2\pi}} \exp\left(\left(\frac{y_1}{\sqrt{2}}\right)^2/2\right) \frac{1}{\sqrt{2}\sqrt{2\pi}} \exp\left(\left(\frac{y_2}{\sqrt{2}}\right)^2/2\right).
\end{aligned}
$$

Therefore, the random variables y_1 and y_2 are independent. Both are normally distributed with mean 0 and standard deviation $\sqrt{2}$. Or in other words the variance is 2, which is the sum of the variances of x_1 and x_2.

More generally, if y is a linear combination of independent normally distributed variables, i.e. $y = c_1 x_1 + \cdots c_K x_K$, with $x_k \sim \mathcal{N}(\mu_k, \sigma_k^2)$, then y is normally distributed with mean

$$\mu = c_1 \mu_1 + \cdots + c_K \mu_K$$

and variance

$$\sigma^2 = c_1^2 \sigma_1^2 + \cdots + c_K^2 \sigma_K^2.$$

So far we have looked at different probability distributions and how to derive the probability density function for functions of random variables. In the next section we look at how to link this to data.

2.6 Conjugate Probability Distributions

The previous sections were concerned with probabilities and how to describe them. However, we are interested in describing data with probabilities. To this end, let us return to the experiment of flipping a coin, or more generally an experiment with two possible outcomes following a Bernoulli distribution $\text{Bern}(x|\mu)$, where μ is unknown. Let $x_1, \ldots, x_N$ be N observations of the experiment. The probability to observe this data $\mathcal{D} = \{x_1, \ldots, x_N\}$ is

$$p(\mathcal{D}|\mu) = \prod_{n=1}^{N} \text{Bern}(x_n|\mu) = \prod_{n=1}^{N} \mu^{x_n}(1-\mu)^{1-x_n} = \mu^{\sum_{n=1}^{N} x_n}(1-\mu)^{\sum_{n=1}^{N}(1-x_n)}.$$

This is known as the *likelihood* of the data $\mathcal{D}$. Now, x_n only takes the values 0 and 1. If m is the number of times we have $x_n = 1$ in the N trials, then

$$p(\mathcal{D}|\mu) = \mu^m (1-\mu)^{N-m}.$$

The value of μ can be estimated by maximizing the likelihood with respect to μ, since we want to choose the μ which explains the data best.

The maximization is equivalent to maximizing the logarithm, which is valid, since the logarithm is a monotonically increasing function. This is the *log likelihood*:

$$\log p(\mathcal{D}|\mu) = m \log \mu + (N - m) \log(1 - \mu).$$

The location of the maximum is where the derivative with respect to μ vanishes. The derivative is

$$\frac{d}{d\mu} p(\mathcal{D}|\mu) = \frac{m}{\mu} - \frac{N - m}{1 - \mu}.$$

Setting to zero, multiplying through by $\mu(1 - \mu)$ and solving for μ gives the *maximum likelihood estimator*

$$\mu_{\mathrm{ML}} = \frac{m}{N}.$$

Returning to flipping the coin, if we happen to get tails ($x_n = 0$) three times in a row in three flips, then the maximum likelihood estimator is $\mu_{\mathrm{ML}} = 0$. The interpretation is that in all future flips we would continue to see tails. Equally, if we happen to flip heads ($x_n = 1$) three times in a row, the maximum likelihood estimator is $\mu_{\mathrm{ML}} = 1$. Both cases are realistic, but the conclusion is not and caused by the small number of trials. In general, to get a good maximum likelihood estimator for any parameter, many observations are necessary.

However, in real world applications such abundance of data might be a luxury. If we describe what we "believe" about μ, mathematically and then use Bayes' rule to modify our "belief", less data is necessary to estimate the parameter μ. It is the probability of success, and therefore $\mu \in [0, 1]$. Thus, the beta distribution $\mathrm{Beta}(\mu|\alpha, \beta)$ lends itself to describe μ, where α and β are known as *hyperparameters* and set to some reasonable initial values. This is the *prior* distribution. In the following, we will see how these parameters get updated with data becoming available.

The joint probability density function of μ and m is the product of the likelihood and the prior

$$
\begin{aligned}
f(\mu, m|N, \alpha, \beta) &= p(\mathcal{D}|\mu)\mathrm{Beta}(\mu|\alpha, \beta) \\
&= \mu^m (1 - \mu)^{N-m} \frac{\Gamma(\alpha + \beta)}{\Gamma(\alpha)\Gamma(\beta)} \mu^{\alpha-1}(1 - \mu)^{\beta-1}.
\end{aligned}
\tag{2.21}
$$

The posterior density function of μ, $f(\mu|m, N, \alpha, \beta)$, is proportional to the joint distribution following Bayes. The constant factors can be dropped when using the proportionality $\propto$, and hence

$$f(\mu|m, N, \alpha, \beta) \propto \mu^{m+\alpha-1}(1 - \mu)^{N-m+\beta-1}.$$

It has the functional form of another, different beta distribution. Inserting the appropriate factor to ensure the integral over the distribution is one gives

$$f(\mu|m, N, \alpha, \beta) = \frac{\Gamma(N + \alpha + \beta)}{\Gamma(m + \alpha)\Gamma(N - m + \beta)}\mu^{m+\alpha-1}(1 - \mu)^{N-m+\beta-1}.$$

Observing m successes and $N - m$ failures changes the shape of the prior distribution to arrive at the posterior distribution. The order of successes and failures is irrelevant.

What is the probability of success after taking the data $\mathcal{D}$ into account, $p(x = 1|\mathcal{D})$? The data is sufficiently specified by N and m. There is no longer a single value for μ, but a posterior distribution over μ. All possible values are considered by taking the integral

$$\begin{aligned} p(x = 1|\mathcal{D}) &= \int_0^1 p(x = 1|\mu)\, f(\mu|m, N, \alpha, \beta)d\mu \\ &= \int_0^1 \mu\, f(\mu|m, N, \alpha, \beta)d\mu = \mathbb{E}(\mu|m, N, \alpha, \beta) \qquad (2.22) \\ &= \frac{m + \alpha}{N + \alpha + \beta}. \end{aligned}$$

Because of the inclusion of α and β, the probability can never become 0, even if $m = 0$, or 1, even if $m = N$, contrary to the maximum likelihood estimator. It incorporates the uncertainty caused by seeing only a finite set of data.

The above result no longer depends on μ. It is an example of a *latent* or *hidden* variable.

Figure 2.31 shows how the beta distribution changes from a prior distribution with $\alpha = \beta = 2$ as more data is observed. If 0 stands for tails when flipping a coin, the second to seventh plot illustrates a sequence of three times tails followed by three times heads. As more tails are observed, the mean of the distribution moves to the left, but with more observations of heads, it moves back right towards the centre. The last plot is after 100 coin flips, where an equal number of heads and tails are observed. The posterior distribution becomes more peaked around $1/2$. As N increases, α and β become relatively small compared to it, and $p(x = 1|\mathcal{D})$ approaches the maximum likelihood estimator μ_{ML} for this $\mathcal{D}$.

As can be seen in Figure 2.31, this approach can be used in real-time learning, where data arrives sequentially, and predictions are updated when more knowledge is gained. It can also be used for large data sets, where it is not possible to load all data into memory at the same time.

The beta distribution is known as the *conjugate* distribution to the Bernoulli and binomial distribution. It was motivated by choosing a suitable prior distribution describing an unknown parameter (μ). There is also a mathematical motivation, since the posterior distribution arising from Bayes' rule in this case has the same functional form as the prior and therefore can be determined in closed form. This is not always the case. When choosing appropriate distributions to describe the data, mathematical convenience should play a subordinate role to the data and expert knowledge.

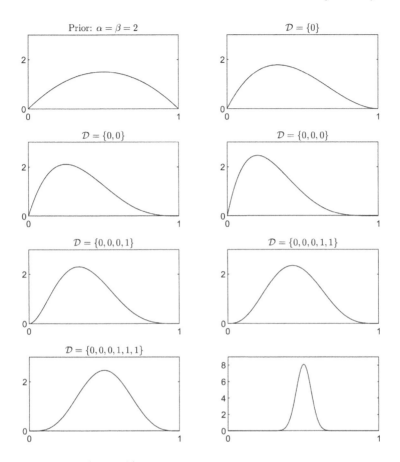

Figure 2.31: Prior (top left) and posterior beta probability distributions after some observations. The bottom right shows the beta distribution after an equal number of 0 and 1.

If the data is categorical in nature, the outcomes are described using a 1-of-K representation and follow the categorical distribution as given in (2.9). Let $\mathcal{D} = \{\mathbf{x}_1, \ldots, \mathbf{x}_N\}$ be the results of N experiments. The likelihood of this data is then

$$p(\mathcal{D}|\boldsymbol{\mu}) = \prod_{n=1}^{N} \mathrm{Cat}(\mathbf{x}_n|\boldsymbol{\mu}) = \prod_{n=1}^{N}\prod_{k=1}^{K} \mu_k^{x_{nk}} = \prod_{k=1}^{K} \mu_k^{\sum_{n=1}^{N} x_{nk}}.$$

Since each $\mathbf{x}_n$ is a 1-of-K representation, let m_k be the number of times we had $x_{nk} = 1$ for $n = 1, \ldots, N$. The likelihood then becomes

$$p(\mathcal{D}|\boldsymbol{\mu}) = \prod_{k=1}^{K} \mu_k^{m_k}.$$

Again an estimator for μ can be found by maximizing the likelihood, or equivalently its logarithm. However, this task is not as straight-forward as before, since μ has to satisfy the constraints

$$\sum_{k=1}^{K} \mu_k = 1 \qquad \text{and} \qquad 0 \le \mu_k \le 1, k = 1, \ldots, K.$$

hence the task is a *constraint optimization*.

Let i be any index in $\{1, \ldots, K\}$. We incorporate the first constraint by noting

$$\mu_i = 1 - \sum_{\substack{k=1 \\ k \ne i}}^{K} \mu_k.$$

With this modification, the log likelihood is

$$\log p(\mathcal{D}|\mu) = \sum_{\substack{k=1 \\ k \ne i}}^{K} m_k \log \mu_k + m_i \log \left(1 - \sum_{\substack{k=1 \\ k \ne i}}^{K} \mu_k \right).$$

Let $j \ne i$ be any other index. The parameter μ_j appears in both sums, and the derivative with respect to μ_j is

$$\frac{d}{d\mu_j} \log p(\mathcal{D}|\mu) = \frac{m_j}{\mu_j} - \frac{m_i}{1 - \sum_{\substack{k=1 \\ k \ne i}}^{K} \mu_k} = \frac{m_j}{\mu_j} - \frac{m_i}{\mu_i}.$$

Setting to zero, we see that the ratio of μ_i to μ_j has to be the same as the ratio of m_i to m_j. Since i and j were freely chosen, the ratio of any pair of components of μ must be the same as the ratio of the corresponding pair in $\mathbf{m} = (m_1, \ldots, m_k)^T$. The only possible choice for μ_k, which also satisfies the second constraint, is the maximum likelihood estimator

$$\mu_k = \frac{m_k}{m_1 + \cdots + m_K} = \frac{m_k}{N}, \qquad k = 1, \ldots, K.$$

As before, this can be a poor estimate, if N is not large enough.

Constraint optimization problems occur often in machine learning. A general technique uses *Lagrange multipliers*, which we will encounter in Section 4.6. For more background knowledge on optimization see [2].

The vector μ of the probabilities of different outcomes needs to be estimated. The Dirichlet distribution $\text{Dir}(\mu|\alpha)$ is chosen as conjugate prior for μ, where α is suitably chosen. The posterior density function is proportional to the product of the likelihood and the prior (which is the joint distribution):

$$f(\mu|\mathbf{m}, \alpha) \propto \prod_{k=1}^{K} \mu_k^{m_k + \alpha_k - 1},$$

where all constant factors were dropped due to the proportional sign. The posterior has again the functional form of a Dirichlet distribution. Incorporating the factor such that the posterior integrates to one leads to

$$f(\boldsymbol{\mu}|\mathbf{m},\boldsymbol{\alpha}) = \frac{\Gamma(\alpha_0 + N)}{\Gamma(\alpha_1 + m_1)\cdots\Gamma(\alpha_K + m_K)}\prod_{k=1}^{K}\mu_k^{m_k+\alpha_k-1} = \mathrm{Dir}(\boldsymbol{\mu}|\mathbf{m}+\boldsymbol{\alpha}),$$

where $\alpha_0 = \alpha_1 + \cdots \alpha_K$.

As before with the beta distribution, the knowledge gained from the experiments is given by $\mathbf{m}$ and modifies the location of the mean and the shape of the distribution. The parameters α and β in the prior beta distribution and $\boldsymbol{\alpha}$ in the prior Dirichlet distribution can be viewed as initial guesses of the true distribution or fictional trial runs informed by expert knowledge and experience.

The probability of a particular outcome $x_k = 1$ is

$$p(x_k = 1|\mathcal{D}) = \int p(x_k = 1|\boldsymbol{\mu})f(\boldsymbol{\mu}|\mathbf{m},\boldsymbol{\alpha})d\boldsymbol{\mu},$$

where the integral is over the simplex, where $\boldsymbol{\mu}$ must lie. Since $p(x_k = 1|\boldsymbol{\mu}) = \mu_k$, this becomes

$$p(x_k = 1|\mathcal{D}) = \mathbb{E}[\mu_k|\mathbf{m},\boldsymbol{\alpha}] = \frac{m_k + \alpha_k}{N + \alpha_0}. \tag{2.23}$$

As N grows, $p(x_k = 1|\mathcal{D})$ approaches the maximum likelihood estimator. Again, the above result no longer depends on $\boldsymbol{\mu}$ and therefore $\boldsymbol{\mu}$ is a latent variable.

We have seen two examples of conjugate probability distributions and will see more. For example, if the mean of a normal distribution is the unknown parameter, the conjugate prior is also a normal distribution. For a more thorough treatment of *conjugacy* see [16].

2.7 Graphical Representations

In the previous section, the joint probability distribution followed by Bayes' rule was used, to achieve better predictions for seeing a particular outcome. In the following, we illustrate how the dependencies between random variables and parameters can be visualized.

The elements of a graph are *nodes*, also known as *vertices*, and *edges*, also called *links* or *arcs*. In a graphical representation, each node represents a random variable, which could be a vector, group of random variables, or parameter or group of parameters. If there is a relationship between nodes, they are joined by an edge. If there is a causal relationship between nodes, the edge has a direction indicated by an arrow. This is then a *directed* graph.

Let $\mathcal{D} = \{x_1, \ldots, x_n\}$ be a set of observed, binary variables. In (2.21), the data $\mathcal{D}$ is governed by the latent variable μ, which in turn is governed by the parameters α and β. The corresponding graph is:

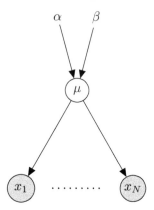

The nodes of random variables are circles, while the nodes of parameters are clear. Random variables which are observed are shaded, while latent ones, which are not observed, are not shaded. The observations $x_1, \ldots, x_N$ are of the same nature. The graph can become very cluttered when writing these out as individual nodes. They are summarized as a representative node x_n surrounded by a box, called a *plate*, labeled with N showing there are N of these.

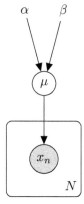

Equation (2.22) gives the probability of $x = 1$ having taken the data $\mathcal{D}$ into account. Graphically, this is depicted as

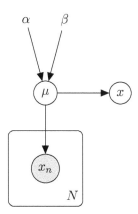

Note that the final equality in (2.22) does not involve μ, since μ is a latent variable, but does involve N and m, which specify $\mathcal{D}$. Graphically, the information from the parameters α and β and the data $\mathcal{D}$ are combined in the node μ to inform node x.

The graph for the categorical case is very similar, except we now have a parameter vector $\boldsymbol{\alpha}$, $\boldsymbol{\mu}$ is vector valued and $\mathbf{x}_1, \ldots, \mathbf{x}_N$ and $\mathbf{x}$ are 1-of-K representations:

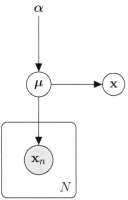

The other main difference is that the underlying prior and posterior distributions are no longer beta distributions, but Dirichlet distributions.

Following the Bayesian paradigm, α, β, and $\boldsymbol{\alpha}$ would not be parameters to be chosen, but would also be treated as random variables with their own prior distributions, called *hyper-priors* as discussed in [16]. This will grow the graphs. Complex inter-dependencies between various random variables and parameters can be visualized graphically. For a good introductory text consult [24].

CHAPTER 3

Sampling

Sampling is necessary, since in machine learning often a closed form describing the probability distribution is not accessible, but the distribution can be sampled. The chapter explains standard sampling techniques such as inverse transform sampling, rejection sampling and importance sampling as well as Markov chains and Markov Chain Monte Carlo in an intuitive way.

In the previous chapter, we have seen how data is used to estimate the parameters of the assumed underlying distribution generating the data $\mathcal{D}$ using Bayes rule. For example, let $\boldsymbol{\theta}_1$ be the vector of all parameters governing the distribution and $\boldsymbol{\theta}_2$ the vector of all hyperparameters governing the prior distributions for the parameters $\boldsymbol{\theta}_1$. Then the posterior is given by

$$f(\boldsymbol{\theta}_1|\mathcal{D}, \boldsymbol{\theta}_2) = \frac{p(\mathcal{D}|\boldsymbol{\theta}_1)f(\boldsymbol{\theta}_1|\boldsymbol{\theta}_2)}{\int p(\mathcal{D}|\boldsymbol{\theta}_1)f(\boldsymbol{\theta}_1|\boldsymbol{\theta}_2)d\boldsymbol{\theta}_1}. \tag{3.1}$$

Unless we are dealing with conjugate probability distributions, no closed solution might exist. It might not be possible to evaluate the integral in the denominator.

Note that, by the product rule, the integrand is the joint probability density function of $\mathcal{D}$ and $\boldsymbol{\theta}_1$ conditioned on $\boldsymbol{\theta}_2$, i.e. $f(\mathcal{D}, \boldsymbol{\theta}_1|\boldsymbol{\theta}_2)$. This means the denominator is the marginal distribution $f(\mathcal{D}|\boldsymbol{\theta}_2)$.

Another way of interpreting the integral is that it is the expectation of the function $p(\mathcal{D}|\boldsymbol{\theta}_1)$ with respect to the distribution described by the probability density function $f(\boldsymbol{\theta}_1|\boldsymbol{\theta}_2)$. We already encountered a similar situation in Equation (2.19), when attempting to derive the probability density function of a random variable which is a function of two other random variables. The outer integral there can be interpreted as the expectation of the function given by the inner integral with respect to the probability density function in the outer integral.

More generally, the situation, where the expectation

$$\mathbb{E}[h] = \int h(x)f(x)dx$$

for some function h of x has to be calculated, arises often, but integrating analytically is too complex. Sampling methods are used to to estimate the expectation. Let $x_1, \ldots, x_N$ be samples independently drawn from the distribution described by the probability density function f. The expectation is approximated by

$$\mathbb{E}[h] \approx \frac{1}{N} \sum_{n=1}^{N} h(x_n).$$

It needs to be ensured that all samples are independent. If some are dependent, the sample size N needs to be increased to counter the effect that some samples might not give any new information.

The relative sizes of $h(x_n)$ to $f(x_n)$ can also skew the result. This happens, for example, if $h(x_n)$ is large in areas where $f(x_n)$ is small and vice versa. Then the expectation is dominated by the few large values of h, even though the probability density is small in this region. Again, it is necessary to have a sufficiently large sample size N to counter act this.

In this chapter, we introduce some general sampling methods. For a more thorough treatment see [10].

3.1   Inverse Transform Sampling

We need to be able to generate samples of the probability density function $f(x)$. We assume that we have an algorithm available that generates uniformly distributed random numbers in the interval $(0, 1)$. The cumulative probability density function is given by

$$F(x) = \int_{-\infty}^{x} f(t)dt.$$

We assume that it is continuous and strictly monotonically increasing on the interval (a, b) and 0 for $x \leq a$ and 1 for $x \geq b$. It takes values in the interval $[0, 1]$. Let y be drawn from the uniform distribution over $(0, 1)$. Then there exists a unique number in (a, b) such that $F(x) = y$, or in other words $x = F^{-1}(y)$. Then x is a continuous random variable with cumulative distribution function $F(x)$ and hence probability density function $f(x)$. This technique is known as *inverse transform sampling*. It is also known as *inverse transformation*, *inversion sampling*, *inverse probability integral transform*, and *Smirnov transform*.

We need to show that for $x = F^{-1}(y)$, we have indeed $p(x \leq \hat{x}) = F(\hat{x})$ for a given value $\hat{x}$. Firstly,

$$p(x \leq \hat{x}) = p(F^{-1}(y) \leq \hat{x})$$

by the definition for x. Next, it follows from F and subsequently F^{-1} being strictly monotonically increasing that $F^{-1}(y) \leq \hat{x}$ if and only if $y \leq F(\hat{x})$. Lastly, since y is from the uniform distribution on $(0, 1)$, the probability of y being less than or equal to $F(\hat{x})$ is in fact $F(\hat{x})$ itself and the assertion follows.

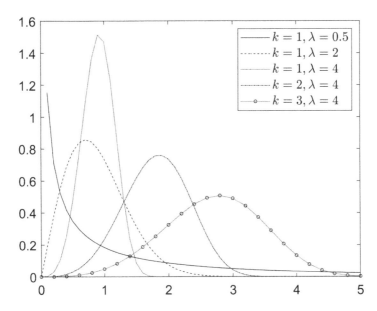

Figure 3.1: The probability density function of the Weibull distribution for various choices of k and λ. For $k = 1$, it is the exponential probability distribution.

As an example, consider the *Weibull distribution*, $x \sim \text{Weibull}(\lambda, k)$. Its probability density function is

$$
\text{Weibull}(x|\lambda, k) = \begin{cases} \dfrac{k}{\lambda} \left(\dfrac{x}{\lambda}\right)^{k-1} \exp\left(-\left(\dfrac{x}{\lambda}\right)^{k}\right) & \text{for} \quad x \geq 0, \\ 0 & \text{for} \quad x < 0, \end{cases}
$$

where $k > 0$ is the *shape parameter* and $\lambda > 0$ is the scale parameter. Its mean is $\lambda\Gamma(1+1/k)$, while the variance is $\lambda^2[\Gamma(1+2/k) - (\Gamma(1+1/k))^2]$. Figures 3.1 and 3.2 give examples of the probability density and cumulative distribution function respectively. For some choices of parameters the shape becomes close to a normal distribution.

For $k = 1$, the Weibull distribution becomes the exponential distribution. The Weibull distribution is used in reliability engineering and failure analysis. There, the parameter k is interpreted in the following way. If $k > 1$, then the failure rate increases with time as parts are more likely to fail as time goes on. If $k = 1$, the failure rate is constant, the system is stable and there is no aging process. If $k < 1$, the failure rate decreases with time. Colloquially, this means that if it has not failed by now, it is less likely to fail in the future. For more information on statistical quality control see [31].

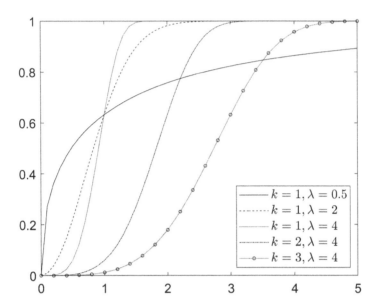

Figure 3.2: The Weibull cumulative distribution function for various choices of k and λ. For $k = 1$, it is the exponential probability distribution.

Since the probability density function is only nonzero for $x \geq 0$, the lower limit of the integral is 0, when calculating the cumulative distribution function

$$
\begin{aligned}
F(x) &= \int_0^x \frac{k}{\lambda} \left(\frac{t}{\lambda} \right)^{k-1} \exp\left(-\left(\frac{t}{\lambda} \right)^k \right) dt \\
&= \left[-\exp\left(-\left(\frac{t}{\lambda} \right)^k \right) \right]_0^x = 1 - \exp\left(-\left(\frac{x}{\lambda} \right)^k \right).
\end{aligned}
$$

Setting this equal to y and solving for x leads to

$$
x = \lambda \left[-\log(1 - y) \right]^{1/k}
$$

which is a random variable drawn from the Weibull distribution, if y is drawn from the uniform distribution over $(0, 1)$. Now, $z = 1 - y$ is equally drawn from the uniform distribution over $(0, 1)$. So the transformation is often stated as

$$
x = \lambda \left[-\log z \right]^{1/k} .
$$

This technique can be used as long as the cumulative distribution function is known and can be inverted. This is not always the case, the normal distribution being a prime example. In the following, we derive the *Box—Muller transform* to generate a pair of random variables following the standard normal distribution from a pair of uniformly distributed random variables.

We combine Equations (2.5) and (2.6) to arrive at

$$
\int_{-\infty}^{\infty} \int_{-\infty}^{\infty} \exp\left(-\frac{x_1^2 + x_2^2}{2} \right) dy\,dz = \int_0^{2\pi} \int_0^{\infty} r \exp\left(\frac{-r^2}{2} \right) dr\,d\theta,
$$

where x_1 and x_2 are two standard normal random variables and $x_1 = r\cos\theta$, $x_2 = r\sin\theta$. From the above we see that θ follows the uniform distribution on the interval $(0, 2\pi)$ which is the circumference of the unit circle, while r has the probability density function $r\exp\left(-r^2/2\right)$ on the interval $(0, \infty)$. To generate r from a uniform random variable, we use the above method. First, we calculate

$$F(r) = \int_0^r t\exp\left(\frac{-t^2}{2}\right)dt = \left[-\exp\left(\frac{-t^2}{2}\right)\right]_0^r = 1 - \exp\left(\frac{-r^2}{2}\right).$$

Let q be a random variable drawn from the uniform distribution on $(0, 1)$. Setting $F(r) = q$ and solving for r, gives $r = \sqrt{-2\log(1-q)}$. Since $1 - q$ is also uniformly distributed on $(0, 1)$, it can be rewritten as

$$r = \sqrt{-2\log q}. \tag{3.2}$$

Summarizing, we generate random variables x_1 and x_2 drawn from the standard normal distribution by drawing two variables y_1 and y_2 from the uniform distribution on $(0, 1)$ and letting

$$x_1 = \sqrt{-2\log y_1}\cos(2\pi y_2),$$

$$x_2 = \sqrt{-2\log y_1}\sin(2\pi y_2).$$

This is the *basic* form of the *Box—Muller transform*.

```
function [y1,y2] = BoxMullerTrig(x1,x2)
% Transforms a bivariate uniform random variable in [0,1] x [0,1] to a
% bivariate normally distributed variable.
if nargin≠2
    error('Two input arguments needed.')
elseif x1<0 || x1>1 || x2<0 || x2 >1
    error('Input arguments must be in [0,1].')
end
y1 = sqrt( - 2 * log(x1) ) * cos( 2 * pi * x2 );
y2 = sqrt( - 2 * log(x1) ) * sin( 2 * pi * x2 );
end
```

Listing 3.1: Basic Box—Muller transform.

To avoid the sine and cosine, we note that $\cos\theta = x_1/r$ and $\sin\theta = x_2/r$. This gives rise to the *polar* form of the *Box—Muller transform*. First, we generate uniformly distributed random numbers $z_1, z_2 \in (-1, 1)$ by letting $z_i = 2y_i - 1$ for variables y_i uniformly distributed in $(0, 1)$, until $s = z_1^2 + z_2^2 < 1$. We then set

$$\cos\theta = \frac{z_1}{\sqrt{s}},$$

$$\sin\theta = \frac{z_2}{\sqrt{s}}.$$

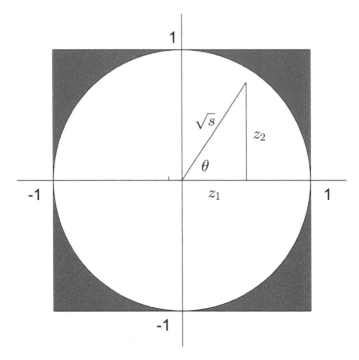

Figure 3.3: The polar Box—Muller method.

The probability $p(s \leq \hat{s})$ is the area of the circle with radius $\sqrt{\hat{s}}$, which is $\pi\hat{s}$, divided by the area of the unit circle which is π. Therefore $p(s \leq \hat{s}) = \hat{s}$, and s follows a uniform distribution on $(0, 1)$. Therefore, s can be used to generate a random variable from the distribution of r. To summarize,

$$x_1 = \sqrt{-2\log s}\frac{z_1}{\sqrt{s}} = z_1\sqrt{\frac{-2\log s}{s}},$$

$$x_2 = \sqrt{-2\log s}\frac{z_2}{\sqrt{s}} = z_2\sqrt{\frac{-2\log s}{s}}$$

follow the required distribution. Another way of viewing this technique is that a random point (z_1, z_2) within the unit circle is mapped to the point $(z_1/\sqrt{s}, z_2/\sqrt{s})$ on the unit circle. This point is then multiplied by a radius following the required distribution in Equation (3.2). Figure 3.3 illustrates this.

The advantage of this method is that the calculation of the sine and cosine is avoided. On the other hand, more random numbers need to be generated, since a fraction of $1 - \pi/4$ are rejected as they lie outside the unit circle. Most software packages include implementations of random number generators for various underlying distributions. Nevertheless, Listings 3.1 and 3.2 are example implementations.

```
function Y = BoxMuller(N)
% Generates N bivariate normally distributed variables avoiding the
% use of the trigonometric functions.
Y = zeros(N,2);
for n = 1:N
    w = 1;
    % Generate random number within the unit circle.
    while w >= 1
        x1 = 2.0 * rand - 1.0;
        x2 = 2.0 * rand - 1.0;
        w = x1 * x1 + x2 * x2;
    end
    w = sqrt( (-2.0 * log( w ) ) / w );
    Y(n,1) = x1 * w;
    Y(n,2) = x2 * w;
end
```

Listing 3.2: Polar Box—Muller transform.

3.2 Rejection Sampling

We now consider the case where the probability density function f is available and can be evaluated, but the cumulative distribution function F is not, and inverse transform sampling is not an option. Let g be a probability density function such that $f(x) \leq cg(x)$ for all x for some finite constant $c > 1$. Let G be the associated cumulative distribution function. We also assume that we have a method to draw samples from the distribution given by G. That is $g(x)$ is a simpler probability density function and $cg(x)$ is an envelope to $f(x)$. Figure 3.4 shows three choices for $cg(x)$ for a probability density function f which has two modes. The first choice is a uniform distribution over the *support* of f. That is the range of x where $f(x)$ is non-zero. For the other two, g is normally distributed. For the second choice the mean of g is at the same location as one of the modes of f, while for the third choice the mean lies between the two modes. We see that there are infinitely many possibilities for g.

Rejection sampling, also known as the *acceptance-rejection* method, proceeds as follows:

1. Draw a random variable x following the distribution given by g;

2. Draw a random variable u from the uniform distribution over $(0, 1)$;

3. If
$$u \leq \frac{f(x)}{cg(x)}, \tag{3.3}$$

   accept x as a sample from the distribution given by f. Otherwise return to 1.

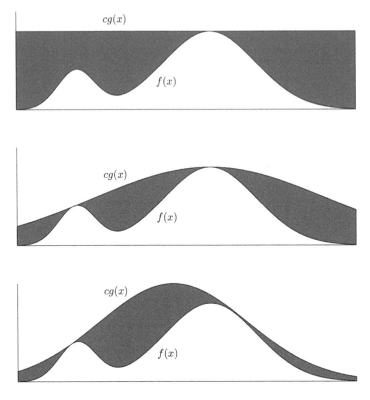

Figure 3.4: A probability density function $f(x)$ with two modes and three choices of enveloping functions $cg(x)$.

Intuitively, the method samples uniformly points from the area under cg and discards those which fall in the shaded area between the curves of cg and f. The x-position of the retained points are samples from the distribution governed by f. To keep the number of discarded points low, the shaded area has to be as small as possible. Or in other words, the ratio of the areas under cg and f needs to be as close to 1 as possible. This ratio is c, since both g and f integrate to 1. If the area under cg is twice the size of the area under f, we expect to discard half the samples.

This process generates samples following the probability distribution defined by g conditioned on inequality (3.3). The cumulative distribution function is given by

$$p\left(x \le \hat{x} \,\middle|\, u \le \frac{f(x)}{cg(x)}\right) = \frac{p\left(u \le \frac{f(x)}{cg(x)}, x \le \hat{x}\right)}{p\left(u \le \frac{f(x)}{cg(x)}\right)}.$$

For the denominator, we know that when fixing $x = \tilde{x}$

$$p\left(u \leq \frac{f(x)}{cg(x)} \Big| x = \tilde{x}\right) = \frac{f(\tilde{x})}{cg(\tilde{x})}.$$

Using the product rule and marginalizing, we obtain

$$\begin{aligned}
p\left(u \leq \frac{f(x)}{cg(x)}\right) &= \int_{-\infty}^{\infty} p\left(u \leq \frac{f(x)}{cg(x)} \Big| x = \tilde{x}\right) p(x = \tilde{x}) d\tilde{x} \\
&= \int_{-\infty}^{\infty} \frac{f(\tilde{x})}{cg(\tilde{x})} g(\tilde{x}) d\tilde{x} = \frac{1}{c}.
\end{aligned}$$

In the numerator, the joint probability distribution can be calculated as

$$\begin{aligned}
p\left(u \leq \frac{f(x)}{cg(x)}, x \leq \hat{x}\right) &= \int_{-\infty}^{\hat{x}} p\left(u \leq \frac{f(x)}{cg(x)}, x = t\right) dt \\
&= \int_{-\infty}^{\hat{x}} p\left(u \leq \frac{f(x)}{cg(x)} \Big| x = t\right) g(t) dt \\
&= \int_{-\infty}^{\hat{x}} \frac{f(t)}{cg(t)} g(t) dt = \frac{F(\hat{x})}{c},
\end{aligned}$$

where we used the product rule. Combining these two results, we see that

$$p\left(x \leq \hat{x} \Big| u \leq \frac{f(x)}{cg(x)}\right) = F(\hat{x})$$

and the samples follow the required distribution.

How many attempts are necessary to draw a sample which we accept? This is governed by the geometric distribution with $\mu = p\left(u \leq \frac{f(x)}{cg(x)}\right) = 1/c$. The expectation is $1/\mu = c$, which agrees with the ratio of the areas under cg and f. We therefore expect to need c draws from the distribution given by g to generate one sample from the distribution given by f. In the examples of Figure 3.4, the values of c are approximately 2.45, 2.09 and 1.67 from top to bottom. Therefore the envelope at the bottom is preferable over the other two.

This process to generate samples also works in the situation arising in Equation (3.1), where the denominator, which is the normalizing constant, is not known. More generally, we assume that a probability density function is given by

$$f(x) = \frac{1}{c_f} \hat{f}(x),$$

where

$$c_f = \int_{-\infty}^{\infty} \hat{f}(x) dx$$

is the normalizing constant. Instead of finding c and g such that $cg(x)$ is an envelope of f, we require $\hat{c}g(x)$ to be an envelope of $\hat{f}$. The normalizing

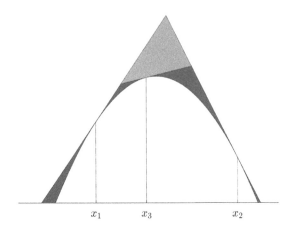

Figure 3.5: Illustration of adaptive rejection sampling of a concave
probability density function.

constant c_f is absorbed in $\hat{c}$. The number of necessary attempts could be
estimated as $\hat{c}/c_f$, would it not be for the fact that c_f is unknown.

If f is *concave*, an envelope can be constructed iteratively as a piecewise
linear function. A function f is concave, if for any points $f(x)$ and $f(y)$ on
its graph the line connecting them lies under the graph. As a consequence
all tangent lines lie above the graph as shown in Figure 3.5, where three
tangent lines are drawn. The tangent lines through $(x_1, f(x_1))$ and $(x_2, f(x_2))$
form a triangle. Its area is approximately 1.2 and thus $c \approx 1.2$. This already
means not many samples are rejected. However, it can be improved upon.
If x_3 happens to be a sample which is rejected, then another tangent line is
drawn at $(x_3, f(x_3))$ and the area where samples are rejected is reduced by the
lightly shaded area. With every new tangent line the probability of rejection
decreases. This method is called *adaptive rejection sampling (ARS)*.

This approach can also be taken, if the probability density function f
is *log concave*, i.e. $\log(f)$ is a concave function. Many probability density
functions are log concave. Assume that the function depicted in Figure 3.5
is $\log(f(x))$. The piecewise linear envelope can be transformed back into the
space of f, by applying the exponential function. The result is shown in Figure
3.6. The envelope is a piecewise exponential function. Samples from this can be
obtained by using the *exponential distribution* and inverse transform sampling.

Rejection sampling is not suitable for high-dimensional problems due to the
curse of dimensionality. The number of attempts necessary to accept a sample
increases exponentially with the number of dimensions, since there simply is
more space to explore. In the following section a technique is introduced which
concentrates on the regions of space considered important.

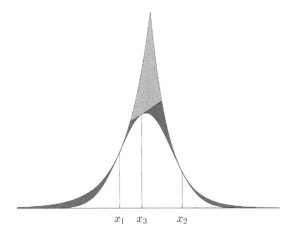

Figure 3.6: Illustration of adaptive rejection sampling of a log concave probability density function.

3.3 Importance Sampling

One reason for sampling is to approximate a multidimensional expectation of some function h of the random variable vector $\mathbf{x}$ by

$$\mathbb{E}[h] = \int h(\mathbf{x})f(\mathbf{x})d\mathbf{x} \approx \frac{1}{N}\sum_{n=1}^{N} h(\mathbf{x}_n),$$

where $\mathbf{x} \in \mathbf{R}^d$ and $\mathbf{x}_n \in \mathbf{R}^d$, $n = 1,\ldots,N$ are samples drawn from the probability density f. However, this is not suitable when sampling is not possible or computationally too expensive. A possible solution is not to sample, but to weigh each evaluation $h(\mathbf{x}_n)$ by $f(\mathbf{x}_n)$,

$$\mathbb{E}[h] \approx \sum_{n=1}^{N} f(\mathbf{x}_n)h(\mathbf{x}_n),$$

where now $\mathbf{x}_n$, $n = 1,\ldots,N$, are selected in some way. A naive approach would be to select them uniformly. Firstly, in this case N grows exponentially with the number of dimensions d to achieve a good approximation. Secondly, this neglects the fact that in many cases the regions of interest, that is where the probability density function concentrates its mass, are relatively small.

Importance sampling uses a proposal distribution g which focuses attention onto regions of interest and which can be easily sampled. It does not need to be an envelope of f. The samples $\mathbf{x}_1, \ldots, \mathbf{x}_n$ are drawn from the proposal distribution. Then

$$
\begin{aligned}
\mathbb{E}_f[h] &= \int h(\mathbf{x}) f(\mathbf{x}) d\mathbf{x} = \int h(\mathbf{x}) \frac{f(\mathbf{x})}{g(\mathbf{x})} g(\mathbf{x}) d\mathbf{x} \\
&= \mathbb{E}_g[hf/g] \approx \frac{1}{N} \sum_{n=1}^{N} \frac{f(\mathbf{x}_n)}{g(\mathbf{x}_n)} h(\mathbf{x}_n),
\end{aligned}
$$

where we used subscripts to indicate with respect to which probability density function the expectation is taken. All generated samples, unlike rejection sampling, are used. The factors $f(\mathbf{x}_n)/g(\mathbf{x}_n)$ are known as *importance weights*. They are a correction to sampling from the wrong distribution. For example, if g follows f in a region, this factor is one and no correction is necessary. If, however, g is large in a region where f is small, the contribution of evaluating h there needs to be reduced. On the other hand, if g is small in a region where f is large and we happen to sample there, the value of h there needs to be magnified.

In the case when

$$
f(\mathbf{x}) = \frac{1}{c_f} \hat{f}(\mathbf{x}) \text{ and } g(\mathbf{x}) = \frac{1}{c_g} \hat{g}(\mathbf{x})
$$

and the normalizing constants

$$
c_f = \int \hat{f}(\mathbf{x}) d\mathbf{x} \text{ and } c_g = \int \hat{g}(\mathbf{x}) d\mathbf{x}
$$

are unknown, the calculation becomes

$$
\begin{aligned}
\mathbb{E}_f[h] &= \int h(\mathbf{x}) f(\mathbf{x}) d\mathbf{x} = \frac{c_g}{c_f} \int h(\mathbf{x}) \frac{\hat{f}(\mathbf{x})}{\hat{g}(\mathbf{x})} g(\mathbf{x}) d\mathbf{x} \\
&= \frac{c_g}{c_f} \mathbb{E}_g[h\hat{f}/\hat{g}] \approx \frac{c_g}{c_f} \frac{1}{N} \sum_{n=1}^{N} \frac{\hat{f}(\mathbf{x}_n)}{\hat{g}(\mathbf{x}_n)} h(\mathbf{x}_n).
\end{aligned}
$$

Even though the normalizing constants are unknown, we can estimate their ratio. Using $c_g = \hat{g}(\mathbf{x})/g(\mathbf{x})$ wherever $g(\mathbf{x})$ is nonzero, we have

$$
\frac{c_f}{c_g} = \frac{1}{c_g} \int \hat{f}(\mathbf{x}) d\mathbf{x} = \int \frac{\hat{f}(\mathbf{x})}{\hat{g}(\mathbf{x})} g(\mathbf{x}) d\mathbf{x} = \mathbb{E}_g[\hat{f}/\hat{g}] \approx \frac{1}{N} \sum_{n=1}^{N} \frac{\hat{f}(\mathbf{x}_n)}{\hat{g}(\mathbf{x}_n)}.
$$

With this result, the expectation is estimated as

$$
\mathbb{E}_f[h] \approx \frac{1}{\sum_{k=1}^{N} \hat{f}(\mathbf{x}_k)/\hat{g}(\mathbf{x}_k)} \sum_{n=1}^{N} \frac{\hat{f}(\mathbf{x}_n)}{\hat{g}(\mathbf{x}_n)} h(\mathbf{x}_n).
$$

The importance weights are

$$
w_n = \frac{\hat{f}(\mathbf{x}_n)/\hat{g}(\mathbf{x}_n)}{\sum_{k=1}^{N} \hat{f}(\mathbf{x}_k)/\hat{g}(\mathbf{x}_k)} = \frac{\hat{f}(\mathbf{x}_n)/g(\mathbf{x}_n)}{\sum_{k=1}^{N} \hat{f}(\mathbf{x}_k)/g(\mathbf{x}_k)}, \tag{3.4}
$$

since g and $\hat{(g)}$ only differ by a multiplicative constant. Note that if the proposal distribution coincides with f, then we have $w_n = 1/N$.

As with rejection sampling, the choice of the proposal distribution is crucial. Most importantly, it should not be small in regions where f is large. The weighting can only make a correction if an actual sample is drawn from this region.

Sampling-importance-resampling (SIR) combines ideas from rejection sampling and importance sampling. First, N samples $\mathbf{x}_1, \ldots, \mathbf{x}_N$ are drawn from the proposal distribution g. Instead of rejecting some of them, weights are calculated according to (3.4). Then N samples are drawn from the discrete set $\{\mathbf{x}_1, \ldots, \mathbf{x}_N\}$ according to the probabilities given by $w_1, \ldots, w_N$. Thus a sample can feature several times in the final set, if its weight is larger than the weights of the other samples. If $g = f$, then all weights will equal $1/N$ and all samples are equally likely to be drawn.

3.4 Markov Chains

Markov chains are an important building block for a class of sampling algorithms which can be applied to many different distributions and also scale well with dimensionality.

A *Markov chain* is a series of random variables $\mathbf{x}_1, \ldots, \mathbf{x}_N$. The indices are associated with a sequence in time. A Markov chain of order m satisfies the following property

$$p(\mathbf{x}_n | \mathbf{x}_{n-1}, \mathbf{x}_{n-2}, \ldots, \mathbf{x}_1) = p(\mathbf{x}_n | \mathbf{x}_{n-1}, \mathbf{x}_{n-2}, \ldots, \mathbf{x}_{n-m})$$

for $n = m + 1, \ldots, M$. This means the nth random variable only depends on the m previous variables. In particular, a first-order Markov chain satisfies

$$p(\mathbf{x}_n | \mathbf{x}_{n-1}, \mathbf{x}_{n-2}, \ldots, \mathbf{x}_1) = p(\mathbf{x}_n | \mathbf{x}_{n-1}).$$

Each random variable only depends on its predecessor. We will only consider first-order Markov chains, since any Markov chain of order m can be transcribed into a first-order Markov chain by letting $\mathbf{y}_{n-m+1}$ be the tuple $(\mathbf{x}_n, \ldots, \mathbf{x}_{n-m+1})$, since then $p(\mathbf{y}_k | \mathbf{y}_{k-1}, \ldots, \mathbf{y}_1) = p(\mathbf{y}_k | \mathbf{y}_{k-1})$.

The probability distribution of the initial variable $\mathbf{x}_1$ needs to be specified. The probabilities $T_n(\mathbf{x}_{n-1}, \mathbf{x}_n) = p(\mathbf{x}_n | \mathbf{x}_{n-1})$ are called *transition probabilities*. If they are the same for all n, i.e. $T_n(\mathbf{x}_{n-1}, \mathbf{x}_n) = T(\mathbf{x}_{n-1}, \mathbf{x}_n)$ for all n, then the Markov chain is called *homogeneous*.

Graphically, a first-order Markov chain can be represented as

The possible values of $\mathbf{x}_n$ form a *state* space. We distinguish between a discrete and countable, possibly finite state space, and a continuous state space, which is sometimes called general state space. Summing or integrating over all possible states $\mathbf{x}_{n-1}$ gives the marginal probability of $\mathbf{x}_n$,

$$p(\mathbf{x}_n) = \sum_{\mathbf{x}_{n-1}} p(\mathbf{x}_n|\mathbf{x}_{n-1})p(\mathbf{x}_{n-1}) \quad \text{or} \quad p(\mathbf{x}_n) = \int p(\mathbf{x}_n|\mathbf{x}_{n-1})p(\mathbf{x}_{n-1})d\mathbf{x}_{n-1}.$$

To do so the marginal probability of $\mathbf{x}_{n-1}$ needs to be known. Starting from the initial distribution $p(\mathbf{x}_1)$, all marginal distributions can be calculated.

For a homogeneous Markov chain, the transition probabilities can be completely described by noting how T acts on different elements of the state space, $T(\mathbf{x}, \hat{\mathbf{x}})$ for all states $\mathbf{x}$ and $\hat{\mathbf{x}}$.

A Markov chain is called *irreducible*, if the probability of reaching state $\mathbf{x}$ from state $\hat{\mathbf{x}}$ in a finite number of steps is non-zero for all states $\mathbf{x}$ and $\hat{\mathbf{x}}$. In other words, any state can be reached from any other state in a finite number of steps.

A distribution specified by the probability mass function or probability density function f is called *invariant* or *stationary* with respect to the Markov chain; if $\mathbf{x}_{n-1}$ follows the distribution, then so does $\mathbf{x}_n$ for all n. In other words, if $\mathbf{x}_0$ follows the distribution, then so does the complete chain. The distribution remains the same after each step. Note that several distributions can be invariant with respect to a given Markov chain. For example, in the degenerate case where

$$T(\mathbf{x}, \hat{\mathbf{x}}) = \begin{cases} 1 & \text{if } \hat{\mathbf{x}} = \mathbf{x}, \\ 0 & \text{otherwise,} \end{cases}$$

we have $\mathbf{x}_1 = \mathbf{x}_2 = \cdots = \mathbf{x}_N$. The transformation in each step is the identity. In this case, any distribution is invariant.

For a homogeneous Markov chain, a distribution specified by f is invariant, if

$$f(\hat{\mathbf{x}}) = \sum_{\mathbf{x}} T(\mathbf{x}, \hat{\mathbf{x}})f(\mathbf{x}) \quad \text{or} \quad f(\hat{\mathbf{x}}) = \int T(\mathbf{x}, \hat{\mathbf{x}})f(\mathbf{x})d\mathbf{x},$$

where the sum or integral is over the state space.

It is possible to construct a Markov chain with transition probabilities such that a given f is invariant. To achieve this, the transition probabilities need to satisfy the property of *detailed balance*, for all pairs of states $\mathbf{x}$ and $\hat{\mathbf{x}}$:

$$T(\mathbf{x}, \hat{\mathbf{x}})f(\mathbf{x}) = T(\hat{\mathbf{x}}, \mathbf{x})f(\hat{\mathbf{x}}).$$

Note that for $\mathbf{x} = \hat{\mathbf{x}}$, the above equation is naturally true. If detailed balance is satisfied, the Markov chain is called *reversible*. Invariance is shown by

$$\sum_{\mathbf{x}} T(\mathbf{x}, \hat{\mathbf{x}})f(\mathbf{x}) = \sum_{\mathbf{x}} T(\hat{\mathbf{x}}, \mathbf{x})f(\hat{\mathbf{x}}) = f(\hat{\mathbf{x}})\sum_{\mathbf{x}} p(\mathbf{x}|\hat{\mathbf{x}}) = f(\hat{\mathbf{x}}),$$

since the sum over the state space of the conditional probabilities is one. The calculation for a continuous probability density function f uses integrals and is analogous.

The transition probabilities can be constructed as linear combinations of base transition probabilities $B_1, \ldots, B_M$,

$$T(\mathbf{x}, \hat{\mathbf{x}}) = \sum_{m=1}^{M} b_m B_m(\mathbf{x}, \hat{\mathbf{x}}),$$

where $b_1, \ldots, b_M$ are known as *mixing coefficients* and satisfy $b_m \geq 0$ and $\sum_{m=1}^{M} b_m = 1$. If each of the base transitions satisfies detailed balance, then so does the linear combination. Often, the base transitions are chosen such that each only changes a subset of components in $\mathbf{x}$.

If the state space is finite, then it can be represented by a 1-of-K vector. That is the random variable $\mathbf{x}_n$ taking on the kth state is a K-dimensional vector, where exactly one element, $x_{n,k}$, equals one and all the others are zero. The transition probabilities $p_{kl} = p(x_{n+1,l} = 1 | x_{n,k} = 1)$, that is that $\mathbf{x}_n$ in state k generates $\mathbf{x}_{n+1}$ in state l, can be represented in matrix form

$$P = \begin{pmatrix} p_{11} & \cdots & p_{1K} \\ \vdots & \ddots & \vdots \\ p_{K1} & \cdots & p_{KK} \end{pmatrix},$$

which is called the *transition matrix*. Each row of P sums to one. It is therefore a *right stochastic matrix* as opposed to a *left stochastic matrix* where each column sums to one. A *doubly stochastic matrix* is one where both columns and rows sum to one. This would for example be the case, if P is symmetric, that is the probability of transitioning from state k to l is the same as from state l to k.

The probability of transitioning from stage k to l in m steps is the (k, l) entry in P^m. A state k has *period* m, if any return to state k occurs in multiples of m time steps. More formally the period m is the greatest common denominator of all numbers n such that the probability $p(x_{n+1,k} = 1 | x_{1,k} = 1)$ is non-zero. If $m = 1$, the state is called *aperiodic*. This happens, for example, if the probability of transitioning to itself is non-zero. A Markov chain is *aperiodic*, if every state is aperiodic. An irreducible Markov chain, i.e. where every state can be reached from any other state, only needs one aperiodic state to be aperiodic. This is because using this one aperiodic state, one can return any state in a number of steps which is any prime number and the greatest common denominator of several prime numbers is 1.

If the initial distribution $p(\mathbf{x}_1)$ is given by $\mathbf{p} = (p_1, \ldots, p_K)$, then the distribution of $\mathbf{x}_{n+1}$ is given by $\mathbf{p}P^n$.

The transition matrix can be depicted in a *state diagram*. For example, if there are $K = 3$ states, then the diagram is

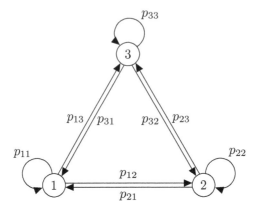

The state diagram should not be confused with the graphical representation of the Markov chain.

If the transition probability from one state to another is zero, this arrow is left off the state diagram. For example,

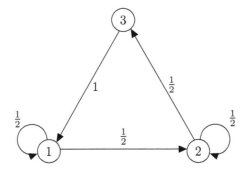

has the transition matrix

$$P = \begin{pmatrix} \frac{1}{2} & \frac{1}{2} & 0 \\ 0 & \frac{1}{2} & \frac{1}{2} \\ 1 & 0 & 0 \end{pmatrix}.$$

State 1 remains state 1 with probability $1/2$ or changes to state 2 with probability $1/2$. Similarly, state 2 remains state 2 with probability $1/2$ or changes to state 3 with probability $1/2$. State 3 always changes to state 1.

The Markov chain generated by this transition matrix is irreducible, since all states can be reached by every other state. States 1 and 2 are obviously aperiodic, making the Markov chain aperiodic.

To find the probability mass function f which is invariant, let f_i be the probability of state i and $\mathbf{f} = (f_1, f_2, f_3)$. We need to solve $\mathbf{f}P = \mathbf{f}$,

$$\frac{1}{2}f_1 + f_3 = f_1,$$
$$\frac{1}{2}f_1 + \frac{1}{2}f_2 = f_2,$$
$$\frac{1}{2}f_2 = f_3,$$

which gives $f_1 = f_2 = 2f_3$. The probabilities also need to sum to one, i.e. $f_1 + f_2 + f_3 = 1$. With this, $\mathbf{f} = (2/5, 2/5, 1/5)$ describes the invariant distribution.

The other direction, i.e. constructing a Markov chain which is invariant for a given distribution f, has more degrees of freedom, since the transition matrix has nine entries under the constraint that all rows have to sum to one. Let $\mathbf{f} = (f_1, f_2, f_3)$ define the probabilities of the random variable being in state $1, 2$ or 3 respectively. The property of detailed balance leads to three equations,

$$p_{12}f_1 = p_{21}f_2$$
$$p_{13}f_1 = p_{31}f_3$$
$$p_{23}f_2 = p_{32}f_3.$$

The diagonal elements p_{11}, p_{22} and p_{33} can be determined, once the off-diagonal elements are chosen by using

$$\sum_{j=1}^{3} p_{ij} = 1,$$

for $i = 1, 2, 3$.

More specifically, let $\mathbf{f} = (1/10, 2/5, 1/2)$. The detailed balance equations are

$$\frac{1}{10}p_{12} = \frac{2}{5}p_{21}$$
$$\frac{1}{10}p_{13} = \frac{1}{2}p_{31}$$
$$\frac{2}{5}p_{23} = \frac{1}{2}p_{32}.$$

Because we have some freedom of choice, we can set $p_{12} = 1$. This implies $p_{21} = 1/4$. It is not possible to set $p_{21} = 1$, since this would lead to a value greater than 1 for p_{12}. Since the probabilities in a row have to sum to one, we have $p_{11} = p_{13} = 0$. The second equation then leads to $p_{31} = 0$ as well. This has determined the first row and column of the transition matrix. Again using our freedom of choice, we set $p_{23} = 3/4$, which implies $p_{22} = 0$. The last equation determines $p_{32} = 2(3/4)(2/5) = 3/5$. Lastly $p_{33} = 2/5$, since the last row has to sum to one, and thus

$$P = \begin{pmatrix} 0 & 1 & 0 \\ \frac{1}{4} & 0 & \frac{3}{4} \\ 0 & \frac{3}{5} & \frac{2}{5} \end{pmatrix}.$$

The state diagram is

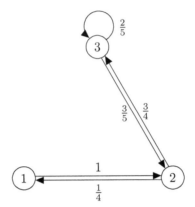

We can check the distribution of $\mathbf{x}_{n+1}$ after transitioning n times from the initial state $\mathbf{x}_1$ with distribution $\mathbf{p} = (p_1, p_12, p_33)$. Recall that this is given by $\mathbf{p}P^n$. For $n = 20$, in our example, we have

$$
P^{20} \approx \begin{pmatrix} 0.1006 & 0.3984 & 0.5011 \\ 0.0996 & 0.4012 & 0.4992 \\ 0.1002 & 0.3994 & 0.5004 \end{pmatrix}.
$$

Let's assume that P^n converges to

$$
F = \begin{pmatrix} f_1 & f_2 & f_3 \\ f_1 & f_2 & f_3 \\ f_1 & f_2 & f_3 \end{pmatrix}
$$

as n converges to infinity. Then the distribution of $\mathbf{x}_{n+1}$ converges to

$$
\mathbf{p}F = (f_1(p_1 + p_2 + p_3), f_2(p_1 + p_2 + p_3), f_3(p_1 + p_2 + p_3)) = \mathbf{f},
$$

since $p_1 + p_2 + p_3 = 1$. Thus we can generate approximate samples for a given distribution f, by generating samples using a Markov chain for which f is invariant irrespective of the starting distribution.

Convergence is due to the *ergodic theorem*, which is beyond the scope of this text, but [33] gives a thorough treatment of Markov chains. If a finite state Markov chain is irreducible and aperiodic, then the the distribution of $\mathbf{x}_n$ will tend to the invariant distribution f irrespective of the initial distribution. The invariant distribution is then called the *equilibrium*.

More generally, any matrix of the form

$$
P = \begin{pmatrix} 1 - (\alpha + \beta) & \alpha & \beta \\ \frac{\alpha}{4} & 1 - \frac{\alpha + 4\gamma}{4} & \gamma \\ \frac{\beta}{5} & \frac{4\gamma}{5} & 1 - \frac{\beta + 4\gamma}{5} \end{pmatrix}
$$

satisfies $\mathbf{f}P = \mathbf{f} = (1/10, 2/5, 1/2)$. All entries have to be valid probabilities and thus $\alpha, \beta, \gamma \in [0,1]$ and $\alpha + \beta \leq 1, \alpha + 4\gamma \leq 4$ and $\beta + 4\gamma \leq 5$.

If any two of α, β and γ are both zero, we have a 1 on the diagonal, meaning that it is impossible to leave that state. This state is called *absorbing*. If there is a non-zero probability of every state to reach that state, then the Markov chain is an *absorbing Markov chain*. This is not the case here, since the other two states change between each other, but not to the absorbing state. This also means that the chain is not irreducible, since the absorbing state cannot be reached from the other two states. Thus in this case, the distribution of $\mathbf{x}_n$ will not converge to the invariant distribution, and the Markov chain cannot be used to generate approximate samples.

If on the other hand at most one of α, β and γ is zero, the Markov chain is irreducible and aperiodic and can be used to generate approximate samples for the invariant distribution f.

An example of a Markov chain on the state space of integers is the *drunkard's walk*, where the drunkard starts at the pub denoted by 0 and whenever taking a step, either steps forward ($x_{n+1} = x_n + 1$) with probability $1/2$, or steps backward ($x_{n+1} = x_n - 1$) also with probability $1/2$, independent of how he got here in the previous step. The probabilities of all other integers are zero. This is an example of a *random walk*. The transition matrix would be infinite with zero on the diagonal and $1/2$ on the subdiagonal and superdiagonal. The state diagram is

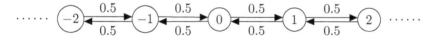

The expectation $\mathbb{E}[x_n]$ is 0, since the expectation of each individual step is that the drunkard stays put. On the other hand, $\mathbb{E}[x_n^2] = n$, which implies that the distance traveled from the pub is of the order of $\sqrt{n}$. The drunkard's partner will find them near the pub. It also shows that this random walk is relatively ineffective in exploring the state space of the integers.

However, if the random walk carries on indefinitely, it will reach each integer an infinite number of times. This property is known as the *level-crossing phenomenon*, *recurrence* or *gambler's ruin*. The last is due to the fact that a gambler with a finite amount of money playing a fair game against a bank with an infinite amount of money will eventually lose, because the amount he has follows a random walk, and will at some point cross zero and the game is over.

The aim is to construct a Markov chain, for which a given distribution with probability mass function or probability density function f is invariant, such that the elements of the Markov chain for n large enough are samples of the given distribution.

3.5 Markov Chain Monte Carlo

Let

$$f(\mathbf{x}) = \frac{1}{c_f}\hat{f}(\mathbf{x})$$

describe the distribution we want to sample from, where the value of the normalizing constant c_f is not necessarily known. We construct a Markov chain by using a proposal distribution $g(\mathbf{x}, \hat{\mathbf{x}})$, which describes the probability of drawing $\hat{\mathbf{x}}$ when $\mathbf{x}$ is given. That is, the next element in the Markov chain $\mathbf{x}_{n+1}$ is drawn from the distribution $g(\mathbf{x}_n, \mathbf{x})$ where $\mathbf{x}_n$ is the current element in the Markov chain. The Markov chain is only homogeneous, if $g(\mathbf{x}, \hat{\mathbf{x}})$ is independent of $\mathbf{x}$.

If, in addition, the proposal distribution is symmetric, i.e. $g(\mathbf{x}, \hat{\mathbf{x}}) = g(\hat{\mathbf{x}}, \mathbf{x})$ for all, a candidate $\mathbf{x}^*$ is drawn from the proposal distribution and accepted with probability

$$\min\left(1, \frac{f(\mathbf{x}^*)}{f(\mathbf{x}_n)}\right) = \min\left(1, \frac{\hat{f}(\mathbf{x}^*)}{\hat{f}(\mathbf{x}_n)}\right).$$

Note that, if $\hat{f}(\mathbf{x}^*) \geq \hat{f}(\mathbf{x}_n)$, the candidate is certain to be kept. If the candidate sample is accepted, then $\mathbf{x}_{n+1} = \mathbf{x}^*$, otherwise $\mathbf{x}_{n+1} = \mathbf{x}_n$. This is known as the *Metropolis* algorithm.

Duplicating the sample is in contrast to rejection sampling. In practical implementations, no extra copies of the sample are stored. Instead a counter of how often the sample was drawn is updated. This counter acts as weighting when, for example, the expectation is calculated. As long as $\hat{f}$ is non-zero over the entire state space, there is always a non-zero probability of $\mathbf{x}_{n+1} = \mathbf{x}_n$ and thus the Markov chain is aperiodic. If, in addition, $g(\mathbf{x}, \hat{\mathbf{x}})$ is non-zero over the entire state space, the Markov chain is obviously also irreducible.

The elements of the Markov chain, $\mathbf{x}_1, \mathbf{x}_2, \ldots$, are *not* independent samples in most cases, because successive samples are highly correlated, if $g(\mathbf{x}, \hat{\mathbf{x}})$ depends on $\mathbf{x}$. In fact, the chain represents a random walk through the state space. For practical purposes, only taking every m^{th} element from the chain approximates independence for m sufficiently large. This practice is known as *thinning*.

Symmetry of $g(\mathbf{x}, \hat{\mathbf{x}})$ can for example be achieved by letting $g(\mathbf{x}, \hat{\mathbf{x}})$ describe the normal distribution with mean $\mathbf{x}$ and a variance which is σ^2 times the identity matrix, since in this case

$$g(\mathbf{x}, \hat{\mathbf{x}}) = \frac{1}{\sqrt{|2\pi\sigma^2 \mathbf{I}|}} \exp\left(\frac{1}{2\sigma^2}(\hat{\mathbf{x}} - \mathbf{x})^T(\hat{\mathbf{x}} - \mathbf{x})\right) = g(\hat{\mathbf{x}}, \mathbf{x}).$$

This choice for the proposal distribution is common, since techniques to generate samples from the standard normal distribution such as the the Box—Muller transform are readily available, and these samples only have to be scaled by σ^2 and translated by $\mathbf{x}$.

```
target_mean = [2 3];
target_variance = [1 1.5; 1.5 3];
% Create function handle to unnormalized probability density.
target = @(x) exp(-0.5*(x-target_mean)/target_variance * ...
    (x-target_mean)');

% The variances of the components are given by the eigenvalues
% of the variance matrix.
e = eig(target_variance);
figure;
hold on;

% Set variance of the proposal distribution to the identity matrix
% times the smallest variance of the target distribution.
proposal_variance = e(1) * eye(2);

N = 300; % Length of Markov chain.

% Set first element of the Markov chain.
start = target_mean;
samples = target_mean;
% Evaluate the target distribution for the last element.
last = feval(target,start);
counter = [1];
total = 1;
accepted = 1;
while total < N
    % Generate candidate from normal distribution with mean
    % samples(accepted) and variance proposal_variance.
    candidate = mvnrnd(samples(accepted,:), proposal_variance);
    %  Evaluate the target distribution for the candidate.
    new =  feval(target,candidate);
    % Evaluate acceptance probability.
    accept = min(1, new/last);
    if accept > rand
        % Draw random walk.
        line([samples(accepted,1) candidate(1)],...
            [samples(accepted,2) candidate(2)],...
            'Color','k');
        % Update Markov chain.
        samples = [samples; candidate];
        counter = [counter; 1];
        accepted = accepted +1;
        % Update evaluation of target distribution for the last element.
        last = new;
    else
        % Plot rejected candidate.
        plot(candidate(1), candidate(2),'k+');
        % Update counter for this state.
        counter(accepted) = counter(accepted)+1;
    end
    total = total +1;
end
% Calculate acceptance rate.
rate = accepted/total
axis equal
```

Listing 3.3: Metropolis sampling example.

We illustrate with an example generating samples of a target distribution which is normal with mean and variance

$$\mu = \begin{pmatrix} 2 \\ 3 \end{pmatrix}, \qquad \Sigma = \begin{pmatrix} 1 & 3/2 \\ 3/2 & 3 \end{pmatrix}.$$

Listing 3.3 implements the Metropolis algorithm for this example. The question is, how should σ^2 be chosen? Here the univariate variances of the distribution differ a lot depending on which direction is considered. The smallest and largest variances are given by the eigenvalues of Σ, which are $\sigma_{\min}^2 \approx 0.2$ and $\sigma_{\max}^2 \approx 3.8$. The variance of the proposal distribution, σ^2, should be of the magnitude of the smallest variance. Or in other words the standard deviation of the proposal distribution should be approximately the same as the smallest standard deviation of the target distribution, $\sigma = \sigma_{\min}$. This choice ensures that not too many candidates are rejected by being too far away in the direction of the smallest standard deviation. The direction of the largest standard deviation is explored by means of a random walk. Listing 3.3 makes this choice for σ.

Also the start of the Markov chain needs to be chosen carefully to ensure the state space is explored in relatively few steps. The technique known as *burn-in* discards the first elements of a Markov chain and restarts the chain from the state it reached. Strictly speaking, this is not necessary. The elements of a Markov chain will be a good set of samples as long as the chain is long enough. Listing 3.3 sets the first element to the mean μ of the target distribution. Of course, these choices for σ and x_1 are only possible if this is known about the target distribution. In any case, good estimates for these are necessary.

These considerations are illustrated in Figure 3.7 which depicts the generated random walks along with the candidates which were rejected (+) and the 95% *confidence ellipse* of the target distribution for different choices. The confidence ellipse gives the area where 95% of samples of the target distribution are expected to lie.

Figure 3.7a used $x_1 = \mu$ and $\sigma = \sigma_{\min}$. The state space is well explored. About two thirds of candidates are accepted. This means that one third of samples coincide with another sample. In Figure 3.7b the starting position was moved to $(0,0)^T$. The acceptance ratio is about the same as before, but the state space is less well explored. The algorithm needs to run for longer. Figure 3.7c used $\sigma = 2 * \sigma_{\min}$. About 87% of candidates are accepted, but the algorithm moves slowly through the state space. Again it needs to run longer to explore the state space. In Figure 3.7d the standard deviation of the proposal distribution was doubled as compared to Figure 3.7a. The acceptance rate dropped to 43%. This means that more than half of samples share their location. This does not matter, if the expectation is to be calculated and the appropriate weighting is applied. If the aim is however to explore the state space, a higher acceptance rate is more desirable.

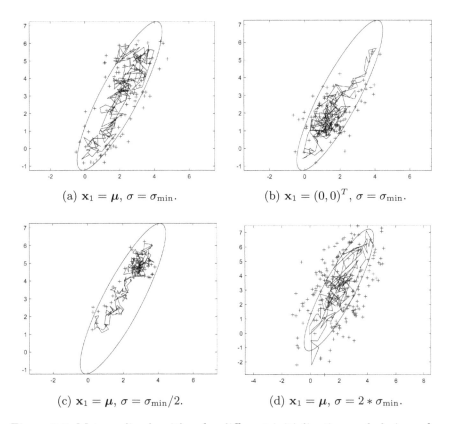

(a) $\mathbf{x}_1 = \boldsymbol{\mu}$, $\sigma = \sigma_{\min}$.

(b) $\mathbf{x}_1 = (0,0)^T$, $\sigma = \sigma_{\min}$.

(c) $\mathbf{x}_1 = \boldsymbol{\mu}$, $\sigma = \sigma_{\min}/2$.

(d) $\mathbf{x}_1 = \boldsymbol{\mu}$, $\sigma = 2 * \sigma_{\min}$.

Figure 3.7: Metropolis algorithm for different initialization and choices of σ.

The *Metropolis–Hastings* algorithm extends the Metropolis method to use proposal distributions which are not symmetric. In this case, the acceptance probability of a candidate $\mathbf{x}^*$ is calculated as

$$\min\left(1, \frac{f(\mathbf{x}^*)g(\mathbf{x}^*, \mathbf{x}_n)}{f(\mathbf{x}_n)g(\mathbf{x}_n, \mathbf{x}^*)}\right) = \min\left(1, \frac{\hat{f}(\mathbf{x}^*)g(\mathbf{x}^*, \mathbf{x}_n)}{\hat{f}(\mathbf{x}_n)g(\mathbf{x}_n, \mathbf{x}^*)}\right).$$

The transition probability from $\mathbf{x}$ to $\hat{\mathbf{x}}$ is given by

$$T(\mathbf{x}, \hat{\mathbf{x}}) = g(\mathbf{x}, \hat{\mathbf{x}}) \min\left(1, \frac{\hat{f}(\hat{\mathbf{x}})g(\hat{\mathbf{x}}, \mathbf{x})}{\hat{f}(\mathbf{x})g(\mathbf{x}, \hat{\mathbf{x}})}\right).$$

We can now check whether f satisfies detailed balance. To this end,

$$
\begin{aligned}
f(\mathbf{x})T(\mathbf{x},\hat{\mathbf{x}}) &= \frac{\hat{f}(\mathbf{x})}{c_f}g(\mathbf{x},\hat{\mathbf{x}})\min\left(1,\frac{\hat{f}(\hat{\mathbf{x}})g(\hat{\mathbf{x}},\mathbf{x})}{\hat{f}(\mathbf{x})g(\mathbf{x},\hat{\mathbf{x}})}\right) \\
&= \frac{1}{c_f}\min\left(\hat{f}(\mathbf{x})g(\mathbf{x},\hat{\mathbf{x}}),\hat{f}(\hat{\mathbf{x}})g(\hat{\mathbf{x}},\mathbf{x})\right) \\
&= \frac{1}{c_f}\min\left(\hat{f}(\hat{\mathbf{x}})g(\hat{\mathbf{x}},\mathbf{x}),\hat{f}(\mathbf{x})g(\mathbf{x},\hat{\mathbf{x}})\right) \\
&= \frac{\hat{f}(\hat{\mathbf{x}})}{c_f}g(\hat{\mathbf{x}},\mathbf{x})\min\left(1,\frac{\hat{f}(\mathbf{x})g(\mathbf{x},\hat{\mathbf{x}})}{\hat{f}(\hat{\mathbf{x}})g(\hat{\mathbf{x}},\mathbf{x})}\right)=f(\hat{\mathbf{x}})T(\hat{\mathbf{x}},\mathbf{x}).
\end{aligned}
$$

Thus detailed balance is satisfied as required to generate approximate samples of the distribution described by f.

The Metropolis–Hastings algorithm is an example of a class of algorithms known as *Markov Chain Monte Carlo (MCMC)*. In general, *Monte-Carlo* methods refer to algorithms, which rely on repeated random sampling to perform calculations, where a closed solution is not possible or prohibitively expensive. They were first developed at the Los Alamos National Laboratory. Since the work was secret, Monte Carlo, referring to the Casino de Monte Carlo in Monaco, was used as code name.

Gibbs sampling is used, when it is easier to sample from the conditional distribution of the components of $\mathbf{x}$ than from the distribution of $\mathbf{x}$ itself. Let $\mathbf{x}_n = (x_{n,1},\ldots,x_{n,D})^T$, where d is the dimension of the state space. The next element $\mathbf{x}_{n+1}$ is then constructed in D steps by drawing $x_{n+1,d}$ sequentially from the conditional probability $p(x|x_{n+1,1},\ldots,x_{n+1,d-1},x_{n,d+1},\ldots,x_{n,D})$ for $d=1,\ldots,D$. This means a component is used in the next draw as soon as it has been drawn. Another version chooses which component to update next randomly.

The distribution described by the probability density function f of $\mathbf{x}$ is invariant of each of the sampling steps for the following reasons. Firstly, the marginal distribution of $\mathbf{x}_{-d} = (x_1,\ldots,x_{d-1},x_{d+1},\ldots,x_D)$ is invariant, since none of these components changes in the dth step. Secondly, the dth component is sampled from the correct conditional distribution, and is therefore invariant. Because the joint distribution (which is the distribution of $\mathbf{x}$) is determined by these marginal and conditional distributions, it is also invariant. The ergodic theorem can be applied, if in addition none of the conditional probabilities is zero anywhere in the state space. Thus the generated Markov chain will produce samples from the desired distribution.

Gibbs sampling can be phrased in terms of the Metropolis–Hastings algorithm. Let the current sample be $\mathbf{x}$. Note, this could be an element of the Markov chain or one of the intermediate steps. The candidate $\mathbf{x}^*$ differs from the previous sample $\mathbf{x}$ in only one component. Let this be the dth component. Hence $\mathbf{x}^*_{-d} = \mathbf{x}_{-d}$. The dth component of $\mathbf{x}^*$ is drawn from $p(x|\mathbf{x}_{-d})$. This defines the proposal distribution

$$
g(\mathbf{x},\mathbf{x}^*) = p(x^*_d|\mathbf{x}_{-d}).
$$

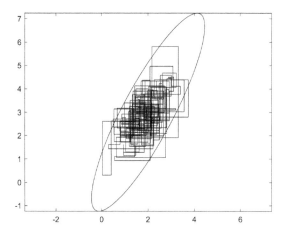

Figure 3.8: Random walk generated by Gibbs sampling.

To determine the acceptance probability, we need to calculate

$$\frac{f(\mathbf{x}^*)g(\mathbf{x}^*,\mathbf{x})}{f(\mathbf{x})g(\mathbf{x},\mathbf{x}^*)} = \frac{f(\mathbf{x}^*)p(x_d|\mathbf{x}^*_{-d})}{f(\mathbf{x})p(x_d^*|\mathbf{x}_{-d})}$$

$$= \frac{f(x_1,\ldots,x_{d-1},x_d^*,x_{d+1},\ldots,x_D)p(x_d|\mathbf{x}_{-d})}{f(\mathbf{x})p(x_d^*|\mathbf{x}_{-d})}.$$

Now using the product rules

$$f(\mathbf{x}) = p(x_d|\mathbf{x}_{-d})p(\mathbf{x}_{-d})$$

$$f(x_1,\ldots,x_{d-1},x_d^*,x_{d+1},\ldots,x_D) = p(x_d^*|\mathbf{x}_{-d})p(\mathbf{x}_{-d})$$

we see that the fraction evaluates to 1. Thus every sample in the intermediate steps is accepted.

Figure 3.8 shows the random walk generated by the steps of the Gibbs algorithm for the same normal distribution as in the examples in Figure 3.7. The samples generated by the Gibbs algorithm are highly correlated. To approximate independence between samples thinning can be used. Another technique is *blocking*, where not the conditional probabilities of individual components are used, but of sets of several components, which can be overlapping.

CHAPTER 4

Linear Classification

The chapter introduces linear classification as simple projections onto a linear subspace. It develops the idea of maximizing the separation deriving Fisher's discriminant analysis and Linear Discriminant Analysis. This idea is extended to multiple classes. The perceptron and online learning are explained as pulling the separation line in different directions according to the seen data sample. It is shown that the perceptron is a simple neural network. Lastly, the Support Vector Machine is derived as maximizing the margin between classes.

4.1 Features

Distinguishing between objects is called *classification*. Already before we are born we learn to distinguish between different sounds in the womb, and we continue to learn and distinguish easily between different objects, different situations and different concepts. In fact, the way we differ or are similar to others is fundamental to our sense of self. Humans have an uncanny ability in the area of classification. For example, despite the variety of dog breeds and their stark visual differences, we still easily identify a dog as a dog. Domestic cats are visually much more similar to each other. A child is not told what is different between a cat or a dog. It is told this is a cat and that is a dog. At most it is told, this goes meow and that goes woof. Based on these examples it has to work out by itself how to distinguish between them. What are cat features and what are dog features?

More formally, there are three different categories of features. *Boolean* or *binary* features can only have two values. The relate to questions answerable by yes or no. For example: Is it red? Is the person tall? Do we have a storm? *Discrete* features take *categorical* also known as *ordinal* values. For example: What colour is it? Is the person short, medium built or tall? What number is the wind force on the Beaufort scale? Lastly there are *continuous*- that is *real-valued* features. Here the questions answered are for example: What is

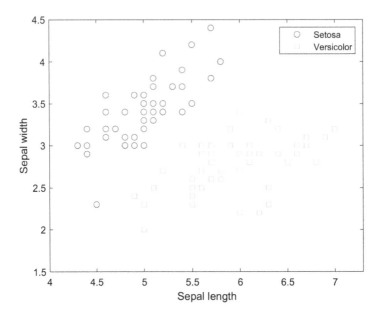

Figure 4.1: Scatter plot of sepal length against sepal width.

the wavelength of the colour? What is the person's height? What is the wind speed?

There are many examples of classification in everyday life. Spam filtering is an application of text categorization. Face detection in images and optical character recognition are examples from the field of machine vision. For businesses, market segmentation is of great importance, while every one of us hopes that the fraud detection algorithms of our credit card company are effective.

Classification is not that new. One of the most used examples dates back to 1936 when Ronald Fisher measured the petal length and width and the sepal length and width of three different species of irises, Iris setosa, iris versicolor, and iris virginica with 50 samples of each species. Thus there were four continuous features.

In the following, we will first only consider two species, that is classes which we label C_0 and C_1. This is also known as *binary classification*. Let N_0 be the number of samples in class C_0 and N_1 the number of samples in class C_1. Hence the overall number of samples is N. It is helpful to look at a visual representation of the data as in Figure 4.1 where we have chosen two features to display, sepal length and sepal width. This is known as a scatter plot.

4.2 Projections onto Subspaces

Each sample has two features and we can write these as *vectors*.

$$\mathbf{v}_1 = \begin{pmatrix} v_{11} \\ v_{12} \end{pmatrix} \qquad \text{and} \qquad \mathbf{v}_2 = \begin{pmatrix} v_{21} \\ v_{22} \end{pmatrix}$$

The *length of a vector* is the square root of the sum of the squares of the coordinates.

$$\|\mathbf{v}_2\| = \sqrt{v_{21}^2 + v_{22}^2}.$$

This is also known as the *Euclidean norm* or L_2 *norm*. Between two vectors a *inner product*, also known as a *scalar product* or *dot product* can be defined:

$$\mathbf{v}_1 \bullet \mathbf{v}_2 = \langle \mathbf{v}_1, \mathbf{v}_2 \rangle = v_{11}v_{21} + v_{12}v_{22}.$$

More generally, this can be viewed as a matrix product between the *transpose* of the first vector and the second vector.

$$\mathbf{v}_1 \bullet \mathbf{v}_2 = \mathbf{v}_1^T \mathbf{v}_2 = \begin{pmatrix} v_{11} & v_{12} \end{pmatrix} \begin{pmatrix} v_{21} \\ v_{22} \end{pmatrix} = v_{11}v_{21} + v_{12}v_{22}.$$

A 1×2 matrix times a 2×1 matrix is a 1×1 matrix.

A common technique is to reduce a higher dimensional problem to a lower dimensional one where it can be solved more easily. In particular, we reduce the two dimensional problem to a one-dimensional one by projecting onto a line going through 0 and in the direction of a vector $\mathbf{w}$. The formula of the *projection* of a vector $\mathbf{v}$ onto $\mathbf{w}$ is given by

$$\text{proj}_{\mathbf{w}}(\mathbf{v}) = \frac{\mathbf{w}^T \mathbf{v}}{\|\mathbf{w}\|^2} \mathbf{w}$$

Figure 4.2 illustrates this.

We then seek a *separation threshold* b, also known as *bias* such that for $\mathbf{v}$

- $\mathbf{w}^T \mathbf{v} < b$ $\qquad \Rightarrow \qquad$ $\mathbf{v} \in C_0$,
- $\mathbf{w}^T \mathbf{v} > b$ $\qquad \Rightarrow \qquad$ $\mathbf{v} \in C_1$.

In Figure 4.2 the bias happens to be 0 for $\mathbf{v}_1$ being in class C_0 and $\mathbf{v}_2$ being in class C_1.

The task is to find the direction $\mathbf{w}$ and separation threshold b. $\mathbf{w}$ is also known as the vector of *weights*. The line vertical to $\mathbf{w}$ through b should separate the classes. In three dimensions, a plane separates the data, while in higher dimensions it is a hyperplane. Naively choosing the coordinate axes is unsuitable as Figure 4.3 shows. One idea could be to use the line through the sample means of each class and the midpoint between sample means as separation threshold. The sample mean of class C_i is

$$\mu_i = \frac{1}{N_i} \sum_{\mathbf{v} \in C_i} \mathbf{v}, \qquad i = 0, 1. \tag{4.1}$$

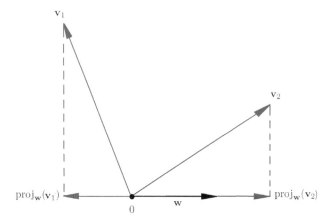

Figure 4.2: Projection of $\mathbf{v}_1$ and $\mathbf{v}_2$ onto $\mathbf{w}$.

The line through the sample means is then parametrized by

$$\mathbf{l}(a) = \boldsymbol{\mu}_0 + a(\boldsymbol{\mu}_1 - \boldsymbol{\mu}_0).$$

For $a \in [0,1]$ we are on the line segment between the sample means and thus the midpoint is given when $a = 1/2$, hence $b = 1/2(\boldsymbol{\mu}_0 + \boldsymbol{\mu}_1)$.

This approach is illustrated in Figure 4.4 where both the projection line and the separation line perpendicular to the projection line through the midpoint between sample means is shown. However, it fails to take into account the *variance* within each class, that is how much and in which way the samples differ from the mean. In Figure 4.4 we have also drawn the 95% *confidence ellipses* and their axes. This means that 95% of the samples lie within the ellipses.

Naively, one could move the separation threshold to the intersection of the confidence ellipses. However, this is subjective and a different value of the confidence, e.g. 99%, would move this point. Also, since the two ellipses have their axes at different angles, it means that the projections of samples of one class are more spread out than the projections of of the samples of the other class.

The goal is to find a line of projection such that:

- The projected means of each class are as far apart as possible,

- The projected samples of each class are as close together as possible.

In the next section we see how Fisher tackled this problem.

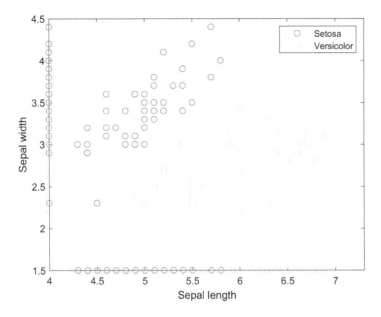

Figure 4.3: Projection onto the coordinate axes.

4.3 Fisher's and Linear Discriminant Analysis

Recall, the *sample mean* of C_i is

$$\boldsymbol{\mu}_i = \frac{1}{N_i} \sum_{\mathbf{v} \in C_i} \mathbf{v}.$$

To take into account the variation of the samples, we define the *sample co-variance* of C_i as

$$\boldsymbol{\Sigma}_i = \frac{1}{N_i} \sum_{\mathbf{v} \in C_i} (\mathbf{v} - \boldsymbol{\mu}_i)(\mathbf{v} - \boldsymbol{\mu}_i)^T. \tag{4.2}$$

The last expression in the above is an *outer product*. In two dimensions it is defined between two vectors $\mathbf{v}_1$ and $\mathbf{v}_2$ as

$$\mathbf{v}_1 \mathbf{v}_2^T = \begin{pmatrix} v_{11} \\ v_{12} \end{pmatrix} \begin{pmatrix} v_{21} & v_{22} \end{pmatrix} = \begin{pmatrix} v_{11}v_{21} & v_{11}v_{22} \\ v_{12}v_{21} & v_{12}v_{22} \end{pmatrix}$$

It can be viewed as a matrix product. A 2×1 matrix times a 1×2 matrix is a 2×2 matrix.

If M is the number of features (in our example $M = 2$), then $\boldsymbol{\mu}_i$ is a vector with M elements and the jth entry is the mean of the jth feature of samples in class C_i. On the other hand, $\boldsymbol{\Sigma}_i$ is an $M \times M$ matrix. The (j, j) diagonal entry is the variance of the jth feature of samples in class C_i. The variance describes how spread the values of a feature are. The (j, k) off-diagonal entry

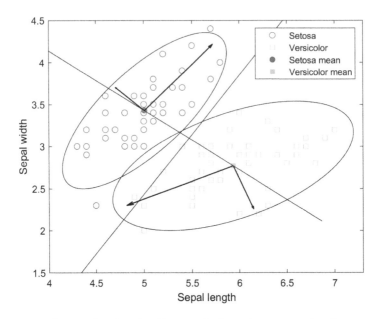

Figure 4.4: Separation by the line vertical to the connection between the sample means and 95% confidence ellipses.

is the covariance between the jth and k^{th} feature of samples in class C_i. The covariance describes how much different features influence each other.

The covariance matrix is in general *positive definite*, since

$$\hat{\mathbf{v}}^T \mathbf{\Sigma}_i \hat{\mathbf{v}} = \frac{1}{N_i} \sum_{\mathbf{v} \in C_i} \hat{\mathbf{v}}^T (\mathbf{v} - \boldsymbol{\mu}_i)(\mathbf{v} - \boldsymbol{\mu}_i)^T \hat{\mathbf{v}} = \frac{1}{N_i} \sum_{\mathbf{v} \in C_i} \left[(\mathbf{v} - \boldsymbol{\mu}_i)^T \hat{\mathbf{v}} \right]^2 \geq 0.$$

It is unlikely that the sum of squares equals zero, since this means that all samples shifted by the mean lie in a space orthogonal to $\hat{\mathbf{v}}$. In this case, one of the dimensions of the feature space is redundant, and dimensionality reduction should be performed first. Positive definiteness implies that the matrix is non-singular and can be inverted, a property which we will use later.

Let $\mathbf{w}$ be the direction of the line of projection. The location of the line of projection can be neglected, since it cancels in the calculations. The mean of the projected samples in class C_i is given by $\mathbf{w}^T \boldsymbol{\mu}_i$, while the variance of the projected samples in class C_i is given by $\mathbf{w}^T \mathbf{\Sigma}_i \mathbf{w}$. We seek $\mathbf{w}$ such that

$$\|\mathbf{w}^T \boldsymbol{\mu}_0 - \mathbf{w}^T \boldsymbol{\mu}_1\|^2 = \mathbf{w}^t (\boldsymbol{\mu}_0 - \boldsymbol{\mu}_1)(\boldsymbol{\mu}_0 - \boldsymbol{\mu}_1)^T \mathbf{w}$$

is as large as possible. That is the projected means are as far apart as possible. This is known as the *between-class scatter* of the projected samples. At the same time, we want

$$\mathbf{w}^T \mathbf{\Sigma}_0 \mathbf{w} + \mathbf{w}^T \mathbf{\Sigma}_1 \mathbf{w} = \mathbf{w}^T (\mathbf{\Sigma}_0 + \mathbf{\Sigma}_1) \mathbf{w}$$

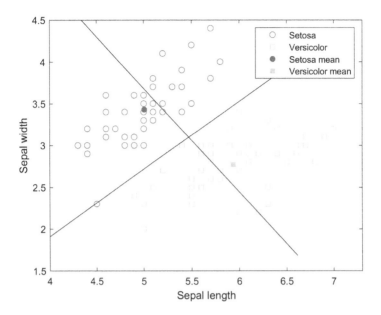

Figure 4.5: Fisher's Discriminant Analysis.

as small as possible, since then the projected samples are close to the projected mean. This is known as the *within-class scatter* of the projected samples.

Depending on $\mathbf{w}$, Fisher defined the separation $\mathbf{w}$ achieves by

$$s(\mathbf{w}) = \frac{\mathbf{w}^T(\boldsymbol{\mu}_0 - \boldsymbol{\mu}_1)(\boldsymbol{\mu}_0 - \boldsymbol{\mu}_1)^T\mathbf{w}}{\mathbf{w}^T(\boldsymbol{\Sigma}_0 + \boldsymbol{\Sigma}_1)\mathbf{w}}.$$

This function needs to be maximized. Or in other words, we maximize the ratio of the variance between projected classes to the variance within projected classes.

The function $s(\mathbf{w})$ takes its maximum, where the *gradient* vector vanishes. The gradient $\nabla s(\mathbf{w})$ is the vector formed by the derivatives of $s(\mathbf{w})$ with respect to each of the components of $\mathbf{w}$. Using the quotient rule of differentiation, the formula given in Appendix A.2.3 and the symmetry of covariance matrices and outer products, the gradient is given by

$$2\frac{(\mathbf{w}^T(\boldsymbol{\Sigma}_0 + \boldsymbol{\Sigma}_1)\mathbf{w})(\boldsymbol{\mu}_0 - \boldsymbol{\mu}_1)(\boldsymbol{\mu}_0 - \boldsymbol{\mu}_1)^T\mathbf{w} - ((\boldsymbol{\mu}_0 - \boldsymbol{\mu}_1)^T\mathbf{w})^2(\boldsymbol{\Sigma}_0 + \boldsymbol{\Sigma}_1)\mathbf{w}}{(\mathbf{w}^T(\boldsymbol{\Sigma}_0 + \boldsymbol{\Sigma}_1)\mathbf{w})^2}$$

$$= 2\frac{(\boldsymbol{\mu}_0 - \boldsymbol{\mu}_1)^T\mathbf{w}}{\mathbf{w}^T(\boldsymbol{\Sigma}_0 + \boldsymbol{\Sigma}_1)\mathbf{w}}(\boldsymbol{\mu}_0 - \boldsymbol{\mu}_1) - 2\left(\frac{(\boldsymbol{\mu}_0 - \boldsymbol{\mu}_1)^T\mathbf{w}}{\mathbf{w}^T(\boldsymbol{\Sigma}_0 + \boldsymbol{\Sigma}_1)\mathbf{w}}\right)^2(\boldsymbol{\Sigma}_0 + \boldsymbol{\Sigma}_1)\mathbf{w}.$$

Setting this to zero and rearranging, we arrive at:

$$\mathbf{w} = \frac{\mathbf{w}^T(\boldsymbol{\Sigma}_0 + \boldsymbol{\Sigma}_1)\mathbf{w}}{(\boldsymbol{\mu}_0 - \boldsymbol{\mu}_1)^T\mathbf{w}}(\boldsymbol{\Sigma}_0 + \boldsymbol{\Sigma}_1)^{-1}(\boldsymbol{\mu}_0 - \boldsymbol{\mu}_1)$$

Since we are only interested in the direction of $\mathbf{w}$, the length can be chosen freely. Thus the separation is maximal, if $\mathbf{w}$ is a multiple of the vector $(\boldsymbol{\Sigma}_0 + \boldsymbol{\Sigma}_1)^{-1}(\boldsymbol{\mu}_0 - \boldsymbol{\mu}_1)$. With this choice for $\mathbf{w}$ the scaling factor in the formula for $\mathbf{w}$ is 1.

The MATLAB code in Listing 4.1 calculates Fisher's discriminant as specified above, and the resulting projection and separation lines are illustrated in Figure 4.5. All samples are correctly separated with the exception of one outlier.

The calculations simplify under the assumption $\boldsymbol{\Sigma}_0 = \boldsymbol{\Sigma}_1 = \boldsymbol{\Sigma}$. This is known as *Linear Discriminant Analysis (LDA)*. The vector $\mathbf{w}$ is set to $\boldsymbol{\Sigma}^{-1}(\boldsymbol{\mu}_0 - \boldsymbol{\mu}_1)$, since scaling can be neglected and the separation threshold is the projection of the midpoint between the means:

$$b = \mathbf{w}^T \frac{1}{2}(\boldsymbol{\mu}_0 + \boldsymbol{\mu}_1) = \frac{1}{2}(\boldsymbol{\mu}_0^T \boldsymbol{\Sigma}^{-1} \boldsymbol{\mu}_0 - \boldsymbol{\mu}_1^T \boldsymbol{\Sigma}^{-1} \boldsymbol{\mu}_1).$$

In the iris example, which we considered so far, the classes do not have the same covariance.

4.4 Multiple Classes

We now consider more than two classes, while we continue to assume that they have the same sample covariance matrix. Let K be the number of classes. For each class C_i we calculate the mean $\boldsymbol{\mu}_i$. Let $\boldsymbol{\mu}$ be the mean of the class means

$$\boldsymbol{\mu} = \frac{1}{K} \sum_{i=1}^{K} \boldsymbol{\mu}_i.$$

The between-class scatter is defined as

$$\boldsymbol{\Sigma}_b = \frac{1}{K} \sum_{i=1}^{K} (\boldsymbol{\mu}_i - \boldsymbol{\mu})(\boldsymbol{\mu}_i - \boldsymbol{\mu})^T.$$

The separation between classes in the direction of $\mathbf{w}$ is given by

$$s(\mathbf{w}) = \frac{\mathbf{w}^T \boldsymbol{\Sigma}_b \mathbf{w}}{\mathbf{w}^T \boldsymbol{\Sigma} \mathbf{w}}.$$

To separate K classes we need to find $K-1$ directions $\mathbf{w}_1, \ldots, \mathbf{w}_{K-1}$ for which $s(\mathbf{w})$ is maximal. This is equivalent to maximizing

$$S(\mathbf{W}) = \frac{|\mathbf{W}^T \boldsymbol{\Sigma}_b \mathbf{W}|}{|\mathbf{W}^T \boldsymbol{\Sigma} \mathbf{W}|},$$

where $\mathbf{W}$ denotes the matrix whose columns are the vectors $\mathbf{w}_1, \ldots, \mathbf{w}_{K-1}$ and $|\cdot|$ denotes the determinant of the matrix. If the number of features M is greater or equal to the number of classes K, it can be shown that $S(\mathbf{W})$

```
load fisheriris;
% meas contains the first species in rows 1 to 50, the second species
% in rows 51 to 100, and the third species in rows 101 to 150, while
% the columns give the measurements of sepal length, sepal width,
% petal length and petal width in this order.
% Extract two species and two attributes.
m = meas(1:100,1:2);
s = species(1:100);
% Calculate the mean for each species.
mean0 = mean( m(1:50,:));
mean1 = mean( m(51:100,:));
% Calculate the covariance for each species.
cov0 = cov( m(1:50,:));
cov1 = cov( m(51:100,:));
% Find the direction of the projection line.
w = ((cov0 + cov1) \ (mean0 - mean1)')';
% Find vector perpendicular to direction.
perp = transpose(null(w));
% Find midpoint between means.
mid = mean0 + 0.5 * (mean1 - mean0);
% Plot data.
figure;
h(1:2) = gscatter(m(:,1), m(:,2), s,'rg','os');
hold on;
% Plot means.
h(3) = scatter(mean0(1),mean0(2),[],'r','filled');
h(4) = scatter(mean1(1),mean1(2),[],'g','filled','s');
% Plot projection line.
temp1 = mid + 0.2 * w;
temp2 = mid - 0.2 * w;
plot([temp1(1) temp2(1)], [temp1(2) temp2(2)], 'k');
% Plot separation line.
temp1 = mid + 1.5 * perp;
temp2 = mid - 2.5 * perp;
plot([temp1(1) temp2(1)], [temp1(2) temp2(2)], 'k');
% Label graph.
legend(h,{'Setosa','Versicolor','Setosa mean','Versicolor mean'},...
    'Location','Northeast')
xlabel('Sepal length');
ylabel('Sepal width');
axis([4 7.3 1.5 4.5]);
```

Listing 4.1: Fisher's discriminant, projection and separation lines.

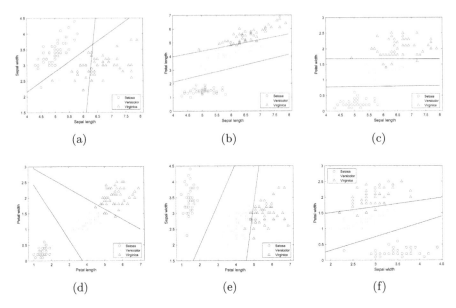

Figure 4.6: Classification using different features.

is maximal if $\mathbf{w}_1, \ldots, \mathbf{w}_{K-1}$ are the generalized eigenvectors corresponding to the $K-1$ largest generalized eigenvalues of the generalized eigenvalue problem

$$\Sigma_B \mathbf{w} = \lambda \Sigma \mathbf{w}.$$

This is implemented in MATLAB as

```
ClassificationDiscriminant.fit
```

There are also other strategies to classify multiple classes. These are based on binary classifiers.

The first one is known as *One vs Rest (OvR)* or *One vs All (OvA)*. K binary classifiers are constructed, each giving a confidence score for a sample belonging to the kth class, $k = 1, \ldots, K$. A sample is labeled with the class with the highest confidence score. There are two disadvantages. Firstly, the confidence scores have to be calibrated between each other. Secondly, the binary classifiers see unbalanced distributions, since the number of samples in the kth class is much smaller than the number of samples NOT in the kth class.

The other one is known as *One vs One (OvO)*. For each pair of classes, $K(K-1)/2$ classifiers are trained. A sample is labeled with the class for which the most classifiers vote. The case where two or more classes receive the same number of votes needs to be handled.

As can be seen in Figure 4.6 some features are more useful for classification than others. We examine this further when discussing feature selection.

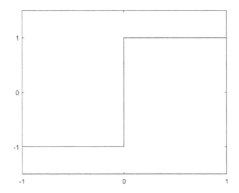

Figure 4.7: Step function sgn.

4.5 Online Learning and the Perceptron

So far we considered learning tasks where the training data is available upfront. This is known as *batch learning* or *offline learning*. In contrast, in *online learning* the training data becomes available sequentially.

Recall that we seek a direction vector $\mathbf{w}$ and separation threshold b such that for sample $\mathbf{v}$

- $\mathbf{w}^T \mathbf{v} < b \qquad \Rightarrow \qquad \mathbf{v} \in C_0,$

- $\mathbf{w}^T \mathbf{v} > b \qquad \Rightarrow \qquad \mathbf{v} \in C_1.$

By extending the vectors $\mathbf{w}$ to $\hat{\mathbf{w}} = (-b, \mathbf{w})$ and $\mathbf{v}$ to $\hat{\mathbf{v}} = (1, \mathbf{v})$ this can be rephrased to

- $\hat{\mathbf{w}}^T \hat{\mathbf{v}} = \mathbf{w}^T \mathbf{v} - b < 0 \qquad \Rightarrow \qquad \mathbf{v} \in C_0$

- $\hat{\mathbf{w}}^T \hat{\mathbf{v}} = \mathbf{w}^T \mathbf{v} - b > 0 \qquad \Rightarrow \qquad \mathbf{v} \in C_1$

This way the bias b, becomes part of the vector of weights.

As depicted in Figure 4.7, we define the step function

$$\operatorname{sgn}(x) = \begin{cases} -1 & \text{if} \quad x < 0 \\ 0 & \text{if} \quad x = 0 \\ 1 & \text{if} \quad x > 0 \end{cases}$$

Then for sample $\mathbf{v}$

- $\operatorname{sgn}(\hat{\mathbf{w}}^T \hat{\mathbf{v}}) = -1 \qquad \Rightarrow \qquad \mathbf{v} \in C_0$

- $\operatorname{sgn}(\hat{\mathbf{w}}^T \hat{\mathbf{v}}) = 1 \qquad \Rightarrow \qquad \mathbf{v} \in C_1$

The *Perceptron* was invented in 1957 by Rosenblatt, and built in 1958 as a physical machine for image recognition. It contained 400 photo cells, and the

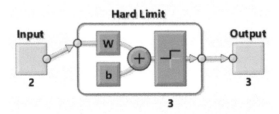

Figure 4.8: Perceptron.

weights were implemented as potentiometers which were updated by electric motors.

Specifically, the Perceptron initializes $\hat{\mathbf{w}}_0 = 0$, and updates the vector $\hat{\mathbf{w}}_{i-1}$ with each new training sample $\mathbf{v}_i$ by

$$\hat{\mathbf{w}}_i = \hat{\mathbf{w}}_{i-1} + \frac{\alpha}{2} \left(c_i - \mathrm{sgn}(\hat{\mathbf{w}}_{i-1}^T \hat{\mathbf{v}}_i)\right) \hat{\mathbf{v}}_i,$$

where c_i is the class label of $\mathbf{v}_i$, that is $c_i = -1$ if $\mathbf{v}_i \in C_0$ and $c_i = 1$ if $\mathbf{v}_i \in C_1$. The parameter $0 < \alpha \le 1$ is the *learning rate* and is set by the user. If the learning rate is chosen too big, any changes are too radical and oscillations occur. Note that $\hat{\mathbf{w}}_i$ is the same as $\hat{\mathbf{w}}_{i-1}$, if the class label was determined correctly. If the class label was determined incorrectly, the separation line is pulled in the direction such that it is more likely that the sample is labeled correctly in future. The sign of $c_i - \mathrm{sgn}(\hat{\mathbf{w}}_{i-1}^T \hat{\mathbf{v}}_i)$ determines in which direction the separation line is pulled. At any stage, the vector of weights, $\hat{\mathbf{w}}_i$, is a linear combination of all samples which caused a change to the separation line so far.

The perceptron is implemented as a single layer neural network within the neural network framework in MATLAB (Figure 4.8 and Listing 4.2). We take a closer look at neural networks, when considering non-linear classification.

The perceptron needs to see enough examples for the separation line to settle. An epoch is one complete presentation of the data to the algorithm. Figure 4.9 shows the different separation lines after a certain amount of data has been seen by the perceptron. The unseen data are marked with ×. The learning rate is set to 1. Because of this large value, the separation line changes position quite dramatically between seeing 90% and the complete data set. After 100 epochs the perceptron classifies all samples correctly apart from one outlier. It takes several hundred more epochs for the perceptron to also classify this sample correctly. Giving a different seed to the random number generator and thus changing the order in which the data is presented leads to different separation lines.

```
load fisheriris;
% meas contains the first species in rows 1 to 50, the second species
% in rows 51 to 100, and the third species in rows 101 to 150, while
% the columns give the measurements of sepal length, sepal width,
% petal length and petal width in this order.
% Extract two species and two attributes.
m = meas(1:100,1:2)';
s = species(1:100);
% Plot data.
figure;
gscatter(m(1,:), m(2,:), s,'rg','os');
hold on;
% Generate binary true/false class labels. First 50 samples belong to
% class zero, next 50 to class 1
s = [zeros(1,50) ones(1,50)];
% Generate random permutation matrix, so that the samples arrive in
% a random order
P = eye(100);
% Create a random stream.
st = RandStream('mt19937ar','Seed',1);
idx = randperm(st,100);
P = P(idx, :);
% Permute data.
m = m*P;
s = s*P;
% Fit a linear separation line using the Perceptron
PerceptronModel = perceptron;
PerceptronModel = configure(PerceptronModel,m(:,1),s(1));
% Use all samples for training.
PerceptronModel.divideFcn = 'dividetrain';
% Set learning rate to 1.
PerceptronModel.trainParam.lr = 1;
% Specify maximum number of epochs.
PerceptronModel.trainParam.epochs = 1000;
% Train the perceptron.
PerceptronModel = train(PerceptronModel,m,s);
% Plot separation line.
w = PerceptronModel.iw{1,1};
b = PerceptronModel.b{1};
h = plotpc(w,b);
set(h, 'Color','k');
% Label graph.
legend('Setosa','Versicolor','Location','Northeast')
xlabel('Sepal length');
ylabel('Sepal width');
axis([4 7.3 1.5 4.5]);
```

Listing 4.2: Perceptron.

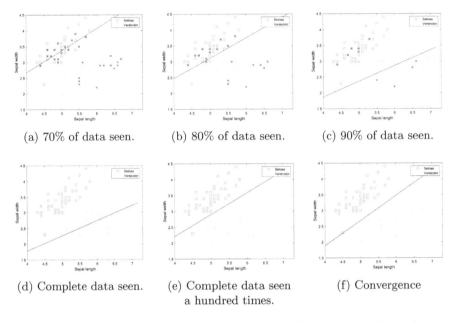

(a) 70% of data seen.   (b) 80% of data seen.   (c) 90% of data seen.

(d) Complete data seen.   (e) Complete data seen   (f) Convergence
a hundred times.

Figure 4.9: Perceptron classification (unseen data is marked by ×).

4.6 The Support Vector Machine

Letting the learning rate α be a choice is crude. All should be determined by the data. When all samples are correctly classified, that is $c_n = \mathrm{sgn}(\hat{\mathbf{w}}^T \hat{\mathbf{v}}_n)$, the *margin* is defined as

$$\min_{n=1,\ldots,N} c_n \frac{\hat{\mathbf{w}}^T \hat{\mathbf{v}}_n}{\|\hat{\mathbf{w}}\|} > 0.$$

The *Support Vector Machine (SVM)* tries to maximize the margin between the two classes.

Note that the margin is independent of any rescaling of $\hat{\mathbf{w}}$, since we divide by the length of $\hat{\mathbf{w}}$. We use the freedom to rescale $\hat{\mathbf{w}}$ such that $c_n \hat{\mathbf{w}}^T \hat{\mathbf{v}}_n - 1 \geq 0$ for all $n = 1, \ldots N$ with equality for at least one sample. Hence we impose some constraints on $\hat{\mathbf{w}}$. In the case of samples, where we have equality, the constraints are said to be *active*. For the other samples, they are *inactive*. Subject to these constraints, maximizing the margin is equivalent to maximizing $\|\hat{\mathbf{w}}\|^{-1}$. Hence, the equivalent problem is: Minimize the *objective function* $\|\hat{\mathbf{w}}\|^2/2$ subject to the constraint $c_n \hat{\mathbf{w}}^T \hat{\mathbf{v}}_n - 1 \geq 0$ for all $n = 1, \ldots N$. This is known as a *quadratic programming problem*. The reasons for which will become clear below, this is called the *primal* optimization problem.

Figure 4.10 illustrates the quadratic programming problem. The circles are the *isolines* of $\|\hat{\mathbf{w}}\|^2/2$, that is lines where it takes a constant value. In the white region, no constraints are satisfied, in the light gray region, one

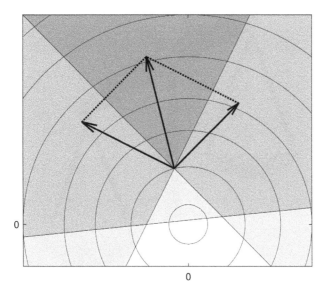

Figure 4.10: Objective function, constraints and gradients.

constraint is satisfied, while in the medium gray region, two constraints are satisfied. In the dark gray region, all constraints are satisfied. This is known as the *feasible set*. The minimum subject to the constraints lies at the point within the feasible set closest to the centre of the circles. In the illustration of Figure 4.10, two constraints are active and one is inactive. At this point, the gradient of $\|\hat{\mathbf{w}}\|^2/2$ is a linear combination of the gradients of the active constraints. In Figure 4.10 the arrows perpendicular to the constraints are the gradients of the constraints multiplied by α_n. The dotted lines illustrate that these add to form the gradient vector of the objective function at the constrained minimum.

The *Lagrangian function* combines the objective function and the constraints into one function and is given by

$$L(\hat{\mathbf{w}}, \boldsymbol{\alpha}) = \frac{1}{2}\|\hat{\mathbf{w}}\|^2 - \sum_{n=1}^{N} \alpha_n \left(c_n \hat{\mathbf{w}}^T \hat{\mathbf{v}}_n - 1 \right),$$

where $\boldsymbol{\alpha} = (\alpha_1, \dots, \alpha_N)^T$ are known as *Lagrange multipliers* which are required to be non-negative. If for a particular $\hat{\mathbf{w}}$, the constraint $c_n \hat{\mathbf{w}}^T \hat{\mathbf{v}}_n - 1 \geq 0$ is satisfied as equality, it does not contribute to the sum. If it is satisfied as a strict inequality, we subtract from the original objective function, but if it is not satisfied, we add to it. Thus, if $\hat{\mathbf{w}}$ lies outside the feasible set, it is penalized.

The function

$$L(\boldsymbol{\alpha}) = \min_{\hat{\mathbf{w}}} L(\hat{\mathbf{w}}, \boldsymbol{\alpha})$$

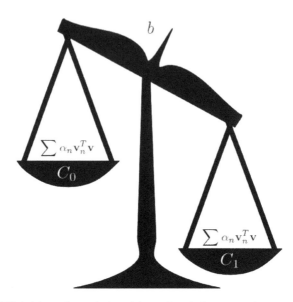

Figure 4.11: Weighing the relationships of training samples to a new sample.

is called the *dual* function of the primal optimization problem. Note that in the feasible set, $L(\boldsymbol{\alpha})$ is always a lower bound of $\|\hat{\mathbf{w}}\|^2/2$. We find the largest lower bound by maximizing with respect to $\boldsymbol{\alpha}$.

We recover the original minimization problem when the multiplier α_n is zero, if the constraint is inactive, and positive otherwise. Formally, this is written as

$$\alpha_n \left(c_n \hat{\mathbf{w}}^T \hat{\mathbf{v}}_n - 1 \right) = 0.$$

This property is known as *complementarity*. If this is the case, the maximum lower bound and the constrained minimum are the same. The solution is a stationary point of $L(\hat{\mathbf{w}}, \boldsymbol{\alpha})$, that is its derivatives vanish. Using Appendices A.2.1 and A.2.2, the derivative of $L(\hat{\mathbf{w}}, \boldsymbol{\alpha})$ with respect to $\hat{\mathbf{w}}$ is

$$\frac{d}{d\hat{\mathbf{w}}} L(\hat{\mathbf{w}}, \boldsymbol{\alpha}) = \hat{\mathbf{w}} - \sum_{n=1}^{N} \alpha_n c_n \hat{\mathbf{v}}_n.$$

Setting this to zero is equivalent to the gradient of the objective function being a linear combination of the gradients of the constraints.

Inserting

$$\hat{\mathbf{w}} = \sum_{n=1}^{N} \alpha_n c_n \hat{\mathbf{v}}_n, \tag{4.3}$$

into the Lagrangian function, we seek $\boldsymbol{\alpha}$ that maximizes the dual function given by

$$L(\boldsymbol{\alpha}) = \sum_{n=1}^{N} \alpha_n - \frac{1}{2} \sum_{i=1}^{N} \sum_{n=1}^{N} \alpha_i \alpha_n c_i c_n \hat{\mathbf{v}}_i^T \hat{\mathbf{v}}_n. \tag{4.4}$$

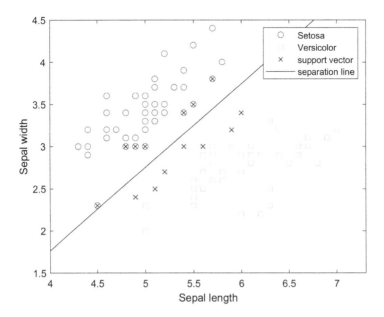

Figure 4.12: Support Vector Machine Classification.

subject to the constraints $\alpha_n \geq 0$ and $\alpha_n \left(c_n \hat{\mathbf{w}}^T \hat{\mathbf{v}}_n - 1\right) = 0$, $n = 1, \ldots, N$. This is the *dual representation* of the maximum margin problem.

The samples for which $\alpha_i \neq 0$ in the linear combination to form the weight vector $\hat{\mathbf{w}}$ are called the *support vectors*. This shaped the name under which the technique is known now. It was invented in 1964 by Vapnik and Lerner as the *Generalized Portrait Method*. The term generalized portrait refers to the centre of a sphere which contains patterns belonging to a certain class, but no other class. That is, it is a representative of this class.

Recall that $\hat{\mathbf{w}} = (-b, \mathbf{w})$ and $\mathbf{v}$ to $\hat{\mathbf{v}} = (1, \mathbf{v})$. Inserting this into (4.3) and (4.4), the bias b is given by $-\sum_{n=1}^{N} \alpha_n c_n$, while the vector of weights is

$$\mathbf{w} = \sum_{n=1}^{N} \alpha_n c_n \mathbf{v}_n.$$

The dual maximization problem becomes

$$L(\boldsymbol{\alpha}) = \sum_{n=1}^{N} \alpha_n - \frac{1}{2} \sum_{i=1}^{N} \sum_{n=1}^{N} \alpha_i \alpha_n c_i c_n \left(1 + \mathbf{v}_i^T \mathbf{v}_n\right).$$

Its solution depends solely on the inner product between samples. The classification is according to the sign of

$$\hat{\mathbf{w}}^T \hat{\mathbf{v}} = \sum_{n=1}^{N} \alpha_n c_n \hat{\mathbf{v}}_n^T \hat{\mathbf{v}} = \sum_{n=1}^{N} \alpha_n c_n \left(1 + \mathbf{v}_n^T \mathbf{v}\right) = -b + \sum_{n=1}^{N} \alpha_n c_n \mathbf{v}_n^T \mathbf{v} = -b + \mathbf{w}^T \mathbf{v}.$$

Again, this only depends on the inner products between the training samples and the new sample weighted by α_n. The inner product can be viewed as measuring the relationship between the two samples. Through the multiplication by c_n, the weighted inner products from one class are added, while the others are subtracted. Depending on where the scales are tipped (Figure 4.11), the new sample is deemed more related to the samples of one class than the other, and classified accordingly.

Since α_n is only non-zero for the support vectors, it is only important how the new sample relates to the support vectors. This is a major advantage of the Support Vector Machine, since it results in a *sparse* model of the classes. Only the support vectors are necessary for future classifications.

We will see the advantages of recasting the problem by examining the relationships between samples when considering non-linear classification and the *kernel trick*. Here the inner product between samples assigns a numeric value to the relationship. The inner product is a counter-intuitive measure. For a given sample vector, all samples lying on the line orthogonal to it are given the value zero in this relation. Samples lying on the same side of this line as the given sample are given a positive value, while samples on the other side are given a negative value. The absolute values increase with the distance from this line.

As we will see later, different measures for the relationship can be chosen.

The MATLAB code in Listing 4.3 uses the Support Vector Machine implementation to calculate the support vectors and plot the separation line. The result can be seen in Figure 4.12.

```
load fisheriris;
% meas contains the first species in rows 1 to 50, the second species
% in rows 51 to 100, and the third species in rows 101 to 150, while
% the columns give the measurements of sepal length, sepal width,
% petal length and petal width in this order.
% Extract two species and two attributes.
m = meas(1:100,1:2);
s = species(1:100);
% Fit a linear separation line using the Support Vector Machine.
SVMModel = fitcsvm(m,s);
% Retrieve support vectors.
sv = SVMModel.SupportVectors;
% Retrieve classes of support vectors.
sl = SVMModel.SupportVectorLabels;
% Retrieve Langrange multipliers.
a = SVMModel.Alpha;
% Calculate vector of weights as linear combination of support vectors
% with the coefficients being the Lagrange multipliers times the class
% labels.
w = (sl.*a)'*sv;
% Retrieve bias.
b = SVMModel.Bias;

figure
% Scatter plot of the data.
gscatter(m(:,1),m(:,2),s,'rg','os');
hold on
% Mark support vectors.
plot(sv(:,1),sv(:,2),'kx')
% Plot separation line.
h = plotpc(w,b);
set(h, 'Color','k');
% Label graph.
legend('Setosa','Versicolor',...
    'support vector','separation line','Location','northeast')
xlabel('Sepal length');
ylabel('Sepal width');
axis([4 7.3 1.5 4.5]);
hold off
```

Listing 4.3: Support Vector Machine.

CHAPTER 5

Non-Linear Classification

The chapter begins by deriving the Quadratic Discriminant analysis as the boundary between two normally distributed classes. This is then shown to be a line in a higher dimensional space leading to the kernel trick. Heuristic methods such as k nearest neighbours and decision trees are covered. For the latter impurity is defined and examples such as the Gini Diversity Index are given. Neural networks are explored further as a set of many conditions of the form to the left or right of a line determined by the network parameters. The chapter concludes with boosting and cascades as a way to build strong classifiers from weak ones and taking the cost into account.

5.1 Quadratic Discriminant Analysis

So far we assumed that the data can be separated by a straight line. However, this assumption of linear classification is very restrictive. Before we go into this, however, we need to recall the subtle difference between two key concepts: *probability* and *likelihood*. Probability answers the question: How probable is it that a sample of class C_i has these features? Likelihood, on the other hand, answers the question: How likely is it that a sample with these features belongs to class C_i?

Let the features of samples in class C_i be normally distributed with mean $\boldsymbol{\mu}_i$ and variance $\boldsymbol{\Sigma}_i$, $i = 0, 1$. Let $\mathbf{v}$ be a feature vector. The probability of $\mathbf{v}$ given that it is in class C_i is:

$$p(\mathbf{v}|\mathbf{v} \in C_i) = \frac{1}{\sqrt{(2\pi)^M |\boldsymbol{\Sigma}_i|}} \exp\left(-\frac{1}{2}(\mathbf{v} - \boldsymbol{\mu}_i)^T \boldsymbol{\Sigma}_i^{-1}(\mathbf{v} - \boldsymbol{\mu}_i)\right).$$

The likelihood of $\mathbf{v} \in C_i$ given that we know its features $\mathbf{v}$ is:

$$\mathcal{L}(\mathbf{v} \in C_i | \mathbf{v}) = \frac{1}{\sqrt{(2\pi)^M |\boldsymbol{\Sigma}_i|}} \exp\left(-\frac{1}{2}(\mathbf{v} - \boldsymbol{\mu}_i)^T \boldsymbol{\Sigma}_i^{-1}(\mathbf{v} - \boldsymbol{\mu}_i)\right)$$

It is the same formula, but a different interpretation.

A classifier can be built by assigning $\mathbf{v}$ to class C_0 if the likelihood of it belonging to C_0 is larger than the likelihood of it belonging to C_1. The boundary between the classes is given by

$$\mathcal{L}(\mathbf{v} \in C_0 | \mathbf{v}) = \mathcal{L}(\mathbf{v} \in C_1 | \mathbf{v}).$$

This is often expressed as likelihood ratio

$$\frac{\mathcal{L}(\mathbf{v} \in C_0 | \mathbf{v})}{\mathcal{L}(\mathbf{v} \in C_1 | \mathbf{v})} = \frac{\sqrt{(2\pi)^M |\boldsymbol{\Sigma}_1|} \exp\left(-\frac{1}{2}(\mathbf{v} - \boldsymbol{\mu}_0)^T \boldsymbol{\Sigma}_0^{-1}(\mathbf{v} - \boldsymbol{\mu}_0)\right)}{\sqrt{(2\pi)^M |\boldsymbol{\Sigma}_0|} \exp\left(-\frac{1}{2}(\mathbf{v} - \boldsymbol{\mu}_1)^T \boldsymbol{\Sigma}_1^{-1}(\mathbf{v} - \boldsymbol{\mu}_1)\right)} = 1.$$

Taking the logarithm, we arrive at

$$\begin{aligned}
\log \frac{\mathcal{L}(\mathbf{v} \in C_0 | \mathbf{v})}{\mathcal{L}(\mathbf{v} \in C_1 | \mathbf{v})} &= \frac{1}{2} \log \frac{|\boldsymbol{\Sigma}_1|}{|\boldsymbol{\Sigma}_0|} - \frac{1}{2}(\mathbf{v} - \boldsymbol{\mu}_0)^T \boldsymbol{\Sigma}_0^{-1}(\mathbf{v} - \boldsymbol{\mu}_0) \\
&\quad + \frac{1}{2}(\mathbf{v} - \boldsymbol{\mu}_1)^T \boldsymbol{\Sigma}_1^{-1}(\mathbf{v} - \boldsymbol{\mu}_1) \\
&= \frac{1}{2}\left(\log \frac{|\boldsymbol{\Sigma}_1|}{|\boldsymbol{\Sigma}_0|} + \boldsymbol{\mu}_1^T \boldsymbol{\Sigma}_1^{-1} \boldsymbol{\mu}_1 - \boldsymbol{\mu}_0^T \boldsymbol{\Sigma}_0^{-1} \boldsymbol{\mu}_0\right) \\
&\quad + \mathbf{v}^T \left(\boldsymbol{\Sigma}_0^{-1} \boldsymbol{\mu}_0 - \boldsymbol{\Sigma}_1^{-1} \boldsymbol{\mu}_1\right) + \frac{1}{2}\mathbf{v}^T \left(\boldsymbol{\Sigma}_1^{-1} - \boldsymbol{\Sigma}_0^{-1}\right) \mathbf{v} = 0.
\end{aligned}$$

Thus the boundary between classes is given by a quadratic of the form

$$\mathbf{v}^T \mathbf{A} \mathbf{v} + \mathbf{v}^T \mathbf{b} + c = 0$$

with

$$\begin{aligned}
\mathbf{A} &= \frac{1}{2}\left(\boldsymbol{\Sigma}_1^{-1} - \boldsymbol{\Sigma}_0^{-1}\right), \\
\mathbf{b} &= \boldsymbol{\Sigma}_0^{-1} \boldsymbol{\mu}_0 - \boldsymbol{\Sigma}_1^{-1} \boldsymbol{\mu}_1, \\
c &= \frac{1}{2}\left(\log \frac{|\boldsymbol{\Sigma}_1|}{|\boldsymbol{\Sigma}_0|} + \boldsymbol{\mu}_1^T \boldsymbol{\Sigma}_1^{-1} \boldsymbol{\mu}_1 - \boldsymbol{\mu}_0^T \boldsymbol{\Sigma}_0^{-1} \boldsymbol{\mu}_0\right).
\end{aligned}$$

This is the *Quadratic Discriminant Analysis (QDA)*.

In the special case of two features, the boundaries are conic sections, that is a line, circle, ellipse, parabola or hyperbola. The MATLAB code in Listing 5.1 applies the Quadratic Discriminant Analysis to Fisher's iris data and the result can be seen in Figure 5.1.

Recall that we arrived at the Linear Discriminant Analysis by assuming $\boldsymbol{\Sigma}_0 = \boldsymbol{\Sigma}_1 = \boldsymbol{\Sigma}$. In this case,

$$\begin{aligned}
\mathbf{A} &= 0, \\
\mathbf{b} &= \boldsymbol{\Sigma}^{-1}(\boldsymbol{\mu}_0 - \boldsymbol{\mu}_1), \\
c &= \frac{1}{2}\left(\boldsymbol{\mu}_1^T \boldsymbol{\Sigma}^{-1} \boldsymbol{\mu}_1 - \boldsymbol{\mu}_0^T \boldsymbol{\Sigma}^{-1} \boldsymbol{\mu}_0\right).
\end{aligned}$$

```
load fisheriris;
% extract two attributes
pl = meas(:,3); % petal length
pw = meas(:,4); % petal width
figure;
h1 = gscatter(pl, pw, species,'rgb','os^');
legend('Setosa','Versicolor','Virginica','Location','best')
hold on;
X = [pl,pw];
cls = ClassificationDiscriminant.fit(X,species,...
    'DiscrimType','quadratic');
% plot the classification boundaries
% retrieve the coefficients for the quadratic boundary between
% the first and second class (setosa and versicolor).
c = cls.Coeffs(1,2).Const;
l = cls.Coeffs(1,2).Linear;
q = cls.Coeffs(1,2).Quadratic;

% plot the curve c + [x1,x2]*l + [x1,x2]*q*[x1,x2]' = 0:
f = @(x1,x2) c + l(1)*x1 + l(2)*x2 + q(1,1)*x1.^2 + ...
    (q(1,2)+q(2,1))*x1.*x2 + q(2,2)*x2.^2;
h2 = ezplot(f,[.9 7.1 0 1]);
set(h2, 'Color','k');
% retrieve the coefficients for the quadratic boundary between
% the second and third class (versicolor and virginica).
c = cls.Coeffs(2,3).Const;
l = cls.Coeffs(2,3).Linear;
q = cls.Coeffs(2,3).Quadratic;
% plot the curve c + [x1,x2]*l + [x1,x2]*q*[x1,x2]' = 0:
f = @(x1,x2) c + l(1)*x1 + l(2)*x2 + q(1,1)*x1.^2 + ...
    (q(1,2)+q(2,1))*x1.*x2 + q(2,2)*x2.^2;
h3 = ezplot(f,[.5 7 0 2.5]);
set(h3, 'Color','k');
xlabel('Petal length');
ylabel('Petal width');
axis([0.8 7 0 3])
title('');
```

Listing 5.1: Quadratic Discriminant Analysis.

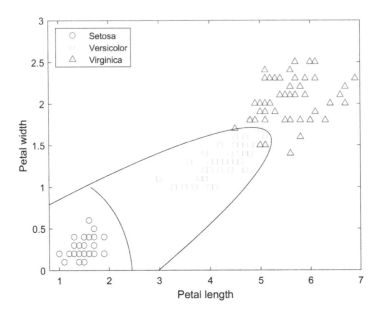

Figure 5.1: Quadratic Discriminant Analysis.

This is exactly the equation of the line in the Linear Discriminant Analysis.

In the following section, we look at the Quadratic Discriminant Analysis from a different angle by introducing artificial features and extending the feature space this way.

5.2 Kernel Trick

In the quadratic

$$\mathbf{v}^T \mathbf{A} \mathbf{v} + \mathbf{v}^T \mathbf{b} + c = 0$$

the matrix $\mathbf{A}$ is symmetric, since A is the difference of covariance matrices which are symmetric. Assuming $M = 2$ and writing

$$\mathbf{A} = \begin{pmatrix} A_{11} & A_{12} \\ A_{12} & A_{22} \end{pmatrix}, \mathbf{b} = \begin{pmatrix} b_1 \\ b_2 \end{pmatrix},$$

the boundary equation can be written as

$$A_{11}x_1^2 + 2A_{12}v_1v_2 + A_{22}v_2^2 + b_1v_1 + b_2v_2 + c = 0.$$

Now we can rearrange this as

$$\begin{pmatrix} A_{11} & 2A_{12} & A_{22} & b_1 & b_2 \end{pmatrix} \begin{pmatrix} v_1^2 \\ v_1v_2 \\ v_2^2 \\ v_1 \\ v_2 \end{pmatrix} + c = 0.$$

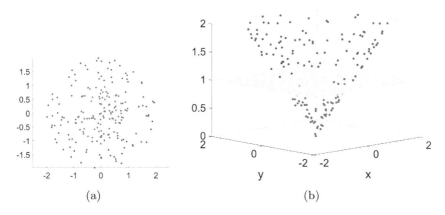

Figure 5.2: Linearly inseparable data and their transformation to a higher dimensional space.

Letting $\mathbf{w}^T = \begin{pmatrix} A_{11} & 2A_{12} & A_{22} & b_1 & b_2 \end{pmatrix}$ and $\check{\mathbf{v}} = \begin{pmatrix} v_1^2 \\ v_1 v_2 \\ v_2^2 \\ v_1 \\ v_2 \end{pmatrix}$, this describes a linear boundary in a five-dimensional feature space.

$$\mathbf{w}^T \check{\mathbf{v}} + c = 0.$$

In other words, we have augmented the feature space by three artificial features

$$v_1^2, \qquad v_1 v_2, \qquad v_2^2.$$

This is known as the *kernel trick*. The features are transformed to a higher dimensional feature space where the classes are linearly separable. We will see below that in practice this transformation is never performed.

Figure 5.2 shows artificial data, which cannot be separated by a line, and their transformation to a higher dimensional space, where they can be separated by a plane. We set $v_3 = \sqrt{v_1^2 + v_2^2}$, and the plane lies at $v_3 = 1$. Going back to two dimensions, $v_3^2 = v_1^2 + v_2^2 = 1$ is the unit circle in the v_1, v_2-plane.

To formalize this approach, recall that the Support Vector Machine classifies according to the sign of

$$-b + \sum_{n=1}^{N} \alpha_n c_n \mathbf{v}_n^T \mathbf{v},$$

where the bias b is given by $-\sum_{n=1}^{N} \alpha_n c_n$ and where $c_i = \pm 1$ is the class label of sample $\mathbf{v}_i$. The $\alpha_i \geq 0$ for $i = 1, \ldots, N$ maximize the function

$$L(\boldsymbol{\alpha}) = \sum_{n=1}^{N} \alpha_n - \frac{1}{2} \sum_{i=1}^{N} \sum_{n=1}^{N} \alpha_i \alpha_n c_i c_n \left(1 + \mathbf{v}_i^T \mathbf{v}_n \right).$$

Let $\phi : \mathbb{R}^M \to \mathbb{R}^{\hat{M}}$ be the mapping from the original M dimensional feature space to the higher $\hat{M}$ dimensional feature space. We now apply the support vector machine to the data in this higher dimensional space; i.e. we need to maximize

$$L(\boldsymbol{\alpha}) = \sum_{n=1}^{N} \alpha_n - \frac{1}{2} \sum_{i=1}^{N} \sum_{n=1}^{N} \alpha_i \alpha_n c_i c_n \left(1 + \phi(\mathbf{v}_i)^T \phi(\mathbf{v}_n)\right).$$

Let $k : \mathbb{R}^M \times \mathbb{R}^M \to \mathbb{R}$ be defined as

$$k(\mathbf{x}, \mathbf{y}) = \phi(\mathbf{x})^T \phi(\mathbf{y}).$$

This is known as the *kernel function* or *kernel* for short. Using this notation, the objective function to be maximized becomes

$$L(\boldsymbol{\alpha}) = \sum_{n=1}^{N} \alpha_n - \frac{1}{2} \sum_{i=1}^{N} \sum_{n=1}^{N} \alpha_i \alpha_n c_i c_n \left(1 + k(\mathbf{v}_i, \mathbf{v}_n)\right),$$

subject to $\alpha_i \geq 0$, $i = 1, \ldots, N$. The classification is according to the sign of

$$-b + \sum_{n=1}^{N} \alpha_n c_n \phi(\mathbf{v}_n)^T \phi(\mathbf{v}) = -b + \sum_{n=1}^{N} \alpha_n c_n k(\mathbf{v}_n, \mathbf{v}).$$

We see the mapping ϕ never has to be evaluated, only the kernel function k in both the maximization and classification.

The kernel trick can be applied to all linear classification methods to separate data which is not linearly separable. All methods reduce to evaluations of the kernel function k.

The kernel function was derived from a higher dimensional inner product. In the previous chapter, we viewed the inner product as a measure of the relationship between different samples. A new sample is classified according to whether it is more related to the support vectors of one class than the other. The kernel function gives a numeric value describing the relationship between two samples.

Such functions describing relationships between samples can be constructed. A necessary and sufficient condition that such a function is suitable for our purposes is that the $N \times N$ matrix $\mathbf{K}$ with (m, n) entry equal to $k(\mathbf{x}_m, \mathbf{x}_n)$ is positive semi-definite for any N and for any set $\{\mathbf{x}_1, \ldots, \mathbf{x}_N\}$. It is known as *Gram* or *Gramian matrix*. The constant function $k(\mathbf{x}, \mathbf{y}) = c$ for positive c is a valid kernel, since then for any vector $\mathbf{v} = (v_1, \ldots, v_n)^T$

$$\mathbf{v}^T \mathbf{K} \mathbf{v} = c \left(\sum n = 1^N v_n \right)^2 \geq 0.$$

We also have positive semi-definiteness of $\mathbf{K}$, if the kernel function is defined via an inner product in a higher dimensional space, because then

$$
\begin{aligned}
\mathbf{v}^T \mathbf{K} \mathbf{v} &= \sum_{m,n=1}^{N} v_m \phi(\mathbf{x}_m)^T \phi(\mathbf{x}_n) v_n \\
&= \left(\sum_{m=1}^{N} v_m \phi(\mathbf{x}_m) \right)^T \left(\sum_{n=1}^{N} v_n \phi(\mathbf{x}_n) \right) = \| \sum_{n=1}^{N} v_n \phi(\mathbf{x}_n) \|^2 \geq 0.
\end{aligned}
$$

Therefore the inner product itself is a valid kernel.

If $\mathbf{A}$ is a symmetric, positive semi-definite, $M \times M$ matrix, then the function defined by

$$
k(\mathbf{x}, \mathbf{y}) = \mathbf{x}^T \mathbf{A} \mathbf{y}
$$

is a valid kernel function, since in this case

$$
\mathbf{v}^T \mathbf{K} \mathbf{v} = \sum_{m,n=1}^{N} v_m \mathbf{x}_m^T \mathbf{A} \mathbf{x}_n v_n = \left(\sum_{m=1}^{N} v_m \mathbf{x}_m \right)^T \mathbf{A} \left(\sum_{n=1}^{N} v_n \mathbf{x}_n \right) \geq 0.
$$

Similarly, if $f : \mathbb{R}^M \to \mathbb{R}$ is any function and $k : \mathbb{R}^M \times \mathbb{R}^M \to \mathbb{R}$ is a valid kernel, then

$$
\sum_{m,n=1}^{N} v_m f(\mathbf{x}_m) k(\mathbf{x}_m, \mathbf{x}_n) f(\mathbf{x}_n) v_n = \mathbf{w}^T K \mathbf{w} \geq 0, \tag{5.1}
$$

where $\mathbf{w} = (f(\mathbf{x}_1) v_1, \ldots, f(\mathbf{x}_N) v_N)^T$. Therefore, the function defined by $f(\mathbf{x}) k(\mathbf{x}, \mathbf{y}) f(\mathbf{y})$ is a kernel function.

Multiplying a valid kernel by a positive constant is obviously a kernel, as is the sum of two valid kernels. Also, the product of two valid kernels is a kernel. Let $\{\mathbf{x}_1, \ldots, \mathbf{x}_N\}$ be an arbitrary set. Let $\mathbf{K}$ denote the Gram matrix of the product of kernels. Since the Gram matrices of the two kernels, which are the factors, are symmetric and positive semi-definite, they can be written as products $\mathbf{A}^T \mathbf{A}$ and $\mathbf{B}^T \mathbf{B}$ respectively. Thus, their (m, n) entries are given by

$$
\sum_{i=1}^{N} A_{mi} A_{ni} \quad \text{and} \quad \sum_{j=1}^{N} B_{mj} B_{nj}
$$

respectively. Now, for an arbitrary vector $\mathbf{v} = (v_1, \ldots, v_n)^T$,

$$
\begin{aligned}
\mathbf{v}^T \mathbf{K} \mathbf{v} &= \sum_{m,n=1}^{N} v_m K_{mn} v_n = \sum_{m,n=1}^{N} v_m \left(\sum_{i=1}^{N} A_{mi} A_{ni} \right) \left(\sum_{j=1}^{N} B_{mj} B_{nj} \right) v_n \\
&= \sum_{i,j=1}^{N} \left(\sum_{m=1}^{N} v_m A_{mi} B_{mj} \right) \left(\sum_{n=1}^{N} v_n A_{ni} B_{nj} \right) \\
&= \sum_{i,j=1}^{N} \left(\sum_{m=1}^{N} v_m A_{mi} B_{mj} \right)^2 \geq 0.
\end{aligned}
$$

Hence, $\mathbf{K}$ is positive semi-definite, and the product of two kernels is a kernel. From this we can deduce that, if p is a polynomial with non-negative coefficients, then $p(k(\mathbf{x}, \mathbf{y}))$ is a valid kernel, if $k(\mathbf{x}, \mathbf{y})$ is a valid kernel function. Also $\exp(k(\mathbf{x}, \mathbf{y}))$ is a valid kernel as the limit of a polynomial with positive coefficients.

Since the positive semi-definiteness has to hold for any set of points, it has to also hold for $\{\psi(\mathbf{a}_1), \ldots, \psi(\mathbf{a}_N)\}$, where $\psi : \mathbb{R}^a \rightarrow \mathbb{R}^M$ is any mapping and $\{\mathbf{a}_1, \ldots, \mathbf{a}_N\}$ is any set in $\mathbb{R}^a$. Therefore, if $k(\mathbf{x}, \mathbf{y})$ is a valid kernel, so is $k(\psi(\mathbf{a}), \psi(\mathbf{b}))$ on $\mathbb{R}^a \times \mathbb{R}^a$. In particular, ψ can be the restriction onto a subset of components of $\mathbf{x}$. Let $\mathbf{x}_a, \mathbf{y}_a \in \mathbb{R}^a$ and $\mathbf{x}_b, \mathbf{y}_b \in \mathbb{R}^b$ be such restrictions. They do not necessarily need to be disjoint or use all components of $\mathbf{x}$. Let $k_a(\mathbf{x}_a, \mathbf{y}_a)$ and $k_b(\mathbf{x}_b, \mathbf{y}_b)$ be the kernels obtained by the restrictions; then

$$k(\mathbf{x}, \mathbf{y}) = k_a(\mathbf{x}_a, \mathbf{y}_a) + k_b(\mathbf{x}_b, \mathbf{y}_b) \quad \text{and} \quad k(\mathbf{x}, \mathbf{y}) = k_a(\mathbf{x}_a, \mathbf{y}_a) k_b(\mathbf{x}_b, \mathbf{y}_b)$$

are valid kernels on $\mathbf{R}^M \times \mathbf{R}^M$.

Thus, there is a multitude of possibilities to construct kernels. Some popular examples of kernels are:

Linear (trivial) kernel $k(\mathbf{x}, \mathbf{y}) = \theta_1^2 + \theta_2^2 \mathbf{x}^T \mathbf{y}$.

Quadratic kernel $k(\mathbf{x}, \mathbf{y}) = (\theta_1^2 + \theta_2^2 \mathbf{x}^T \mathbf{y})^2$.

Polynomial kernel (of degree d) $k(\mathbf{x}, \mathbf{y}) = (\theta_1^2 + \theta_2^2 \mathbf{x}^T \mathbf{y})^d$.

Hyperbolic tangent (Sigmoid) kernel $k(\mathbf{x}, \mathbf{y}) = \tanh(\theta_1^2 + \theta_2^2 \mathbf{x}^T \mathbf{y})$.

Additive and multiplicative constants have to be positive. To ensure this, squared parameters are chosen. This also helps, when the parameters are optimized, since then this is an unconstrained optimization problem, while otherwise it would be a constrained optimization problem.

Considering the inner product $\mathbf{x}^T \mathbf{y}$ for a given vector $\mathbf{x}$, vectors $\mathbf{y}$ lying on the line orthogonal to $\mathbf{x}$ are given the value zero for this relationship. Samples lying on the same side of this line as $\mathbf{x}$ are given a positive value, while samples on the other side are given a negative value. The absolute values increase with the distance from this line. The additive constant θ_1 shifts the zero line, while the multiplicative constant θ_2 controls the rate of this increase. For the linear kernel the shift is absorbed in the bias. Hence setting θ_1 to zero in this context suffices.

The hyperbolic tangent kernel addresses the problem that absolute values of the inner product increase with the distance from the zero line. As the arguments of the hyperbolic tangent tend to plus or minus infinity, the result tends to $+1$ or -1 respectively. The values of the hyperbolic tangent are illustrated in Figure 5.3 for different multiplicative constants θ_2, but fixed $\mathbf{y} = (1, 1)^T$ and $\theta_1 = 0$. For smaller θ_2 the change from -1 to $+1$ is less rapid. The additive and multiplicative constants are parameters which need to be optimized or tuned for the specific application. When tuning, a validation set

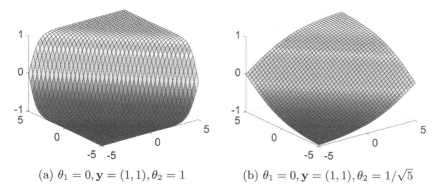

(a) $\theta_1 = 0, \mathbf{y} = (1,1), \theta_2 = 1$ (b) $\theta_1 = 0, \mathbf{y} = (1,1), \theta_2 = 1/\sqrt{5}$

Figure 5.3: Hyperbolic tangent kernel.

is used to ensure the resulting classifier generalizes well to unseen data. It is important to note that for certain choices of θ_1 and θ_2, the hyperbolic tangent is not a valid kernel. For example, for $\theta_1 = \theta_2 = 1$ and $\mathbf{x}_1 = (0.5, 0)^T$ and $\mathbf{x}_2 = (1, 0)$, the Gram matrix is

$$\mathbf{K} = \left(\begin{array}{cc} \tanh(1.25) & \tanh(1.5) \\ \tanh(1.5) & \tanh(2) \end{array} \right)$$

and has a negative eigenvalue and therefore is not positive semi-definite. Nevertheless, the hyperbolic tangent kernel is used in practice. Burges in [7] derives necessary conditions for functions based on the inner product to be valid kernels, as well as giving insight into the construction of kernels.

In some cases, it is possible to recover the mapping ϕ. To illustrate this let $M = 2$ and we consider the quadratic kernel:

$$
\begin{aligned}
k(\mathbf{x}, \mathbf{y}) &= (\theta_1^2 + \theta_2^2 \mathbf{x}^T \mathbf{y})^2 \\
&= \left(\theta_1^2 + \theta_2^2 x_1 y_1 + \theta_2^2 x_2 y_2 \right)^2 \\
&= \theta_1^4 + \theta_2^4 x_1^2 y_1^2 + \theta_2^4 x_2^2 y_2^2 + 2\theta_2^4 x_1 y_1 x_2 y_2 + 2\theta_1^2 \theta_2^2 x_1 y_1 + 2\theta_1^2 \theta_2^2 x_2 y_2 \\
&= \left(\theta_1^2, \theta_2^2 x_1^2, \theta_2^2 x_2^2, \sqrt{2}\theta_2^2 x_1 x_2, \sqrt{2}\theta_1 \theta_2 x_1, \sqrt{2}\theta_1 \theta_2 x_2 \right) \left(\begin{array}{c} \theta_1^2 \\ \theta_2^2 y_1^2 \\ \theta_2^2 y_2^2 \\ \sqrt{2}\theta_2^2 y_1 y_2 \\ \sqrt{2}\theta_1 \theta_2 y_1 \\ \sqrt{2}\theta_1 \theta_2 y_2 \end{array} \right) \\
&= \phi(\mathbf{x})^T \phi(\mathbf{y}).
\end{aligned}
$$

Hence, in this case, the quadratic kernel is an implicit mapping to a six-dimensional space.

It is more intuitive to let the relationship between samples depend on the distance between them, that is the *Euclidean norm* (L_2 *norm*) of their difference, $\|\mathbf{x} - \mathbf{y}\|$. This leads to the class of *radial basis function (RBF)* kernels.

Gaussian kernel $k(\mathbf{x}, \mathbf{y}) = \theta_1^2 \exp\left(-\dfrac{\|\mathbf{x} - \mathbf{y}\|^2}{2\theta_2^2}\right).$

Exponential kernel $k(\mathbf{x}, \mathbf{y}) = \theta_1^2 \exp\left(-\dfrac{\|\mathbf{x} - \mathbf{y}\|}{\theta_2}\right).$

Multiquadric kernel $\theta_1^2 \left(1 + \dfrac{\|\mathbf{x} - \mathbf{y}\|^2}{2\theta_2^2}\right)^{1/2}.$

Inverse multiquadric kernel $\theta_1^2 \left(1 + \dfrac{\|\mathbf{x} - \mathbf{y}\|^2}{2\theta_2^2}\right)^{-1/2}.$

Rational quadratic kernel $\theta_1^2 \left(1 + \dfrac{\|\mathbf{x} - \mathbf{y}\|^2}{2\alpha\theta_2^2}\right)^{-\alpha}, \alpha > 0.$

Thin plate spline kernel $\theta_1^2 \|\mathbf{x} - \mathbf{y}\|^2 \log(\|\mathbf{x} - \mathbf{y}\| + \theta_2^2).$

As an example, we show that the Gaussian kernel is a valid kernel. Firstly,

$$\|\mathbf{x} - \mathbf{y}\|^2 = \mathbf{x}^T\mathbf{x} - 2\mathbf{x}^T\mathbf{y} + \mathbf{y}^T\mathbf{y}$$

and the kernel can be written as

$$
\begin{aligned}
k(\mathbf{x}, \mathbf{y}) &= \theta_1 \exp\left(-\frac{1}{2\theta_2^2}\mathbf{x}^T\mathbf{x}\right) \exp\left(\frac{1}{\theta_2^2}\mathbf{x}^T\mathbf{y}\right) \theta_1 \exp\left(-\frac{1}{2\theta_2^2}\mathbf{y}^T\mathbf{y}\right) \\
&= f(\mathbf{x}) \exp\left(\frac{1}{\theta_2^2}\mathbf{x}^T\mathbf{y}\right) f(\mathbf{y}).
\end{aligned}
$$

This is a valid kernel, since it is the exponential of the inner product multiplied by a positive constant and because of Equation (5.1).

The Gaussian kernel is also known as a *squared exponential kernel* or *exponentiated quadratic kernel*. It is an intuitive similarity measure. If the sample $\mathbf{y}$ happens to be the same as sample $\mathbf{x}$, the Gaussian kernel evaluates to θ_1^2, its maximum value. θ_1 is known as the *signal standard deviation*. With the Euclidean distance between samples increasing, the similarity measure given by the Gaussian kernel decreases, the rate of decrease being determined by θ_2, which is the *characteristic length scale*.

Figure 5.4 gives two examples of Gaussian kernels, for different choices of θ_2. The choice of θ_2 is crucial in the performance of the kernel, since it determines the width of the kernel. The larger the θ_2, the wider the base of the kernel. The kernels are centred at the support vectors. If the width is chosen too narrow, the kernels evaluate to nearly zero for all samples except the ones close to support vectors. If it is chosen too wide, the kernels evaluate to values similar to each other, since the slope is at a low angle. It is then hard to make a distinction between samples.

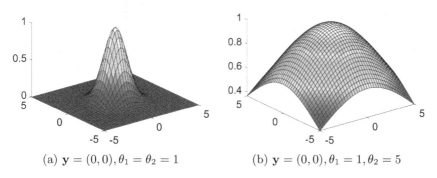

(a) $\mathbf{y} = (0,0), \theta_1 = \theta_2 = 1$ (b) $\mathbf{y} = (0,0), \theta_1 = 1, \theta_2 = 5$

Figure 5.4: Gaussian kernel.

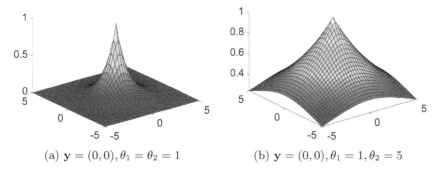

(a) $\mathbf{y} = (0,0), \theta_1 = \theta_2 = 1$ (b) $\mathbf{y} = (0,0), \theta_1 = 1, \theta_2 = 5$

Figure 5.5: Exponential kernel.

The exponential kernel is also an intuitive similarity measure. Equality of $\mathbf{x}$ and $\mathbf{y}$ is, however, more emphasized than by the Gaussian kernel with the exponential kernel being sharply peaked at zero. Figure 5.5 gives two examples for different choices of θ_2.

Some of the other kernels are shown in Figure 5.6. The multiquadric and thin plate spline kernels are contra-intuitive as similarity measures, since their value increases as the distance between $\mathbf{x}$ and $\mathbf{y}$ increases. Also their Gram matrices are not necessarily positive semi-definite. With the addition of a polynomial kernel, they can be made to be positive definite (see for example [6]).

The parameter α in the rational quadratic kernel

$$k(\mathbf{x}, \mathbf{y}) = \theta_1^2 \left(1 + \frac{\|\mathbf{x} - \mathbf{y}\|^2}{2\alpha\theta_2^2} \right)^{-\alpha},$$

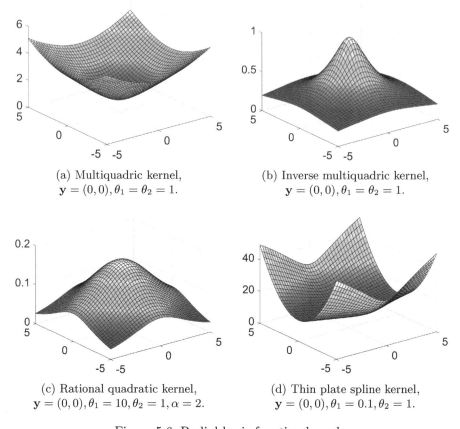

(a) Multiquadric kernel,
$\mathbf{y} = (0,0), \theta_1 = \theta_2 = 1$.

(b) Inverse multiquadric kernel,
$\mathbf{y} = (0,0), \theta_1 = \theta_2 = 1$.

(c) Rational quadratic kernel,
$\mathbf{y} = (0,0), \theta_1 = 10, \theta_2 = 1, \alpha = 2$.

(d) Thin plate spline kernel,
$\mathbf{y} = (0,0), \theta_1 = 0.1, \theta_2 = 1$.

Figure 5.6: Radial basis function kernels.

is known as *scale-mixture parameter*. As α increases, the error in the approximation

$$1 + \frac{\|\mathbf{x} - \mathbf{y}\|^2}{2\alpha\theta_2^2} = \exp\left(\frac{\|\mathbf{x} - \mathbf{y}\|^2}{2\alpha\theta_2^2}\right) + O(\alpha^2)$$

decreases for fixed $\mathbf{x}$ and $\mathbf{y}$. Therefore, as α approaches infinity, the rational quadratic kernel becomes the Gaussian kernel. Using expansions, the rational quadratic kernel can be viewed as an infinite sum of Gaussian kernel with different characteristic length scales. Radial basis functions have many theoretical properties which can influence their performance; see for example [6].

All feature dimensions are treated equally in the kernel definitions so far. However, one feature might be more spread out than another. One possibility to tackle this is, instead of using the characteristic length scale θ_2, to have a separate length scale θ_{m+1} for each feature dimension $m = 1, \ldots, M$. That is, instead of $\|\mathbf{x} - \mathbf{y}\|^2/\theta_2^2$, we have

$$\left(\frac{x_1 - y_1}{\theta_2}\right)^2 + \ldots + \left(\frac{x_M - y_M}{\theta_{M+1}}\right)^2.$$

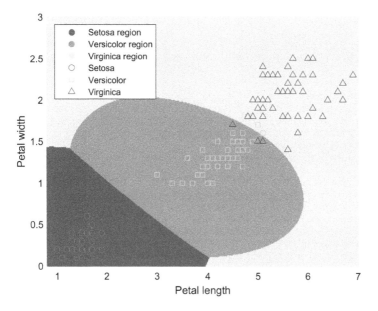

Figure 5.7: Support Vector Machine classification with Gaussian kernel.

Increasing the number of parameters to be chosen is, however, better avoided. Instead, the training data is *standardized*. For each feature dimension, the training data, $\mathbf{v}_1, \ldots, \mathbf{v}_N$, is shifted by the mean and scaled by the standard deviation. Writing $\mathbf{v}_n = (v_{n1}, \ldots, v_{nM})^T$, for $m = 1, \ldots, M$ we calculate

$$\mu_m = \frac{1}{N} \sum_{n=1}^{N} v_{nm} \text{ and } \sigma_m^2 = \frac{1}{N} \sum_{n=1}^{N} (v_{nm} - \mu_m)^2.$$

The standardized training data is then $\tilde{\mathbf{v}}_1, \ldots, \tilde{\mathbf{v}}_N$ with

$$\tilde{v}_{nm} = \frac{v_{nm} - \mu_m}{\sigma_m}.$$

The research on kernels is an active field with kernels being developed for specific applications. Their usefulness lies in the fact that they do not need to be defined over $\mathbb{R}^M \times \mathbb{R}^M$, but can be defined over different categories of objects. An example are string kernels [37] and graph kernels [32]. It all comes down to defining a suitable similarity measure between objects.

The MATLAB code in Listing 5.2 applies a kernel Support Vector Machine to Fisher's iris data using the Gaussian kernel. The results can be seen in Figure 5.7.

```
load fisheriris
% extract two attributes
pl = meas(:,3); % petal length
pw = meas(:,4); % petal width
X = [pl,pw];

% determine classes
classes = unique(species);
% classifiers are constructed on the principle of One versus All
% as many classifier as classes are needed
SVMModels = cell(numel(classes),1);
rng(1); % seeding the random number generator for reproducibility

for j = 1:numel(classes)
    % create binary classes for each classifier
    indx = strcmp(species,classes(j));
    % create classifier
    SVMModels{j} = fitcsvm(X,indx,'ClassNames',[false true],...
        'Standardize',true,...        % standardize data
        'KernelFunction','gaussian'); % specifying the kernel

end
% lay grid over the region
d = 0.01;
[x1Grid,x2Grid] = meshgrid(0.8:d:7,0:d:3);
xGrid = [x1Grid(:),x2Grid(:)];
N = size(xGrid,1);
Scores = zeros(N,numel(classes));

% for each grid point calculate the score of each classifier
for j = 1:numel(classes)
    % predict both returns the predicted class labels as well as a
    % score indicating the likelihood of the negative class (false)
    % and positive class (true)
    [¬,score] = predict(SVMModels{j},xGrid);
    Scores(:,j) = score(:,2); % second column contains positive
                              % class scores
end
% classify according to the maximum score
[¬,maxScore] = max(Scores,[],2);

% plot classifier regions
figure
h(1:3) = gscatter(xGrid(:,1),xGrid(:,2),maxScore,...
    [0.5 0.5 0.5; 0.7 0.7 0.7; 0.9 0.9 0.9]);
hold on
% plot data
h(4:6) = gscatter(pl, pw, species,'rgb','os^');
xlabel('Petal length');
ylabel('Petal width');
legend(h,{'Setosa region','Versicolor region','Virginica region',...
    'Setosa','Versicolor','Virginica'},...
    'Location','Northwest');
axis([0.8 7 0 3])
hold off
```

Listing 5.2: Support Vector Machine using Gaussian kernels.

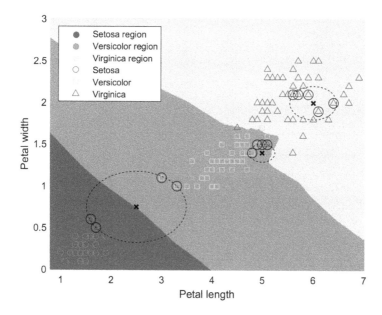

Figure 5.8: k Nearest Neighbour classification.

5.3 k Nearest Neighbours

In the previous chapter, we considered classifying samples according to their relationship to samples in the training set. This relation was quantified using kernels. A simple idea is to classify a new sample according to the majority of classes to which its neighbours in the training set belong. Here neighbour is defined as the samples closest with regards to the *Euclidean norm*, also known as the L_2 *norm*. The number of neighbours considered is denoted by k. The classifier depends on the choice of k. If a small number of neighbours is chosen, then the boundaries between classes are more irregular, while they are smoother, if k is chosen larger. The technique is known as k *Nearest Neighbours (k-NN)*. Listing 5.3 creates a k Nearest Neighbour classifier and uses it to classify the entire region. Adjusting the parameter specifying the number of neighbours shows the different outcomes. Three new sample points are also generated, their k nearest neighbours marked and ellipses drawn around the nearest neighbours. Figure 5.8 shows the results.

5.4 Decision Trees

Another intuitive way to classify are *decision trees*. These are binary trees consisting of *nodes* and *leaves*, also known as *terminal nodes*. Each non-terminal node has two branches. A sample enters the tree at the *root* node at the top. At each node, a decision is made based on the value of a single feature. If

```
load fisheriris
% Extract two attributes.
pl = meas(:,3); % petal length
pw = meas(:,4); % petal width
X = [pl,pw];

% Create classifier.
k = 5;
kNNModel = fitcknn(X,species,...
    'NumNeighbors',k,...            % number of neighbours
    'Standardize',true);           % standardize data

% Lay grid over the region.
d = 0.01;
[x1Grid,x2Grid] = meshgrid(0.8:d:7,0:d:3);
xGrid = [x1Grid(:),x2Grid(:)];
N = size(xGrid,1);

% For each grid point calculate the score of each class.
% 'predict' returns the predicted class labels corresponding to the
% minimum misclassification cost, the score (posterior probability)
% for each class as well as the expected classification cost for
% each class
[¬,score,¬] = predict(kNNModel,xGrid);

% Classify according to the maximum score.
[¬,maxScore] = max(score,[],2);

% Plot classifier regions.
figure
h(1:3) = gscatter(xGrid(:,1),xGrid(:,2),maxScore,...
    [0.5 0.5 0.5; 0.7 0.7 0.7; 0.9 0.9 0.9]);
hold on
% Plot data.
h(4:6) = gscatter(pl, pw, species,'rgb','os^');
xlabel('Petal length');
ylabel('Petal width');
legend(h,{'Setosa region','Versicolor region','Virginica region',...
    'Setosa','Versicolor','Virginica'},...
    'Location','Northwest');
axis([0.8 7 0 3])

% Plot new points with ellipses containing nearest neighbours.
newpoints = [2.5 .75;...
    5 1.4;...
    6 2];
plot(newpoints(:,1),newpoints(:,2),'xk','linewidth',1.5);
% Find nearest neighbours.
[idx,d] = knnsearch(X,newpoints,...
    'k',k,...                       % number of neighbours
    'Distance','seuclidean');% Euclidean distance on standardized data
% Mark neighbours.
plot(X(idx,1),X(idx,2),'ok','markersize',10);
% Plot ellipses.
for i=1:3
    s = kNNModel.Sigma *d(i,end); % scale standardized coordinates
    c = newpoints(i,:) − s;   % corner of rectangle containing ellipse
    % Draw an ellipse around the nearest neighbours.
    h = rectangle('position',[c,2*s(1),2*s(2)],...
        'curvature',[1 1],'Linestyle','—');
end
hold off
```

Listing 5.3: k Nearest Neighbours.

the feature is continuous, the decision is based on the question of whether the value is larger or smaller than a threshold. The sample traverses the tree down to a leaf. Each leaf is associated with a class. Hence, the sample is assigned the class of the leaf it ends up in.

Decision trees are grown recursively from a training set. There are many possible ways the tree can grow. At each node, the set of training samples which reached that node is split in two. Each of the two subsets is passed down one of the branches to the child nodes. At each node, we need to choose the feature on which to base the split, and the threshold. To do so all possible splits for every feature are considered, and the optimal split is selected. The optimality is based on different possible criteria.

Before we discuss the criteria, we need to introduce some notation. Let $t = P, L$, or R refer to quantities relating to the parent, left or right child node respectively. For example, N_P is the number of samples reaching the parent node, while N_L and N_R are the number of samples reaching the left and right child nodes. Let N_{tk}, be the number of samples in class k, where $k = 1, \ldots, K$, reaching node t. We define p_{tk} as the portion of samples belonging to class k at node t. That is

$$p_{tk} = \frac{N_{tk}}{N_t},$$

and we have

$$\sum_{k=1}^{K} p_{tk} = 1.$$

The proportions of the parent node are related to the proportions of the child nodes in the following way:

$$p_{Pk} = \frac{N_{Pk}}{N_P} = \frac{N_{Lk} + N_{Rk}}{N_P} = \frac{N_L p_{Lk} + N_R p_{Rk}}{N_P}. \tag{5.2}$$

A node is called *pure*, if it only contains samples of one class. This naturally becomes a leaf. However, not all leaves are pure nodes, since a tree where the number of samples per leaf is low is likely to over-fit the data and not generalize well to unseen data. Therefore, in a leaf with samples from a mixture of classes, some training samples will be misclassified.

We define the node *impurity* $i(t)$ as a function of class proportions

$$i(t) = \phi(p_{t1}, \ldots, p_{tk}).$$

We will discuss later the specific choices for ϕ. A good decision split should optimize the purity, or in other words minimize the impurity. Since the two child nodes can differ in size, we define the change in impurity Δi as the impurity of the parent node minus the weighted average of the impurities of the child nodes:

$$\Delta i = i(P) - \left(\frac{N_L}{N_P} i(L) + \frac{N_R}{N_P} i(R) \right).$$

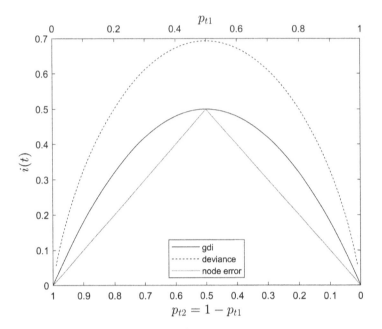

Figure 5.9: Different node impurities for two classes.

The best possible split maximizes the change in impurity over all possible splits for all features. If no splits, where the change is positive, are possible, the node is not split and it becomes a leaf.

Before considering specific choices for ϕ, it needs to have certain properties. The impurity should be zero, if only one class is present. That is $i(t) = 0$, if and only if $p_{tk} = 1$ for some k, and zero for all others. On the other hand, the impurity should be maximal, when all classes are mixed in equal proportions, i.e. $p_{t1} = \ldots = p_{tK} = 1/K$. It should also be symmetric, if the classes are re-labeled.

One popular choice for the impurity $i(t) = \phi(p_{t1}, \ldots, p_{tK})$ is the *Gini Diversity Index (gdi)*

$$i(t) = \sum_{k=1}^{K} p_{tk}(1 - p_{tk}) = 1 - \sum_{k=1}^{K} p_{tk}^2,$$

since $\sum_{k=1}^{K} p_{tk} = 1$. It vanishes, if $p_{tk} = 1$ for some k, and zero for all others. Figure 5.9 shows, for two classes, $i(t)$ as function of p_{t1} (or equivalently $p_{t2} = 1 - p_{t1}$).

With this choice, the change in impurity is

$$
\begin{aligned}
\Delta i &= \left(1 - \sum_{k=1}^{K} p_{Pk}^2\right) - \frac{N_L}{N_P}\left(1 - \sum_{k=1}^{K} p_{Lk}^2\right) - \frac{N_R}{N_P}\left(1 - \sum_{k=1}^{K} p_{Rk}^2\right) \\
&= \sum_{k=1}^{K} \frac{N_R}{N_P} p_{Rk}^2 + \frac{N_L}{N_P} p_{Lk}^2 - p_{Pk}^2,
\end{aligned}
$$

where we used $N_L + N_R = N_P$. Inserting (5.2), gives

$$
\begin{aligned}
\Delta i &= \sum_{k=1}^{K} \frac{N_R}{N_P} p_{Rk}^2 + \frac{N_L}{N_P} p_{Lk}^2 - \frac{(N_L p_{Lk} + N_R p_{Rk})^2}{N_P^2} \\
&= \frac{1}{N_P^2} \sum_{k=1}^{K} N_R (N_L + N_R) p_{Rk}^2 + N_L (N_L + N_R) p_{Lk}^2 \\
&\qquad\qquad - N_L^2 p_{Lk}^2 - 2 N_L N_R p_{Lk} p_{Rk} - N_R^2 p_{Rk}^2 \\
&= \frac{N_R}{N_P} \frac{N_L}{N_P} \sum_{k=1}^{K} (p_{Lk} - p_{Rk})^2.
\end{aligned}
$$

We see that, if the proportions in the child nodes are similar to each other, then the change in impurity will be small. If, however, the proportions are quite different, then Δi will be large.

Similar to this, another possibility known as *twoing* chooses the split which maximizes

$$
\Delta i = \frac{N_L}{N_P} \frac{N_R}{N_P} \left(\sum_{k=1}^{K} |p_{Lk} - p_{Rk}|\right)^2.
$$

It is commonly used if there are many classes, i.e. K is large.

Another choice for $i(t)$ is the *deviance*, also known as *cross-entropy*

$$
i(t) = -\sum_{k=1}^{K} p_{tk} \log p_{tk}.
$$

It is also plotted in Figure 5.9. In this case, we have

$$
\Delta i = -\sum_{k=1}^{K} p_{Pk} \log p_{Pk} + \frac{N_L}{N_P} \sum_{k=1}^{K} p_{Lk} \log p_{Lk} + \frac{N_R}{N_P} \sum_{k=1}^{K} p_{Rk} \log p_{Rk}.
$$

Using $p_{tk} = N_{tk}/N_t$ and $N_{Pk} = N_{Lk} + N_{Rk}$, this can be rewritten as

$$
\Delta i = \frac{1}{N_P} \sum_{k=1}^{K} N_{Lk} \log \frac{p_{Lk}}{p_{Pk}} + N_{Rk} \log \frac{p_{Rk}}{p_{Pk}}.
$$

If the proportions in the child nodes remain similar to the proportions in the parent node, then the change in impurity will be small, since the fractions are close to one and the logarithms evaluate to zero there. If, however, the proportions are dissimilar to the ones in the parent node, and especially if they are close to zero or one, then Δi is larger.

One might consider the node error as a possibility for $i(t)$. If a node is assigned the class with the largest proportion of training samples at that node, the node error is the fraction of misclassified samples:

$$i(t) = 1 - \max_k p_{tk}.$$

Figure 5.9 illustrates this graphically. Comparing the graphs of the *gdi, deviance* and *node error*, nodes where the classes have similar parity are considered as more impure under *gdi* and *deviance* than under the *node error*. This has the effect that the *node error* does not result in a preference to create purer child nodes. To see this, consider

$$
\begin{aligned}
\Delta i &= (1 - \max_k p_{Pk}) - \frac{N_L}{N_P}(1 - \max_k p_{Lk}) - \frac{N_R}{N_P}(1 - \max_k p_{Rk}) \\
&= 1 - \max_k \frac{N_{Pk}}{N_P} - \frac{N_L}{N_P} + \frac{N_L}{N_P}\max_k \frac{N_{Lk}}{N_L} - \frac{N_R}{N_P} + \frac{N_R}{N_P}\max_k \frac{N_{Rk}}{N_R} \\
&= \frac{1}{N_P}\left(\max_k N_{Lk} + \max_k N_{Rk} - \max_k N_{Pk}\right),
\end{aligned}
$$

where we used $N_L + N_R = N_P$. Thus, the change in impurity is solely dependent on the sum of the numbers of largest classes in the child nodes. This number can be kept constant for different splits, while the purity of the child nodes is very different. [5] gives a number example: Suppose there are only two classes and at the parent node they have equal parity, $N_{P1} = N_{P2} = 400$. One possible split is $N_{L1} = 100, N_{L2} = 300, N_{R1} = 300, N_{R2} = 100$, while another one is $N_{L1} = 200, N_{L2} = 400, N_{R1} = 200, N_{R2} = 0$. Both result in $\Delta i = (600 - 400)/800 = 1/4$. However, the second split is preferable, since there the right node is pure, and thus a leaf. For the first split, both child nodes are impure and need to grow branches.

Figure 5.10a shows our previous example where the region is this time classified by a decision tree. Figure 5.10b displays the corresponding decision tree. Each node corresponds to a split of a certain region. The first node splits the entire region along a vertical line at 2.45. The region to the right of this line is then split along a horizontal line at 1.75. The last node splits the region below this line and to the right of the first line vertically at 4.95. It can be seen that this decision tree could be improved by letting the horizontal split be slightly lower, at 1.65 say. This would result in the one misclassified virginica sample being classified correctly.

The depth of decision trees is controlled by three parameters: the maximum number of allowed branch node splits, the minimum number of samples needed in a node which is split, and the minimum number of samples per leaf node. In the example in Figure 5.10 the minimum number of samples in a leaf node

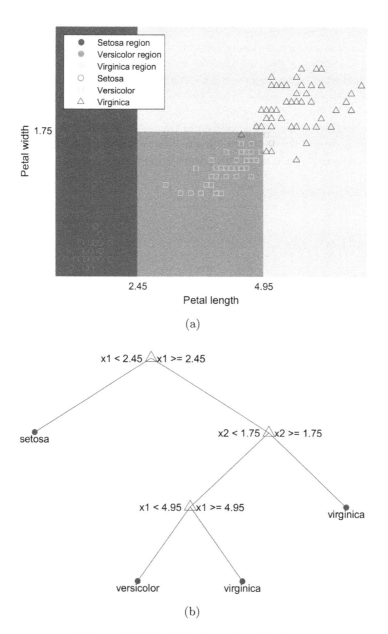

Figure 5.10: Decision tree classification.

was first set to 2. Setting it to one results in one further split. In fact, the last region is split horizontally at 1.65 which results in the misclassified virginica sample being classified correctly. This increases the depth of the decision tree. If the depth of a decision tree is not controlled, it generalizes less well to unseen data, since it is too closely fitted to the training data.

Listing 5.4 grows a decision tree for the more challenging case when the data is interleaved. The region is classified according to the decision tree. Figure 5.11a shows the result when the minimum number of samples in a node which is being split is set to 10, while for Figure 5.11b this is set to 5. The latter results in more sub regions and the resulting tree which is shown in Figure 5.13 is deeper than the tree resulting from when the minimum number of samples in a split node is larger (Figure 5.12).

```
load fisheriris
% Extract two attributes.
sl = meas(:,1); % sepal length
sw = meas(:,2); % sepal width
X = [sl,sw];

% Create classifier.
% The depth of a decision tree is governed by three arguments:
% Maximum number of branch node splits; a large value results in a
% deep tree.
MaxNumSplits = size(X,1) − 1;
% Minimum number of samples per branch node; a small number results in
% a deep tree.
MinParentSize = 5;
% Minimum number of samples per leaf; a small number results in a deep
% tree.
MinLeafSize = 1;
treeModel = fitctree(X,species,...
    'MaxNumSplits',MaxNumSplits,...
    'MinLeafSize',MinLeafSize,...
    'MinParentSize',MinParentSize);
view(treeModel,'mode','graph') % visualization

% Lay grid over the region
d = 0.01;
[x1Grid,x2Grid] = meshgrid(4:d:8.2,1.5:d:4.5);
xGrid = [x1Grid(:),x2Grid(:)];
N = size(xGrid,1);

% For each grid point calculate the score of each class.
% 'predict' returns the predicted class labels corresponding to the
% minimum misclassification cost, the score (posterior probability)
% for each class as well as the predicted node number and class
% number.
[¬,score,¬,¬] = predict(treeModel,xGrid);

% Classify according to the maximum score.
[¬,maxScore] = max(score,[],2);

% Plot classifier regions.
figure
h(1:3) = gscatter(xGrid(:,1),xGrid(:,2),maxScore,...
    [0.5 0.5 0.5; 0.7 0.7 0.7; 0.9 0.9 0.9]);
hold on
% Plot data.
h(4:6) = gscatter(sl, sw, species,'rgb','os^');
xlabel('Sepal length');
ylabel('Sepal width');
legend(h,{'Setosa region','Versicolor region','Virginica region',...
    'Setosa','Versicolor','Virginica'},...
    'Location','Southeast');
axis([4 8.2 1.5 4.5])
```

Listing 5.4: Decision tree.

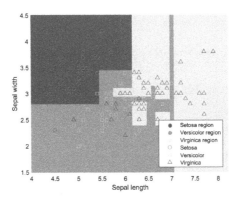

(a) Minimum number of samples per
split node set to 10.

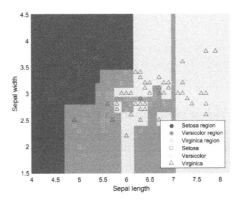

(b) Minimum number of samples per
split node set to 5.

Figure 5.11: Decision tree classification.

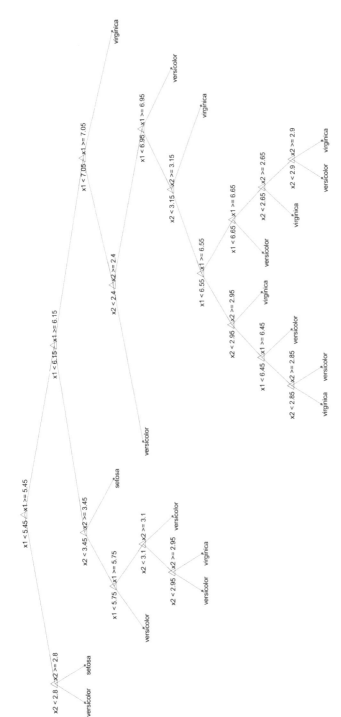

Figure 5.12: Decision tree with the minimum number of samples per split node set to 10.

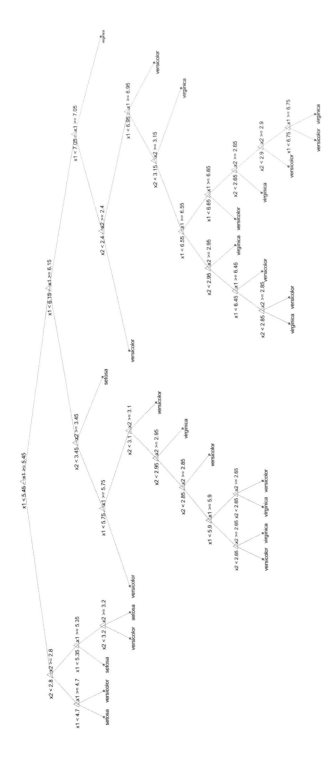

Figure 5.13: Decision tree with the minimum number of samples per split node set to 5.

5.5   Neural Networks

It was already mentioned that the perceptron is in fact a simple, single layer neural network. In the following, we describe neural networks more generally. Neural networks are dynamic systems characterized by non-linear, distributed, parallel and local processing. A neural network consists of *neurons*, also known as *nodes* or *units* and *synapses* connecting the neurons. The neurons are organized into three types:

- *Input*: Each feature is an input neuron.

- *Hidden*: Each neuron is a (possibly complex) mathematical function creating a predictor.

- *Output*: The neurons gather the predictions and produce the final result.

The synapses not only connect neurons, but also store weights.

As an example we consider data with two features, v_1 and v_2. Recall that the perceptron classifies according to $\text{sgn}(\hat{\mathbf{w}}^T \hat{\mathbf{v}})$. The corresponding neural network is

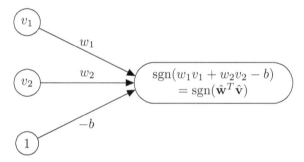

Note that here the bias is implemented as the weight of a synapse connecting a dummy input neuron which always has the value 1.

There are some differences in what is considered a layer in the neural network literature. Sometimes, each set of hidden neurons is considered a layer. In this context, the above has no layers. Sometimes, the sets of input and output neurons as well as the sets of hidden neurons are counted as layers. Then the above is a two layer neural network. However, of importance are the weights of the synapses. We therefore follow the convention that a layer is the connections between two sets of neurons. With this definition, the above is a single layer neural network. The weights are updated by the chosen learning process. We will not go into specific learning processes in this chapter, but leave this to Chapter 9 on feature learning where we will revisit neural networks.

The output neuron

$$
\begin{array}{c}
\text{sgn}(w_1 v_1 + w_2 v_2 - b) \\
= \text{sgn}(\hat{\mathbf{w}}^T \hat{\mathbf{v}})
\end{array}
$$

consists of two elements:

- the *propagation function* given by $\hat{\mathbf{w}}^T \hat{\mathbf{v}}$, which passes the information in a weighted manner to the next neuron.

- and the *activation function*, also called *transfer function*

$$h(x) = \text{sgn}(x).$$

Sometimes the bias is modeled as part of the propagation function. In this case there will be no dummy input neuron with the synapse having the bias as weight. It is important to be aware of these two different conventions when studying neural networks from different sources.

At the inception of neural networks, it was asked whether they can model basic logical operations, since a more complex deductive system can be built out of combinations of basic logical operators. Let 0 stand for the logical *false*, and 1 for the logical *true*.

A neural network is capable of implementing the logical AND

v_1	v_2	AND
0	0	0
0	1	0
1	0	0
1	1	1

,

since this is equivalent to separating the point $(1, 1)$ which is in the true class from the points $(0, 0), (0, 1)$ and $(1, 0)$ which belong to the false class. Any of the red lines in Figure 5.14 will do. Each line is described by $w_1 v_1 + w_2 v_2 - b = 0$. For any of the lines, points to the right should result in output 1, while points to the left should result in output 0.

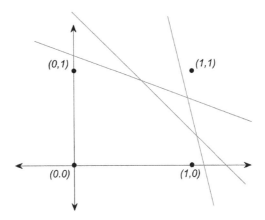

Figure 5.14: Possible implementations of the logical AND.

The *Heaviside step function* is defined by

$$H(x) = \begin{cases} 0 & \text{if} \quad x \leq 0 \\ 1 & \text{if} \quad x > 0 \end{cases}$$

It is also known as *hard-limit transfer function*. Using this as the activation function, the resulting neural network might be:

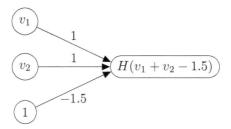

Note, the learning process might have arrived at different choices for w_1, w_2 and b.

Equally, the logical OR

v_1	v_2	OR
0	0	0
0	1	1
1	0	1
1	1	1

,

can be implemented as illustrated in Figure 5.15. Using the Heaviside step function again, one possibility for the resultant neural network is

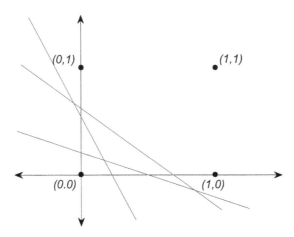

Figure 5.15: Possible implementations of the logical OR.

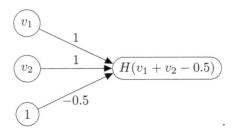

Continuing on this theme, it is possible to implement the logical AND NOT = NAND, for example:

v_1	v_2	NAND
0	0	1
0	1	1
1	0	1
1	1	0

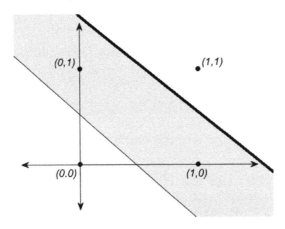

However, a simple neural network is *not* capable of implementing the exclusive OR = XOR (one or the other, but not both).

v_1	v_2	XOR
0	0	0
0	1	1
1	0	1
1	1	0

Figure 5.16 illustrates the region evaluating to true needs to be separated from the region evaluating to false on two sides.

Figure 5.16: Graphically distinguishing XOR.

One possible solution is an activation function with two steps, for example

$$h(x) = \begin{cases} 0 & \text{if} \quad x \le 0.25 \\ 1 & \text{if} \quad 0.25 < x < 0.75 \\ 0 & \text{if} \quad x \ge 0.75 \end{cases} \tag{5.3}$$

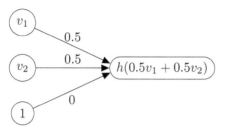

Again, the learning process might have chosen different weights.

Indeed, there are many different activation functions in use. Some of these are:

- Linear

$$h(x) = x \tag{5.4}$$

- Logistic sigmoid

$$h(x) = \frac{1}{1 + \exp(-a(x - c))} \tag{5.5}$$

- Hyperbolic tangent

$$h(x) = \tanh(x)$$

- Gaussian

$$h(x) = \exp\left(-\frac{(x-c)^2}{2a^2}\right)$$

- Multiquadratics

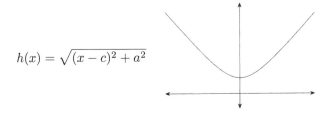

$$h(x) = \sqrt{(x-c)^2 + a^2}$$

- Inverse multiquadratics

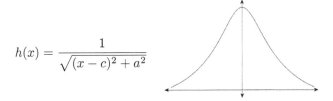

$$h(x) = \frac{1}{\sqrt{(x-c)^2 + a^2}}$$

The parameters a and c have to be tuned for the task at hand.

If the sgn activation function in the perceptron is replaced by the logistic sigmoid function, then the output is interpreted as the probability of the sample belonging to the positive class C_1. The probability of belonging to the negative class C_0 is one minus the output. This is then known as *logistic regression*. While it is solving a classification task, regression is part of the name, since it is trying to model the probability function of belonging to the positive class over the feature space.

An activation function particularly important for classifying multiple classes is the *softmax* function, also known as *normalized exponential* function. It is different from the other activation functions in that it takes into account the result of the propagation functions of other neurons. Let a_j be the result of the propagation function in the jth neuron: $a_j = \hat{\mathbf{w}}_j^T \hat{\mathbf{v}}$. If the number of neurons is K, then the softmax function maps the K-dimensional vector $\mathbf{a} = (a_1, \ldots, a_K)^T$ to a K-dimensional vector $\boldsymbol{\sigma}(\mathbf{a})$ with the jth entry of $\boldsymbol{\sigma}(\mathbf{a})$ being

$$\boldsymbol{\sigma}(\mathbf{a})_j = \frac{\exp(a_j)}{\sum_{k=1}^{K} \exp(a_k)}. \tag{5.6}$$

Obviously, the elements of $\boldsymbol{\sigma}(\mathbf{a})$ sum to 1. The softmax function is most often used in the final layer, where the number of synapses is the same as the number of classes, K. It gives a probabilistic interpretation to which class the sample, which passed through the neural network, belongs.

However, the construction of specific activation functions as in (5.3) is not in the spirit of machine learning, where as many tasks as possible shall be completed by the machine. Looking at Figure 5.16, we see that the region can actually be described as points on the right of the thin line and on the left of the bold line shall output 1. XOR is the combination of simpler building blocks. It can be implemented by introducing a set of *hidden* neurons. We denote the *hidden* variables, also known as *latent* variables, by z_1 and z_2. This is a two layer neural network, since a layer are the synapses from one set of neurons to another.

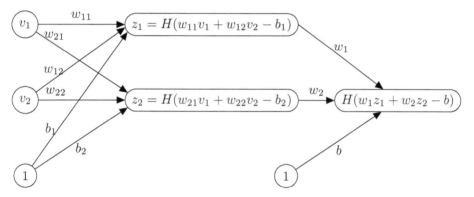

For example:

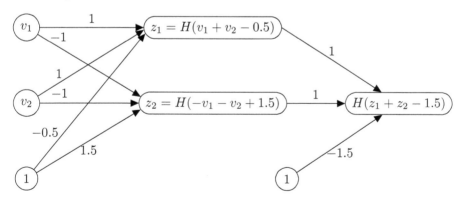

z_1 represents an OR which is equivalent to the right of the thin line in Figure 5.16, while z_2 represents NAND, or equivalently to the left of the bold line. The output neuron combines these two results with an AND, since both have to hold at the same time.

Using several sets of hidden neurons, and therefore several layers if necessary, any region can be described as a combination of on the right or left of several lines (or hyperplanes in higher dimensions). This makes neural networks particularly powerful for classification. In Chapter 9 we see how their performance is enhanced when continuous, non-linear activation functions are used.

The results of neural networks are sensitive to the choice of

- the number of layers,

- the connection pattern,

- the initialization of weights and

- the activation functions and their parameters.

All these are generally chosen by the user. Different choices are validated on validation sets to arrive at a set with which the neural network generalizes best to unseen data.

Listing 5.5 generates a neural network with a set of 10 hidden neurons. The output layer has 3 neurons, since the number of classes is 3. The hidden neurons use the heaviside step function as an activation function while the output neurons use the softmax function. An illustration of the neural network is given in Figure 5.18. The default MATLAB initialization of weights and biases is used. This is random to a certain degree. Therefore separate runs lead to different results. The listing contains code for both using all data for training and using a certain percentage for training, another percentage for validation and a third percentage for testing. The results can be seen in Figure 5.17.

5.6 Boosting and Cascades

We distinguish between *weak* and *strong* learners. A weak learner will only be slightly better than random guessing, while a strong learner provides the right classification most of the time. In this section we consider the question of whether a set of weak learners can create a single strong learner.

To simplify things we consider binary classification. We also assume that both classes are present in equal parity. That is the number of samples in each class is the same. The goal is to generate a classifier $\mathcal{C}$ where the sign of the output determines the class, and the absolute value of the output gives the confidence in this prediction.

The total error E of $\mathcal{C}$ is defined as the sum of *exponential error* at each training sample

$$E(\mathcal{C}) = \sum_{i=1}^{N} \exp\left(-c_i \mathcal{C}(\mathbf{v}_i)\right),$$

where $c_i = \pm 1$ is the class label of $\mathbf{v}_i$. If $\mathcal{C}$ predicts the class label of $\mathbf{v}_i$ *correctly* with great confidence, then c_i and $\mathcal{C}(\mathbf{v}_i)$ will have the same sign. Great confidence means that the absolute value of $\mathcal{C}(\mathbf{v}_i)$ is large, and hence

```
load fisheriris
% Extract two attributes.
pl = meas(:,3); % petal length
pw = meas(:,4); % petal width
% Prepare data.
X = [pl,pw];
X = X';
classes = unique(species);
t = [];
for j = 1:numel(classes)
    indx = strcmp(species,classes(j));
    t = [t indx];
end
t = t';

% Create classifier.
hiddenSizes = [10]; % row vector of one or more hidden layer sizes
netModel = patternnet(hiddenSizes);
% Use all samples for training.
%netModel.divideFcn = 'dividetrain';
% Use 70% of samples for training, 15% for validation, 15% for testing.
netModel.divideFcn = 'dividerand';
netModel.divideParam.trainRatio = 0.7;
netModel.divideParam.valRatio = 0.15;
netModel.divideParam.testRatio = 0.15;
% Set transfer function of the set of hidden neurons to the hard-limit
% transfer function. This is equivalent to a particular hidden neuron
% returning whether its input lies to the left or right of the
% line given by the weights.
netModel.layers{1}.transferFcn = 'hardlim';
% By default the transfer function of the last layer is set to the
% softmax function. Depending upon on which sides of the lines produced
% in the first layer a sample lies, a probability score for each class
% is given.

netModel = train(netModel,X,t);

% Lay grid over the region.
d = 0.01;
[x1Grid,x2Grid] = meshgrid(0.8:d:7,0:d:3);
xGrid = [x1Grid(:),x2Grid(:)];
N = size(xGrid,1);

% For each grid point calculate the score of each class.
score = netModel(xGrid');

% Classify according to the maximum score.
[¬,maxScore] = max(score,[],1);

% Plot classifier regions.
figure
h(1:3) = gscatter(xGrid(:,1),xGrid(:,2),maxScore,...
    [0.5 0.5 0.5; 0.7 0.7 0.7; 0.9 0.9 0.9]);
hold on

% Plot data.
h(4:6) = gscatter(pl, pw, species,'rgb','os^');
xlabel('Petal length');
ylabel('Petal width');
legend(h,{'Setosa region','Versicolor region','Virginica region',...
    'Setosa','Versicolor','Virginica'},...
    'Location','Northwest');
axis([0.8 7 0 3])
```

Listing 5.5: Neural Network.

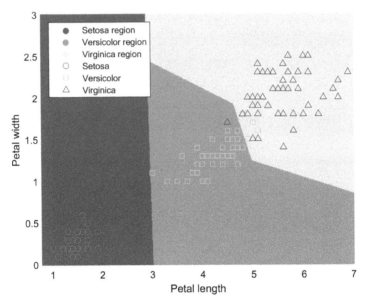

(a) All samples used for training.

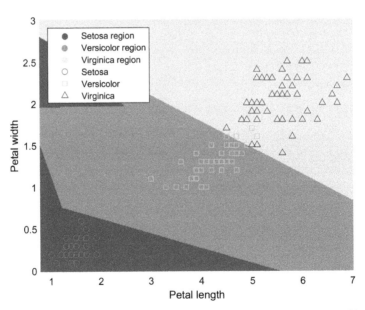

(b) 70% of samples used for training, 15% used for validation, 15% used for testing.

Figure 5.17: Two layer neural network classification with a set of ten hidden neurons using the heaviside step function.

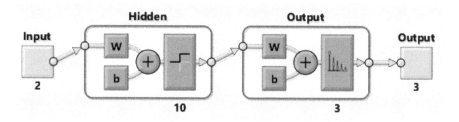

Figure 5.18: Two layer neural network with two input neurons, ten hidden neurons using the heaviside step function, and 3 output neurons using the softmax function.

$\exp\left(-c_i\mathcal{C}(\mathbf{v}_i)\right)$ will contribute very little to the total error, since the exponent is a large negative number. If $\mathcal{C}$ predicts the class label of $\mathbf{v}_i$ *incorrectly* with great confidence, then $\exp\left(-c_i\mathcal{C}(\mathbf{v}_i)\right)$ will contribute a lot to the total error, since c_i and $\mathcal{C}(\mathbf{v}_i)$ will have opposite sign and the exponent will be a large positive number. We consider $\mathcal{C}$ a strong classifier if it predicts correctly with great confidence, and has little confidence in the prediction when making errors.

AdaBoost, short for *Adaptive Boosting* iteratively generates a series of classifiers $\mathcal{C}_m$. Each iteration improves the current classifier by concentrating on the misclassified elements. Each classifier is stronger than its predecessor. It is generated as a linear combination of simple, weak classifiers k_j which only return ± 1. The perceptron could be one of these classifiers. The coefficients in this linear combination are positive. Each coefficient gives the confidence in the weak classifier.

First, $\mathcal{C}_1$ is initialized to the weak classifier $k_1(\mathbf{v})$ which misclassifies the least number of training examples. Its coefficient α_1 is chosen to minimize

$$
\begin{aligned}
E(\mathcal{C}_1) &= \sum_{i=1}^{N} \exp\left(-c_i\alpha_1 k_1(\mathbf{v}_i)\right) \\
&= \sum_{c_i \neq k_1(\mathbf{v}_i)} \exp\left(\alpha_1\right) + \sum_{c_i = k_1(\mathbf{x}_i)} \exp\left(-\alpha_1\right),
\end{aligned}
$$

where we used the fact that both c_i and $k_1(\mathbf{v}_i)$ only have the values $+1$ or -1. Differentiating with respect to α_1 gives

$$
\frac{dE(\mathcal{C}_1)}{d\alpha_1} = \sum_{c_i \neq k_1(\mathbf{v}_i)} \exp\left(\alpha_1\right) - \sum_{c_i = k_1(\mathbf{v}_i)} \exp\left(-\alpha_1\right).
$$

Let N be the number of samples and N_C be the number of correctly classified samples. We then have

$$
\frac{dE}{d\alpha_1} = (N - N_C)\exp\left(\alpha_1\right) - N_C \exp\left(-\alpha_1\right).
$$

To find the extremum, we set this to zero and solve for α_1:

$$\exp(2\alpha_1) = \frac{N_C}{N - N_C}$$

$$\alpha_1 = \frac{1}{2} \log \frac{N_C}{N - N_C}.$$

Thus α_1 is half the logarithm of the ratio of the number of correctly classified samples over the number of misclassified samples. If $N_C = N/2$, then $\alpha_1 = 0$, meaning so that we have no confidence in the classification, since it is equal to random guessing. Note that N_C cannot be less than $N/2$, since the weak classifiers need to be better than random guessing, i.e. they need to classify at least more than $N/2$ samples correctly. The larger the N_C, the higher the confidence.

In the mth iteration we generate $\mathcal{C}_m$ as $\mathcal{C}_{m-1} + \alpha_m k_m$. The total error is

$$
\begin{aligned}
E(\mathcal{C}_m) &= \sum_{i=1}^{N} \exp\left(-c_i \mathcal{C}_{m-1}(\mathbf{v}_i) - c_i \alpha_m k_m(\mathbf{v}_i)\right) \\
&= \sum_{i=1}^{N} \exp\left(-c_i \mathcal{C}_{m-1}(\mathbf{v}_i)\right) \exp\left(-c_i \alpha_m k_m(\mathbf{v}_i)\right) \\
&= \exp(-\alpha_m) \sum_{c_i = k_m(\mathbf{v}_i)} w_{i,m} + \exp(\alpha_m) \sum_{c_i \neq k_m(\mathbf{v}_i)} w_{i,m},
\end{aligned}
\tag{5.7}
$$

where we defined the weights $w_{i,m} = \exp\left(-c_i \mathcal{C}_{m-1}(\mathbf{v}_i)\right)$. Hence the weight $w_{i,m}$ is the exponential error the current classifier makes when classifying sample $\mathbf{v}_i$. The total error can be rewritten as

$$E(\mathcal{C}_m) = \exp(-\alpha_m) \sum_{i=1}^{N} w_{i,m} + (\exp(\alpha_m) - \exp(-\alpha_m)) \sum_{c_i \neq k_m(\mathbf{v}_i)} w_{i,m}.$$

Only the last sum depends on the the weak classifier k_m. To make $E(\mathcal{C}_m)$ small, the weak classifier where the sum of weights over the misclassified elements is smallest should be chosen. That is the classifier which classifies most samples with large weights correctly.

Having chosen k_m, we differentiate the last line of 5.7 with respect to α_m and set to zero, solving for α_m. This results in

$$\alpha_m = \frac{1}{2} \log \frac{\sum_{c_i = k_m(\mathbf{x}_i)} w_{i,m}}{\sum_{c_i \neq k_m(\mathbf{x}_i)} w_{i,m}}.$$

The more samples with large weights k_m classify correctly the larger α_m and the confidence in k_m. On the other hand, k_m classifying a sample with small weight incorrectly has little effect, since a small weight indicates that $\mathcal{C}_{m-1}$ classifies this sample correctly.

The above relies on the fact that the classes are balanced. Imbalanced classes have their own challenges. For example, if there are 9 times more samples of one class than of the other class, a classifier which always assigns the first class will be correct 90% of the time. However, it classifies the second class incorrectly all the time. For more information on tackling imbalanced classes see for example [21].

So far we have not considered the cost of acquiring features. The sources of cost can be quite varied. For example, when considering the computational cost, it can be the low cost associated with a simple algorithm. At the other end of the scale are computationally intensive algorithms which require also large storage. There is also the difference between cheap and expensive diagnostic tools. It is easy and inexpensive to check the oil level in ones car; beyond that however, often a car mechanic is required, whose hourly rate is a considerable cost. Then there is also the human cost. Invasive procedures put much more strain on a patient than a simple medical test.

A *cascade* is a procedural method which takes into account the cost. Each stage of a cascade of classifiers uses features with increasing predictive power and increasing cost. For example, in medical diagnosis a general practitioner first does a simple inexpensive test which can exclude a diagnosis. Should the test, however, reveal the possibility of an underlying health problem, the patient can be referred for further tests.

Let class C_0 be the *negatives* and class C_1 be the *positives*. The number of samples in C_0 is commonly denoted N, while the number of samples in C_1 is denoted P.

A *confusion matrix* visualizes the goodness of a classifier by listing the number of samples of C_0 correctly classified, known as *true negatives*, TN, the number of samples of C_0 misclassified, which are the *false positives*, FP on one hand, and on the other hand the number of samples of C_1 correctly classified, the *true positives*, TP, and the number of samples of C_1 misclassified, the *false negatives*, FP:

$$\begin{pmatrix} TN & FN \\ FP & TP \end{pmatrix}.$$

This concept can be easily extended to multiple classes. The (i, j) entry in the matrix then contains the number of samples in class C_j classified as belonging to class C_i. For example, Figure 5.19 shows the confusion matrices of the neural networks the data depicted in Figure 5.18. The neural network using validation and testing sets misclassifies altogether only four samples.

From these numbers we can calculate the *sensitivity*, also known as *recall*, *true positive rate* (TPR) and *probability of detection*, as the fraction of positive samples correctly classified:

$$TPR = \frac{TP}{P}.$$

The *specificity* or *true negative rate* (TNR) is the fraction of negative samples correctly identified:

$$TNR = \frac{TN}{N}.$$

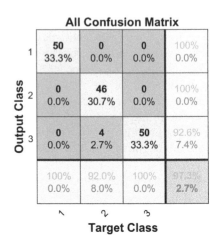

(a) All samples used for training.

(b) 70% of samples used for training, 15% used for validation, 15% used for testing.

Figure 5.19: Confusion matrices of the neural networks of Figure 5.18.

The *false negative rate*, also known as *miss rate* is given by

$$\text{FNR} = \frac{\text{FN}}{\text{P}} = 1 - \text{TPR}.$$

It quantifies the percentage of time a positive sample is overlooked. Not identifying a patient who has an underlying health problem can have disastrous consequences for that patient. Another commonly used term is the *fall-out* or *false positive rate* (FPR):

$$\text{FPR} = \frac{\text{FP}}{\text{N}} = 1 - \text{TNR}.$$

It is the percentage of time a negative sample is registered as positive. In this case the patient worries needlessly. However, this is acceptable in some cases as long as subsequent tests avoid needless procedures.

A perfect classifier would be 100% sensitive (all positives are correctly identified) and 100% specific (no negatives are incorrectly classified). In a cascade we want all stages to have high sensitivity so that as few as possible patients with a health problem are overlooked. High specificity is desirable so that patients are not subjected to costly (both financial and human cost) procedures. However, in early stages a lower specificity is acceptable. Samples classified as negative at a stage are not considered in the following stages, thus decreasing the overall cost.

CHAPTER 6

Clustering

Clustering is introduced as a sorting process where the criteria governing the sort are not known. It builds from the heuristic K means clustering, explaining mixture models in general and specifically Gaussian mixture models. The Expectation-Maximization technique is shown to be a general iterative technique to maximize the data likelihood. Next Bayesian mixture models are explored and illustrated with the Chinese Restaurant Process. The chapter concludes with the Dirichlet process which is explained in detail and illustrated by examples.

So far we have looked at data which had class labels. The task was to learn the class membership from the features of each data sample, in order to classify unseen data in the future. In this chapter, we look at unlabeled data. As before, we denote the vector of features by $\mathbf{v}$. Let $\mathbf{v}_1, \ldots, \mathbf{v}_N$ be the feature vectors of N data samples. The choice of features or combination of features is important.

Figure 6.1: Sorting sweets.

For example, a popular exercise for children to apply their observation and sorting skills is to let them sort sweets. It is immediately obvious that there are numerous distinguishing features. Firstly, there is the size, then the colour, or the sweet might be multicoloured. They come in all sorts of shapes. Next, we might consider the main ingredient which may be chocolate, jelly, licorice or something completely different, or it might be a combination of different ingredients, or have a filling. The parents might be more interested in keeping the overall sugar intake low, while a dentist might be interested in whether the sweets stick to the teeth or not. Depending on their tastes and preferences, different people will put different emphasis on each feature.

Clustering often is the first step when analyzing a new data set. If not much is known yet about the data, it is useful to see whether the data separates naturally into different groups. The aim is to find clusters where data samples within one cluster are more similar to each other than to data samples in other clusters. We will see in the following what similar means for different algorithms.

There are many application areas. For example, clustering is used to find news articles with similar contents or relating to the same topic. Similarly, search engines use clustering algorithms. The suggestions we receive on streaming sites depending on our past preferences are based on clusters as well. Image segmentation is another application area as is lossy data compression, where it is sufficient to represent the data by the cluster it belongs to. It is also used in bioinformatics to analyze biological data of all sorts.

6.1   K Means Clustering

K means clustering decides at the outset on the numbers of clusters denoted by K and each data sample is assigned to exactly one cluster. This is known as *hard clustering*. The aim is to find cluster centres $\boldsymbol{\mu}_1, \ldots, \boldsymbol{\mu}_K$ such that the sum of the squared distances of each data sample to its nearest cluster centre (the one it is assigned to) is minimal. Nearest here is with respect to the *Euclidean norm* (L_2 *norm*). Thus the *objective function* is

$$J = \sum_{n=1}^{N} \min_{k} \|\mathbf{v}_n - \boldsymbol{\mu}_k\|^2.$$

This optimization problem is *NP hard*. That is, there is no known algorithm to solve this in polynomial time. This is because as cluster centres move around, for each sample it changes which is the nearest cluster. The algorithm finds an approximate solution by separating this interdependency with the introduction of *hidden* (*latent*) variables $\mathbf{z}_n$, one for each data sample $\mathbf{v}_n$. The latent variables are binary vectors, i.e. $\mathbf{z}_n \in \{0,1\}^K$, where only one entry can be 1 and the others have to be 0. This is known as a *1-of-K representation*. A 1 in the kth entry indicates that $\boldsymbol{\mu}_k$ is the nearest cluster centre to $\mathbf{v}_n$. Let $z_{n,k}$

be the kth component of $\mathbf{z}_n$. The objective function can then be rewritten as

$$J = \sum_{n=1}^{N} \sum_{k=1}^{K} z_{n,k} \|\mathbf{v}_n - \boldsymbol{\mu}_k\|^2.$$

This is quadratic in $\boldsymbol{\mu}_k$. The minimum is found by differentiating with respect to $\boldsymbol{\mu}_k$ for each k using Appendices A.2.1 and A.2.2 and setting this to zero,

$$2 \sum_{n=1}^{N} z_{n,k} (\mathbf{v}_n - \boldsymbol{\mu}_k) = 0.$$

This gives new centre locations for all k

$$\boldsymbol{\mu}_k = \frac{\sum_{n=1}^{N} z_{n,k} \mathbf{v}_n}{\sum_{n=1}^{N} z_{n,k}}. \tag{6.1}$$

However, now that the centres have moved, the indicator vectors $\mathbf{z_n}$ have to be adjusted. This can be viewed as minimizing J with respect to all $\mathbf{z}_n$. This in turn will cause the centres to move again. The algorithm alternates between moving the centres $\boldsymbol{\mu}_k$ and recalculating $\mathbf{z}_n$. It terminates, when after moving the centres, none of the indicator vectors changes.

Looking at (6.1), note that $\sum_{n=1}^{N} z_{n,k}$ gives the number of data samples for which $\boldsymbol{\mu}_k$ is the closest centre, while $\sum_{n=1}^{N} z_{n,k} \mathbf{v}_n$ is the sum of those samples. Thus $\boldsymbol{\mu}_k$ is the mean of the samples assigned to this particular cluster.

The algorithm arrives at a local minimum of J. However, this is by no means necessarily a global minimum. In fact, the solution is highly dependent on the initialization of the cluster centres. These can be chosen randomly at the start, or the samples can be randomly assigned clusters and the initial cluster centers be calculated as the mean of these random assignments. To counteract this dependency on initial values, the algorithm is run (possibly in parallel) with many different initializations. After convergence, the result with the lowest value of J is chosen.

The squared Euclidean norm of the difference between sample and cluster centre was used. The algorithm can be generalized using any dissimilarity measure between two feature vectors, just as kernels were used as similarity measures for the kernel trick described in Section 5.2. In this case, the method is known as K-medoids algorithm. However, the minimization with regards to $\boldsymbol{\mu}_k$ might be much more involved. It depends on the differentiability of the dissimilarity measure, and whether it is possible to find where the derivative vanishes. If this is not possible, it is common practice to require each cluster centre to be one of the data samples. The minimization with respect to $\boldsymbol{\mu}_k$ is then a search among the data samples assigned to the kth cluster.

Lossy data compression is achieved by choosing a suitable number of clusters and storing the feature vectors of the centres, $\boldsymbol{\mu}_k$. For each data sample only the cluster it belongs to is stored. This can be applied to images. Typically images are represented by storing three values for each pixel, with each

(a) $K = 4$, 8.3% compression ratio.   (b) $K = 16$, 16.7% compression ratio.

Figure 6.2: K mean compression.

value being one byte. Since one byte is 8 bits, $24N$ bits are necessary to store an image with N pixels. If instead we cluster the pixels using K clusters, we need to store the pixel values of each cluster centre which requires $24K$ bits. The cluster number can be encoded in $\log_2 K$ bits, and this needs to be stored for each pixel. For example, for $K = 4$, two bits are necessary. The cluster numbers are stored as $00, 01, 10, 11$. When $K = 16$, four bits are required. Figure 6.2 illustrates this. Using 4 clusters results in an image consisting of lighter and darker shades of brown which means the cluster centres did not stray too far off the axis of shades of gray where the red, green and blue components are equal. 16 clusters seem an adequate representation at a compression ratio of 16.7%. The colours are more realistic.

Listing 6.1 reads an image of tissue stained with hemotoxylin and eosin to make different tissue types distinguishable (Image courtesy of Alan Partin, Johns Hopkins University). Using three clusters, one of these identifies the nuclei.

We used K means clustering on two examples of image processing. This is a very large area of ongoing research with algorithms tailored to specific applications. A comprehensive overview can be found in [18].

6.2 Mixture Models

Let's change our view point and say that the indicator variable $\mathbf{z}_n$ does not indicate to which cluster the sample $\mathbf{v}_n$ belongs, but shows which process generated that data sample, and we assume there are K distinct processes generating the data. Let $p_k(\mathbf{v})$ be the probability distribution of the samples generated by the kth process. Let π_k be the probability that process k

```
% Read image.
original = imread('hestain.png');
% Display image.
figure;
imshow(original);
nrows = size(original,1);
ncols = size(original,2);
% Reshape the image to obtain data vector.
% The features are the red, green and blue colour values
% converted to doubles.
data_vector = reshape(double(original),nrows*ncols,3);
% Choose the number of clusters.
nClusters = 3;
% Choose the number of repetition to avoid local minima
nRepeats = 3;
% Perform the clustering using the Euclidean distance.
[cluster_labels, cluster_centres] = kmeans(data_vector,nClusters,...
    'distance', 'sqEuclidean', 'Replicates', nRepeats);

% Construct segmented image finding all pixels belonging to a specific
% cluster and giving them the colour of the centre of the cluster it
% belongs to.
segmented = zeros(nrows*ncols, 3);
for i = 1:nClusters
    segmented(cluster_labels == i,:) = repmat(cluster_centres(i,:),...
        size(segmented(cluster_labels == i,:),1),1);
end
figure;
segmentedim = uint8(reshape(segmented, nrows, ncols, 3));
imshow(segmentedim);

% Only show specific cluster from image.
for i = 1:nClusters
    removed = data_vector;
    removed(cluster_labels ~= i,:) = 255;
    figure;
    removedim = uint8(reshape(removed, nrows, ncols, 3));
    imshow(removedim);
end
```

Listing 6.1: Biological image processing.

generates a sample. Thus we have

$$0 \leq \pi_k \leq 1 \text{ and } \sum_{k=1}^{K} \pi_k = 1.$$

This can be interpreted as that the latent variables $z_1, \ldots, z_N$ are drawn from a probability distribution $p(\mathbf{z})$. Since $\mathbf{z}$ has a 1-of-K representation, $\mathbf{z}$ is uniquely identified by specifying which component is 1. Therefore it is often written as

$$p(z_k = 1) = \pi_k.$$

Now, the probability of a feature vector $\mathbf{v}$ given an indicator variable $\mathbf{z}$ is the conditional probability distribution

$$p(\mathbf{v}|\mathbf{z}) = p(\mathbf{v}|z_k = 1) = p_k(\mathbf{v}).$$

The joint probability distribution of $\mathbf{v}$ and $\mathbf{z}$ is given by the product rule as $p(\mathbf{v}, \mathbf{z}) = p(\mathbf{v}|\mathbf{z})p(\mathbf{z})$. We obtain the probability distribution for $\mathbf{v}$ by *marginalizing* over $\mathbf{z}$. This means summing the joint distribution over all possible values of $\mathbf{z}$:

$$
\begin{aligned}
p(\mathbf{v}) &= \sum_{\mathbf{z}} p(\mathbf{v}, \mathbf{z}) = \sum_{\mathbf{z}} p(\mathbf{z})p(\mathbf{v}|\mathbf{z}) \\
&= \sum_{k=1}^{K} p(z_k = 1)p(\mathbf{v}|z_k = 1) = \sum_{k=1}^{K} \pi_k p_k(\mathbf{v}).
\end{aligned}
\tag{6.2}
$$

This is called a *mixture* distribution. The probabilities π_k are known as *mixing coefficients*.

For a given feature vector $\mathbf{v}$, we can calculate the probability that it was generated by process k. This is the conditional probability $p(z_k = 1|\mathbf{v})$, whose value can be calculated using *Bayes' Rule*. This is given by

$$P(A|B) = \frac{P(B|A)P(A)}{P(B)},$$

where A and B are events, $P(A)$ and $P(B)$ are the probabilities of A and B without regard to each other, $P(A|B)$ is the *conditional probability* of A given that B is true, and $P(B|A)$ is the conditional probability of B given that A is true. Applying Bayes' rule here leads to

$$p(z_k = 1|\mathbf{v}) = \frac{p(z_k = 1)p(\mathbf{v}|z_k = 1)}{p(\mathbf{v})} = \pi_k \frac{p_k(\mathbf{v})}{p(\mathbf{v})}.$$

We see that for a particular $\mathbf{v}$ the probability that it was generated by process k is the mixing coefficient π_k adjusted by the ratio of the probability $p_k(\mathbf{v})$ to the overall probability $p(\mathbf{v})$. $p(z_k = 1|\mathbf{v})$ is known as the *responsibility* process k takes for explaining the sample $\mathbf{v}$.

Using (6.2), we deduce

$$\sum_{k=1}^{K} p(z_k = 1|\mathbf{v}) = \frac{\sum_{k=1}^{K} \pi_k p_k(\mathbf{v})}{p(\mathbf{v})} = 1.$$

Thus the responsibilities, i.e. the probabilities of the processes generating a particular sample, sum to one.

In particular, for the data samples $\mathbf{v}_n$, $n = 1, \ldots, N$, we can now make cluster assignments according to the values $p(z_{n,k} = 1|\mathbf{v}_n)$ for $k = 1, \ldots, K$. This is called *soft clustering*, since our belief is quantified by probabilities. It is possible that for a particular sample the probabilities are the same for two (or even more) values of k. These are samples which lie between clusters.

Before we can however calculate any probabilities for cluster membership, we need to determine π_k and $p_k(\mathbf{v})$ for $k = 1, \ldots, K$. To this end we consider the joint likelihood under this model of the data samples $\mathbf{v}_1, \ldots, \mathbf{v}_n$ which is given by

$$\prod_{n=1}^{N} p(\mathbf{v}_n) = \prod_{n=1}^{N} \sum_{k=1}^{K} \pi_k p_k(\mathbf{v}_n).$$

We obtain values for π_k and the parameters describing $p_k(\mathbf{v})$, by maximizing this likelihood, or alternatively its logarithm, known as the *log likelihood*

$$\mathcal{L} = \sum_{n=1}^{N} \log p(\mathbf{v}_n) = \sum_{n=1}^{N} \log \left(\sum_{k=1}^{K} \pi_k p_k(\mathbf{v}_n) \right). \tag{6.3}$$

When maximizing with respect to the mixing coefficients π_k, it is important to remember the constraint that they sum to one. This can be incorporated into the maximization problem using a *Lagrange multiplier* λ. That is we maximize

$$\sum_{n=1}^{N} \log \left(\sum_{k=1}^{K} \pi_k p_k(\mathbf{v}_n) \right) + \lambda \left(\sum_{k=1}^{K} \pi_k - 1 \right).$$

Differentiating with respect to π_k and setting to zero gives

$$\sum_{n=1}^{N} \frac{1}{\sum_{k=1}^{K} \pi_k p_k(\mathbf{v}_n)} p_k(\mathbf{v}_n) + \lambda = 0. \tag{6.4}$$

We first solve for λ by multiplying through by π_k and summing over all k. The right hand side remains zero, while the left hand side becomes

$$\sum_{n=1}^{N} \frac{\sum_{k=1}^{K} \pi_k p_k(\mathbf{v}_n)}{\sum_{k=1}^{K} \pi_k p_k(\mathbf{v}_n)} + \sum_{k=1}^{K} \pi_k \lambda = N + \lambda.$$

Hence, we find $\lambda = -N$. Inserting this result into (6.4) and again multiplying by π_k, we have

$$
\begin{aligned}
0 &= \sum_{n=1}^{N} \frac{\pi_k p_k(\mathbf{v}_n)}{\sum_{k=1}^{K} \pi_k p_k(\mathbf{v}_n)} - N\pi_k = \sum_{n=1}^{N} \frac{\pi_k p_k(\mathbf{v}_n)}{p(\mathbf{v}_n)} - N\pi_k \\
&= \sum_{n=1}^{N} p(z_{n,k} = 1|\mathbf{v}_n) - N\pi_k.
\end{aligned}
$$

Therefore,

$$
\pi_k = \frac{1}{N} \sum_{n=1}^{N} p(z_{n,k} = 1|\mathbf{v}_n). \tag{6.5}
$$

Thus the mixing coefficient π_k is the average responsibility that all data samples are generated by process k.

We need to emphasize here that this is not a closed form solution, since the responsibilities $p(z_{n,k} = 1|\mathbf{v}_n)$ depend on π_k itself. However, an iterative scheme can be derived from this, where the mixing coefficients π_k and the parameters of the probabilities $p_k(\mathbf{v})$ are initially estimated, then the mixing coefficients are updated according to (6.5). The next step is to maximize with respect to the parameters. The method alternates between these two steps.

In the following sections, we will see different techniques of maximizing with respect to the parameters. A note of caution is however appropriate at this point, and this is in relation to outliers in the data. Assume that $K = 2$ and that all data samples are roughly grouped together apart from one outlier. In the maximization procedure π_1 will tend to $(N - 1)/N$ while π_2 will tend to $1/N$. $p_1(\mathbf{v})$ will roughly describe the distribution of $N - 1$ samples. The problem comes with $p_2(\mathbf{v})$. The likelihood can be increased again and again by concentrating the probability mass of $p_2(\mathbf{v})$ more and more tightly around the outlier. Hence, the evaluation of $p_2(\mathbf{v})$ at the outlier will tend to infinity, while the area where $p_2(\mathbf{v})$ is non-zero tends to zero. This in turn leads to the likelihood tending to infinity. Heuristics are used to identify such degenerate cases. For example, such outliers could be identified and removed from the data set during the maximization procedure. Or K is reduced such that it is not possible for outliers to be assigned their own cluster.

Something else to be aware of is that any solution actually has $K!$ equivalent solutions. These arise by just relabeling the clusters by permutations of $1 \ldots K$. This equivalence is known as *identifiability*, and is important when comparing and interpreting models. In this context, however, this is of no importance, since just any one of the equivalent solutions needs to be found.

6.3 Gaussian Mixture Models

It is common to choose the distributions $p_k(\mathbf{v})$, $k = 1, \ldots, K$ in the mixture model given in (6.2) from the same family of distributions. As an example we consider normal distributions $\mathcal{N}(\boldsymbol{\mu}_k, \boldsymbol{\Sigma}_k)$. That is

$$p_k(\mathbf{v}) = \frac{1}{\sqrt{|2\pi\boldsymbol{\Sigma}_k|}} \exp\left(-\frac{1}{2}(\mathbf{v} - \boldsymbol{\mu}_k)^T \boldsymbol{\Sigma}_k^{-1}(\mathbf{v} - \boldsymbol{\mu}_k)\right), \tag{6.6}$$

where $|\cdot|$ denotes the matrix determinant. The derivative of $p_k(\mathbf{v})$ with respect to $\boldsymbol{\mu}_k$ is given by

$$\frac{\partial}{\partial \boldsymbol{\mu}_k} p_k(\mathbf{v}) = p_k(\mathbf{v}) \boldsymbol{\Sigma}_k^{-1}(\mathbf{v} - \boldsymbol{\mu}_k),$$

where we used the chain rule, Appendix A.2.3 and the symmetry of $\boldsymbol{\Sigma}_k$. Differentiating (6.3) with respect to $\boldsymbol{\mu}_k$ therefore results in

$$\begin{aligned}
\frac{\partial}{\partial \boldsymbol{\mu}_k} \mathcal{L} &= \sum_{n=1}^{N} \frac{1}{p(\mathbf{v}_n)} \pi_k \frac{\partial}{\partial \boldsymbol{\mu}_k} p_k(\mathbf{v}) \\
&= \sum_{n=1}^{N} \frac{1}{p(\mathbf{v}_n)} \pi_k p_k(\mathbf{v}_n) \boldsymbol{\Sigma}_k^{-1}(\mathbf{v}_n - \boldsymbol{\mu}_k) \\
&= \sum_{n=1}^{N} p(z_{n,k} = 1|\mathbf{v}_n) \boldsymbol{\Sigma}_k^{-1}(\mathbf{v}_n - \boldsymbol{\mu}_k).
\end{aligned}$$

To approximate the extremum, we set this to zero and multiply through by $\boldsymbol{\Sigma}_k$ which is non-singular, since it is a covariance matrix:

$$\sum_{n=1}^{N} p(z_{n,k} = 1|\mathbf{v}_n)(\mathbf{v}_n - \boldsymbol{\mu}_k) = 0.$$

We define

$$N_k = \sum_{n=1}^{N} p(z_{n,k} = 1|\mathbf{v}_n),$$

which is comparable to the expected number of samples in cluster k. With this definition

$$\boldsymbol{\mu}_k = \frac{1}{N_k} \sum_{n=1}^{N} p(z_{n,k} = 1|\mathbf{v}_n)\mathbf{v}_n. \tag{6.7}$$

We see that $\boldsymbol{\mu}_k$ is a weighted average of all samples in the data set where the weights are the responsibilities that the sample was generated by process k. It is reminiscent of formula (4.1) for the sample mean.

To maximize with respect to the covariance $\boldsymbol{\Sigma}_k$, we need the derivative of $p_k(\mathbf{v})$ with respect to $\boldsymbol{\Sigma}_k$. Using the product and chain rule and the differentiation rules given in Appendices A.2.4 and A.2.8, we have

$$\frac{\partial}{\partial \boldsymbol{\Sigma}_k} p_k(\mathbf{v}) = -\frac{1}{2} p_k(\mathbf{v}) \left[\boldsymbol{\Sigma}_k^{-1} - \boldsymbol{\Sigma}_k^{-1}(\mathbf{v} - \boldsymbol{\mu}_k)(\mathbf{v} - \boldsymbol{\mu}_k)^T \boldsymbol{\Sigma}_k^{-1}\right],$$

since $\boldsymbol{\Sigma}_k^{-1}$ is symmetric. From this we derive the derivative of the logarithm of the likelihood given in (6.3) as

$$\frac{\partial}{\partial \boldsymbol{\Sigma}_k} \mathcal{L} = \frac{1}{2} \sum_{n=1}^{N} p(z_{n,k} = 1 | \mathbf{v}_n) \left[\boldsymbol{\Sigma}_k^{-1} - \boldsymbol{\Sigma}_k^{-1} (\mathbf{v}_n - \boldsymbol{\mu}_k)(\mathbf{v}_n - \boldsymbol{\mu}_k)^T \boldsymbol{\Sigma}_k^{-1} \right].$$

As before, we approximate the extremum by setting the derivative to zero and multiplying through with $\boldsymbol{\Sigma}_k$, but this time from both sides:

$$\sum_{n=1}^{N} p(z_{n,k} = 1 | \mathbf{v}_n) \left[\boldsymbol{\Sigma}_k - (\mathbf{v}_n - \boldsymbol{\mu}_k)(\mathbf{v}_n - \boldsymbol{\mu}_k)^T \right] = 0.$$

Solving for $\boldsymbol{\Sigma}_k$, we arrive at

$$\boldsymbol{\Sigma}_k = \frac{1}{N_k} \sum_{n=1}^{N} p(z_{n,k} = 1 | \mathbf{v}_n)(\mathbf{v}_n - \boldsymbol{\mu}_k)(\mathbf{v}_n - \boldsymbol{\mu}_k)^T. \tag{6.8}$$

The formula is very similar to (4.2) for the sample covariance. The main difference is that the sum is over all samples and that each summand is weighted by the responsibility that process k generated this sample.

To summarize, the Gaussian mixture algorithm proceeds as follows

1. Choose K and convergence threshold.

2. Initialize means $\boldsymbol{\mu}_k$, covariances $\boldsymbol{\Sigma}_k$, and mixing coefficients π_k for $k = 1, \ldots, K$, and calculate the initial value of the logarithm of the likelihood according to (6.3).

3. For $n = 1, \ldots, N$ and $k = 1, \ldots, K$, calculate all the responsibilities $p(z_{n,k} = 1 | \mathbf{v}_n)$.

4. Use these responsibilities to update means $\boldsymbol{\mu}_k$, covariances $\boldsymbol{\Sigma}_k$, and mixing coefficients π_k for $k = 1, \ldots, K$ according to equations (6.7), (6.8) and (6.5).

5. Evaluate the change in the logarithm of the likelihood and terminate if this is below the convergence threshold. Otherwise return to step 3.

Alternatively, the algorithm converges if the change in all parameters is below the convergence threshold. It is common to use the K means algorithm to obtain good initializations. The means are initialized to the cluster centres obtained from the K means procedure, while the covariances are set to the sample covariances of the clusters, and the mixing coefficients are initially the fractions of data samples assigned to a particular cluster.

```
% Number of samples to be generated.
N = 100;
% Array of means.
MU = [1 1;-4 -1; 1 -2];
% Concatenation of co-variance matrices.
SIGMA = cat(3,[2 0; 0 .5],[1 0.5; 0.5 1],[1 0; 0 1]);
% Mixing coefficients.
p = [0.4 0.5 0.1];
% Gaussian mixture model.
GMModel = gmdistribution(MU,SIGMA,p);
% Generate and display data.
rng(1); % for reproducibility
[V,idx] = random(GMModel,N);
gscatter(V(:,1),V(:,2),idx,'bgr','...',[10 10 10]);
```

Listing 6.2: Generating data samples from a Gaussian mixture model.

Listing 6.2 uses a two-dimensional Gaussian mixture model with three components with mixing coefficients $\pi_1 = 0.4, \pi_2 = 0.5$ and $\pi_3 = 0.1$ to generate data samples. The means are

$$\mu_1 = \begin{pmatrix} 1 \\ 1 \end{pmatrix}, \mu_2 = \begin{pmatrix} -4 \\ -1 \end{pmatrix}, \mu_3 = \begin{pmatrix} 1 \\ -2 \end{pmatrix},$$

while the covariance matrices are given by

$$\Sigma_1 = \begin{pmatrix} 2 & 0 \\ 0 & 0.5 \end{pmatrix}, \Sigma_2 = \begin{pmatrix} 1 & 0.5 \\ 0.5 & 1 \end{pmatrix}, \Sigma_3 = \begin{pmatrix} 1 & 0 \\ 0 & 1 \end{pmatrix}.$$

The result can be seen in Figure 6.3a with each data sample coloured according to the process which generated it. This is sometimes referred to as the *complete* data set, since both the samples and the latent variables are given. In Figure 6.3b the visual clue of colours has been removed. Without this, it is harder to tell how many clusters are present. Two clusters are another possibility. When the information about the latent variables is not given, then the data set is sometimes called *incomplete*.

Listing 6.3 fits a Gaussian mixture model and gives the relevant means and covariances. When two components are fitted, the estimates for the mixing coefficients are $\pi_1 = 0.49$ and $\pi_2 = 0.51$ with means

$$\mu_1 = \begin{pmatrix} 1.09 \\ 0.43 \end{pmatrix}, \mu_2 = \begin{pmatrix} -3.98 \\ -1.05 \end{pmatrix},$$

and covariances

$$\Sigma_1 = \begin{pmatrix} 1.87 & 0.11 \\ 0.11 & 2.10 \end{pmatrix}, \Sigma_2 = \begin{pmatrix} 0.91 & 0.47 \\ 0.47 & 1.05 \end{pmatrix}.$$

The graphical representation in Figure 6.3c is plausible.

```
load('MixtureData.mat');
%Set number of components.
K = 2;
%For Gaussian mixture distribution using the EM algorithm.
GMModel = fitgmdist(V,K);
% Posterior probabilities of each data sample to be generated by
% each component.
P = posterior(GMModel,V);
% Estimated means, covariances and mixing coefficients of each
% component.
MixtureMeans = GMModel.mu;
MixtureCovariances = GMModel.Sigma;
MixtureProportions = GMModel.PComponents;
for k =1:K
    k
    MixtureMeans(k,:)
    MixtureCovariances(:,:,k)
    MixtureProportions(k)
end
% Colour coordinated plot for K=2 or K=3. The posterior probabilities
% are interpreted as red, green, blue colour proportions.
% If k=2, the blue proportion is set to zero.
if K==2 || K==3
    if K == 2
        P = cat(2,P, zeros(length(P),1));
    end
    figure;
    scatter(V(:,1), V(:,2),10,P,'filled');
end
```

Listing 6.3: Fitting a Gaussian mixture model.

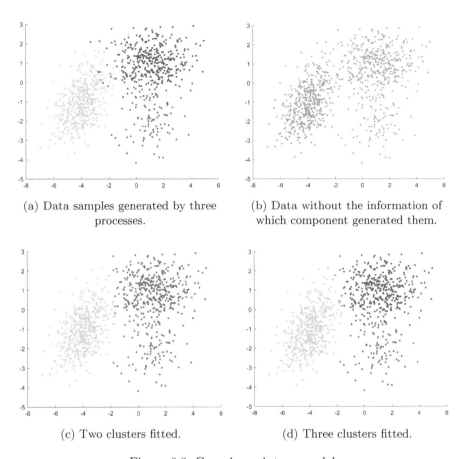

(a) Data samples generated by three processes.

(b) Data without the information of which component generated them.

(c) Two clusters fitted.

(d) Three clusters fitted.

Figure 6.3: Gaussian mixture model.

For three components, we have $\pi_1 = 0.399, \pi_2 = 0.507$ and $\pi_3 = 0.093$ and

$$\boldsymbol{\mu}_1 = \left(\begin{array}{c} 1.07 \\ 1.02 \end{array} \right), \boldsymbol{\mu}_2 = \left(\begin{array}{c} -3.99 \\ -1.06 \end{array} \right), \boldsymbol{\mu}_3 = \left(\begin{array}{c} 1.06 \\ -2.02 \end{array} \right),$$

and covariances

$$\boldsymbol{\Sigma}_1 = \left(\begin{array}{cc} 2.10 & 0.06 \\ 0.06 & 0.64 \end{array} \right), \boldsymbol{\Sigma}_2 = \left(\begin{array}{cc} 0.90 & 0.44 \\ 0.44 & 1.02 \end{array} \right), \boldsymbol{\Sigma}_3 = \left(\begin{array}{cc} 1.16 & 0.11 \\ 0.11 & 0.78 \end{array} \right).$$

This is close to the Gaussian mixture which generated the data. In Figures 6.3c and 6.3d each sample is given a colour which is derived by mixing red, green and blue according to the responsibilities. That is a sample where two responsibilities are zero and the third is one is given a pure colour, while one where two of the responsibilities are 0.5 and the third is 0, is given a mixture of two primary colours, for example purple.

6.4 Expectation-Maximization

In this section we introduce a more general technique to maximize

$$\mathcal{L} = \sum_{n=1}^{N} \log p(\mathbf{v}_n) = \sum_{n=1}^{N} \log \left(\sum_{k=1}^{K} \pi_k p_k(\mathbf{v}_n) \right).$$

We start by considering the update formulae (6.7), (6.8) and (6.5) for $\boldsymbol{\mu}_k$, $\boldsymbol{\Sigma}_k$ and π_k in the case of Gaussian mixtures. Maximizing the function

$$
\begin{aligned}
\widehat{\mathcal{L}} &= \sum_{n=1}^{N}\sum_{k=1}^{K} p(z_{n,k}=1|\mathbf{v}_n)\left[\log \pi_k + \log p_k(\mathbf{v}_n)\right] \\
&= \sum_{n=1}^{N}\sum_{k=1}^{K} p(z_{n,k}=1|\mathbf{v}_n)\left[\log \pi_k - \frac{1}{2}\log|2\pi\boldsymbol{\Sigma}_k|\right. \\
&\qquad\left. -\frac{1}{2}(\mathbf{v}_n - \boldsymbol{\mu}_k)^T \boldsymbol{\Sigma}_k^{-1}(\mathbf{v}_n - \boldsymbol{\mu}_k)\right]
\end{aligned}
$$

with respect to $\boldsymbol{\mu}_k$, $\boldsymbol{\Sigma}_k$ and π_k subject to the constraint that the mixing coefficients sum to one, arrives at the same update formulae.

To see this, we consider first the derivatives

$$
\begin{aligned}
\frac{\partial}{\partial \boldsymbol{\mu}_k}\widehat{\mathcal{L}} &= \sum_{n=1}^{N} p(z_{n,k}=1|\mathbf{v}_n)\boldsymbol{\Sigma}_k^{-1}(\mathbf{v}_n - \boldsymbol{\mu}_k), \\
\frac{\partial}{\partial \boldsymbol{\Sigma}_k}\widehat{\mathcal{L}} &= -\frac{1}{2}\sum_{n=1}^{N} p(z_{n,k}=1|\mathbf{v}_n)\left[\boldsymbol{\Sigma}_k^{-1} - \boldsymbol{\Sigma}_k^{-1}(\mathbf{v}_n - \boldsymbol{\mu}_k)(\mathbf{v}_n - \boldsymbol{\mu}_k)^T\boldsymbol{\Sigma}_k^{-1}\right],
\end{aligned}
$$

employing Appendices A.2.3, A.2.4 and A.2.8. Setting these to zero and solving for $\boldsymbol{\mu}_k$ and $\boldsymbol{\Sigma}_k$ respectively gives the update formulae for those parameters.

To find an optimal value for π_k, we need to introduce a Lagrange multiplier:

$$\frac{\partial}{\partial \pi_k}\left[\widehat{\mathcal{L}} + \lambda\left(\sum_{k=1}^{K}\pi_k - 1\right)\right] = \sum_{n=1}^{N} p(z_{n,k}=1|\mathbf{v}_n)\frac{1}{\pi_k} + \lambda.$$

To find $\lambda = -N$, we set the derivative to zero, multiply through by π_k and sum over k, using the fact that both the responsibilities for a particular sample and the mixing coefficients sum to 1. The update formula for π_k follows directly then.

How are the objective functions $\mathcal{L}$ and $\widehat{\mathcal{L}}$ related and why do the parameters where their derivatives vanish coincide?

Rewriting

$$\widehat{\mathcal{L}} = \sum_{n=1}^{N}\sum_{k=1}^{K} p(z_{n,k}=1|\mathbf{v}_n)\log\left(p(z_{n,k}=1)p(\mathbf{v}_n|z_{n,k}=1)\right),$$

we see that $\widehat{\mathcal{L}}$ is the expectation of the logarithm of the complete data likelihood

$$\sum_{n=1}^{N} \log p(\mathbf{v}_n, \mathbf{z}_n),$$

where the expectation is taken with respect to the responsibilities, that is the posterior probabilities of the latent variables. In the following, we will derive a lower bound which shows the relationship between $\mathcal{L}$ and $\widehat{\mathcal{L}}$.

Using the product rule for probabilities, the logarithm of the complete data likelihood can also be written as

$$\sum_{n=1}^{N} \log p(\mathbf{v}_n, \mathbf{z}_n) = \sum_{n=1}^{N} \log p(\mathbf{v}_n) + \log p(\mathbf{z}_n | \mathbf{v}_n).$$

Both sides can be viewed as functions of the random variables $\mathbf{z}_n$, and the expectation with respect to any distribution $q(\mathbf{z}_n)$ can be taken. Since $\mathbf{z}_n$ is a 1-of-K representation, the expectation is calculated by summing over all possible values for $\mathbf{z}_n$:

$$\sum_{k=1}^{K} \sum_{n=1}^{N} q(z_{n,k} = 1) \log p(\mathbf{v}_n, z_{n,k} = 1)$$

$$= \sum_{k=1}^{K} \sum_{n=1}^{N} q(z_{n,k} = 1) \log p(\mathbf{v}_n) + \sum_{k=1}^{K} \sum_{n=1}^{N} q(z_{n,k} = 1) \log p(z_{n,k} = 1 | \mathbf{v}_n)$$

$$= \underbrace{\sum_{n=1}^{N} \log p(\mathbf{v}_n)}_{\mathcal{L}} + \sum_{k=1}^{K} \sum_{n=1}^{N} q(z_{n,k} = 1) \log p(z_{n,k} = 1 | \mathbf{v}_n),$$

because of $\sum_{k=1}^{K} q(z_{n,k} = 1) = 1$. Rearranging, we see that the the logarithm of the data likelihood is given by

$$\mathcal{L} = \sum_{n=1}^{N} \log p(\mathbf{v}_n) = \sum_{k=1}^{K} \sum_{n=1}^{N} q(z_{n,k} = 1) \log p(\mathbf{v}_n, z_{n,k} = 1)$$

$$- \sum_{k=1}^{K} \sum_{n=1}^{N} q(z_{n,k} = 1) \log p(z_{n,k} = 1 | \mathbf{v}_n).$$

Subtracting and adding the term

$$\sum_{n=1}^{N} \sum_{k=1}^{K} q(z_{n,k} = 1) \log q(z_{n,k} = 1)$$

on the right hand side leaves the equation unchanged, and we obtain

$$
\begin{aligned}
\mathcal{L} \;=\; & \sum_{k=1}^{K}\sum_{n=1}^{N} q(z_{n,k}=1)\log\frac{p(\mathbf{v}_n, z_{n,k}=1)}{q(z_{n,k}=1)}\\
& +\sum_{n=1}^{N}\sum_{k=1}^{K} q(z_{n,k}=1)\log\frac{q(z_{n,k}=1)}{p(z_{n,k}=1|\mathbf{v}_n)},
\end{aligned}
\tag{6.9}
$$

Now, in the last line each sum over $k = 1,\ldots,K$ is the *Kullback–Leibler divergence (KL divergence)* from the discrete distribution $p(\mathbf{z}_n|\mathbf{v}_n)$ to the discrete distribution $q(\mathbf{z}_n)$.

More generally, given two discrete probability distributions P and Q, the Kullback–Leibler divergence from Q to P is defined as

$$
D_{KL}(P\|Q) = \sum_i P(i)\log\frac{P(i)}{Q(i)}.
$$

To avoid a division by zero, it is only defined if $Q(i) = 0$ implies $P(i) = 0$. In this case, that particular term of the sum is interpreted as zero, since $\lim_{x\to 0} x\log x = 0$. If both distributions are the same, then the Kullback–Leibler divergence is zero. It is also non-negative, since $\log x \le x - 1$ and therefore

$$
\begin{aligned}
D_{KL}(P\|Q) \;=\; & -\sum_i P(i)\log\frac{Q(i)}{P(i)} \ge -\sum_i P(i)\left(\frac{Q(i)}{P(i)} - 1\right)\\
\;=\; & -\sum_i Q(i) + \sum_i P(i) = 0,
\end{aligned}
$$

since the probabilities need to sum to 1.

The last line in (6.9) is always non-negative, and therefore the first line is a lower bound for the logarithm of the data likelihood. We can rewrite this lower bound as

$$
\begin{aligned}
\tilde{\mathcal{L}} \;=\; & \underbrace{\sum_{k=1}^{K}\sum_{n=1}^{N} q(z_{n,k}=1)\log\frac{p(\mathbf{v}_n, z_{n,k}=1)}{q(z_{n,k}=1)}}_{\widehat{\mathcal{L}}}\\
\;=\; & \sum_{k=1}^{K}\sum_{n=1}^{N} q(z_{n,k}=1)\log\Big(p(z_{n,k}=1)p(\mathbf{v}_n|z_{n,k}=1)\Big)\\
& -\sum_{k=1}^{K}\sum_{n=1}^{N} q(z_{n,k}=1)\log q(z_{n,k}=1).
\end{aligned}
$$

The lower bound has the same value as the logarithm of the data likelihood $\mathcal{L}$, if $q(\mathbf{z}_n)$ is equal to $p(\mathbf{z}_n|\mathbf{v}_n)$, because then the Kullback–Leibler divergence is zero.

The maximum of (6.9) is estimated by alternating between maximizing the lower bound with respect to $q(\mathbf{z}_n)$ and with respect to $\pi_k = p(z_k = 1)$ and the parameters of $p_k(\mathbf{v}) = p(\mathbf{v}|z_k = 1)$, $k = 1, \ldots, K$. The former means setting $q(\mathbf{z}_n)$ to $p(\mathbf{z}_n|\mathbf{v}_n)$, where the responsibilities are evaluated using the current parameters of $p_k(\mathbf{v})$. With this choice for $q(\mathbf{z}_n)$, however, the first term of the lower bound becomes exactly $\widehat{\mathcal{L}}$. The second term is independent of the parameters of $p_k(\mathbf{v})$ and hence is irrelevant for the maximization. The location of the maximum is the same whether maximizing $\widehat{\mathcal{L}}$ or $\widetilde{\mathcal{L}}$.

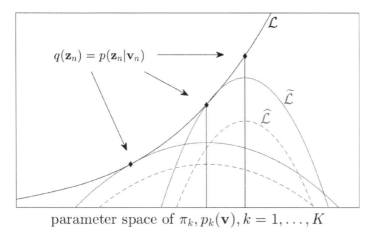

$$q(\mathbf{z}_n) = p(\mathbf{z}_n|\mathbf{v}_n)$$

parameter space of $\pi_k, p_k(\mathbf{v}), k = 1, \ldots, K$

Figure 6.4: Illustration of the maximization steps.

Figure 6.4 illustrates this. The horizontal axis stands for the space formed by $\pi_k = p(z_k = 1)$ and the parameters of $p_k(\mathbf{v}) = p(\mathbf{v}|z_k = 1)$, $k = 1, \ldots, K$. Generally, it has a large dimensionality. The logarithm of the joint likelihood of the data samples, that is the objective function $\mathcal{L}$ is drawn in black. The diamonds pick out particular choices for π_k and the parameters of $p_k(\mathbf{v})$, $k = 1, \ldots, K$. If $q(\mathbf{z}_n)$ is set to the responsibilities $p(\mathbf{z}_n|\mathbf{v}_n)$ calculated for theses choices, $\mathcal{L}$ and $\widehat{\mathcal{L}}$ have the same value. For two different choices of $q(\mathbf{z}_n)$, the lower bound $\widetilde{\mathcal{L}}$ is drawn in blue. Its curve touches the curve of $\mathcal{L}$ at the diamond. This signifies a maximization with respect to $q(\mathbf{z}_n)$. The expectation of the logarithm of the complete data likelihood $\widehat{\mathcal{L}}$ is in red. Note that the locations of the maxima of both $\widehat{\mathcal{L}}$ and $\widetilde{\mathcal{L}}$ are the same, indicated by the vertical lines, where a maximization with respect to π_k and the parameters of $p_k(\mathbf{v})$, $k = 1, \ldots, K$, took place.

The above analysis leads to the following algorithm to maximize the logarithm of the data likelihood:

1. Choose K and convergence threshold.

2. Initialize all parameters of $p_k(\mathbf{v}) = p(\mathbf{v}|z_k = 1)$ and mixing coefficients $\pi_k = p(z_k = 1)$.

3. *E-step*: Evaluate the responsibilities $p(z_{n,k} = 1|\mathbf{v}_n)$ for $n = 1, \ldots, N$ and $k = 1, \ldots, K$.

4. *M-step*: Maximize

$$\sum_{k=1}^{K} \sum_{n=1}^{N} p(z_{n,k} = 1|\mathbf{v}_n) \log \Big(p(z_{n,k} = 1)p(\mathbf{v}_n|z_{n,k} = 1) \Big)$$

with respect to the parameters of $p_k(\mathbf{v}) = p(\mathbf{v}|z_k = 1)$ and mixing coefficients $\pi_k = p(z_k = 1)$.

5. Terminate, if all changes are below the convergence threshold. Otherwise return to step 3

This method is known as the *Expectation-Maximization* algorithm or *EM* algorithm for short. Each step increases the logarithm of the data likelihood. Listing 6.3 uses the internal implementation of the EM algorithm in MATLAB within the call to fit a Gaussian Mixture. We revisit the Expectation-Maximization technique in the context of regression where we consider continuous latent variables. The EM algorithm is widely used and has many extensions. For an overview of those see [29].

6.5 Bayesian Mixture Models

The previous sections have treated the mixing coefficients π_k and the parameters of the probability distribution p_k of the kth process for $k = 1, \ldots, K$ as constant unknown parameters, whose values need to be determined by maximizing the likelihood. We now take a Bayesian approach.

The prior assumption is that the vector $\boldsymbol{\pi} = (\pi_1, \ldots, \pi_K)$ follows a Dirichlet distribution with parameter $\boldsymbol{\alpha} = (\alpha/K, \ldots, \alpha/K)^T$. For example, if $\alpha = K$, then $\boldsymbol{\alpha} = (1, \ldots, 1)^T$ and the probability distribution is the uniform distribution over the simplex in which $\boldsymbol{\pi}$ lies. So all possibilities for $\boldsymbol{\pi}$ are equally likely. As α increases, the resulting Dirichlet probability density functions get more and more peaked at $\boldsymbol{\pi} = (1/K, \ldots, 1/K)^T$ with any other vectors becoming less likely. The effect is illustrated in Figure 6.5. When $\alpha = K$ in Figures 6.5a and 6.5b, the cluster sizes can be very different. As α increases (Figures 6.5c and 6.5d), the clusters become similar in size. Their sizes approach N/K.

Similarly, the probability distributions p_k are drawn themselves from a probability distribution. It is a distribution over distributions, known as *base distribution* G_0. For example, if each process is a normal distribution with its own mean $\boldsymbol{\mu}_k$ and covariance matrix $\boldsymbol{\Sigma}_k$, these can be drawn from the *normal inverse Wishart distribution*, which has four parameters:

- the *location vector* $\mathbf{m}$ lying in the feature space,

- the *mean fraction* λ,

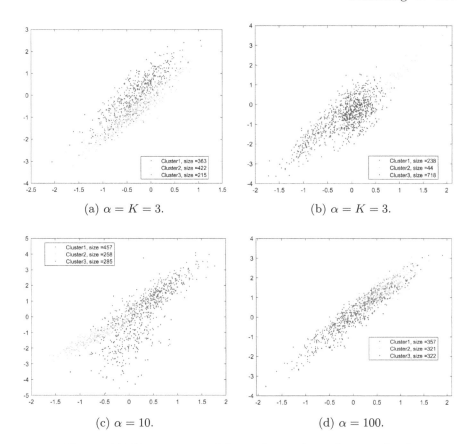

Figure 6.5: Data generated for various values of α and fixed $K = 3, N = 1000$.

- the *inverse scale matrix* $\boldsymbol{\Psi}$, which has to be symmetric and positive definite,

- and ν, which has to be at least the number of dimensions d of the feature space and regulates the degrees of freedom.

The notation is $(\boldsymbol{\mu}_k, \boldsymbol{\Sigma}_k) \sim \text{NIW}(\mathbf{m}, \lambda, \boldsymbol{\Psi}, \nu)$.

This way of generating the data can be represented graphically as

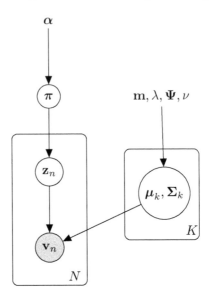

The covariance matrices $\mathbf{\Sigma}_k$ follow the *inverse Wishart distribution* given by $\mathbf{\Psi}$ and ν, $\mathbf{\Sigma}_k \sim \mathcal{W}^{-1}(\mathbf{\Psi}, \nu)$. It has the probability density function

$$\mathcal{W}^{-1}(\mathbf{\Sigma}_k | \mathbf{\Psi}, \nu) = \frac{|\mathbf{\Psi}|^{\nu/2}}{2^{\nu d/2} \Gamma_d\left(\frac{\nu}{2}\right)} |\mathbf{\Sigma}_k|^{-(\nu+d+1)/2} \exp\left(-\frac{1}{2} \mathrm{tr}(\mathbf{\Psi} \mathbf{\Sigma}_k^{-1})\right), \quad (6.10)$$

where $|\cdot|$ denotes the determinant and $\mathrm{tr}(\cdot)$ denotes the trace of a matrix. Γ_d is the multivariate gamma function.

The expectation of the inverse Wishart distribution is given by

$$\mathbb{E}[\mathbf{\Sigma}_k] = \frac{\mathbf{\Psi}}{\nu - d - 1}.$$

The variance of each element Σ_{ij} of $\mathbf{\Sigma}$ is calculated as

$$\mathrm{var}[\Sigma_{ij}] = \frac{(\nu - d + 1)\Psi_{ij}^2 + (\nu - d - 1)\Psi_{ii}\Psi_{jj}}{(\nu - d)(\nu - d - 1)^2(\nu - d - 3)} = \mathrm{var}[\Sigma_{ji}],$$

since $\mathbf{\Psi}$ is symmetric. On the diagonal, that is $i = j$, this simplifies to

$$\mathrm{var}[\Sigma_{ii}] = \frac{2\Psi_{ii}^2}{(\nu - d - 1)^2(\nu - d - 3)}.$$

Since ν appears several times in the denominator, it controls the variability in the generated covariances and in turn in the different data sets which are generated. The variability decreases as ν increases. Figures 6.6 and 6.7 show

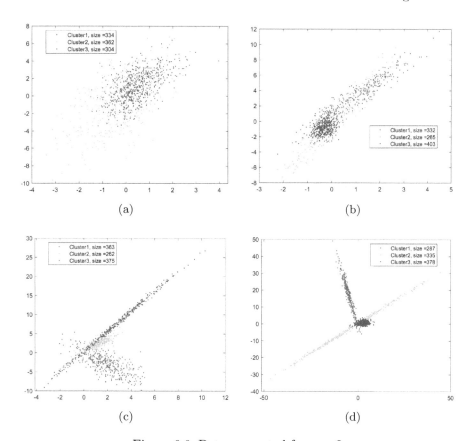

Figure 6.6: Data generated for $\nu = 2$.

the differences in generated data sets for $\nu = 2$ the smallest possible choice, and $\nu = 10$ for $\alpha = 10K$, $\mathbf{m} = (0,0)^T$, $\lambda = 1$ and

$$\boldsymbol{\Psi} = \begin{pmatrix} 1 & 3/2 \\ 3/2 & 1 \end{pmatrix}.$$

Looking at the ranges of the axes in Figure 6.6, the generated data sets are indeed very different in nature. In Figure 6.7, in particular 6.7a and 6.7d, on the other hand, the data seem to come from the base distribution specified by $\mathbf{m}$ and $\boldsymbol{\Psi}$, if it weren't for the distinction by processes.

The inverse Wishart distribution $\mathcal{W}^{-1}(\boldsymbol{\Sigma}|\boldsymbol{\Psi}, \nu)$ is the conjugate prior for the covariance matrix $\boldsymbol{\Sigma}$ of a multivariate normal distribution with a known mean. We can assume that the mean of this distribution is zero, since we can re-centre the data at zero by subtracting the mean from all samples.

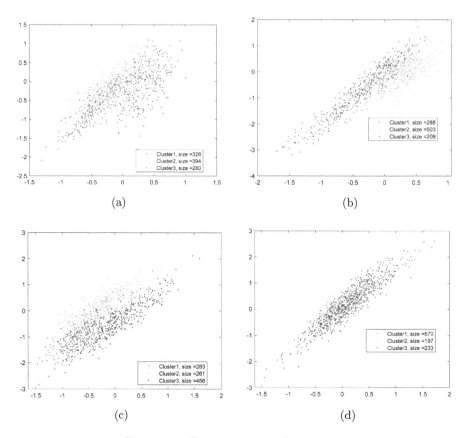

Figure 6.7: Data generated for $\nu = 10$.

Let $\mathcal{D} = \{\mathbf{v}_1, \ldots, \mathbf{v}_N\}$ be draws from $\mathcal{N}(\mathbf{0}, \boldsymbol{\Sigma})$. The likelihood of this data is given by

$$\prod_{n=1}^{N} \frac{1}{\sqrt{(2\pi)^d |\boldsymbol{\Sigma}|}} \exp\left(-\frac{1}{2}\mathbf{v}_n^T \boldsymbol{\Sigma}^{-1} \mathbf{v}_n\right) =$$
$$(2\pi)^{-dN/2} |\boldsymbol{\Sigma}|^{-N/2} \exp\left(-\frac{1}{2}\sum_{n=1}^{N} \mathbf{v}_n^T \boldsymbol{\Sigma}^{-1} \mathbf{v}_n\right).$$

The posterior distribution of $\boldsymbol{\Sigma}$ is proportional to the product of the likelihood and the prior probability density function as given in (6.10) by Bayes' rule. Because of the proportionality, we drop all multiplicative constants in both. These are the factors $(2\pi)^{-dN/2}$, $|\boldsymbol{\Psi}|^{\nu/2}$, $2^{-\nu d/2}$ and $1/\Gamma_d(\nu/2)$. Hence, the posterior is proportional to

$$|\boldsymbol{\Sigma}|^{-N/2} \exp\left(-\frac{1}{2}\sum_{n=1}^{N} \mathbf{v}_n^T \boldsymbol{\Sigma}^{-1} \mathbf{v}_n\right) |\boldsymbol{\Sigma}|^{-(\nu+d+1)/2} \exp\left(-\frac{1}{2}\mathrm{tr}(\boldsymbol{\Psi}\boldsymbol{\Sigma}^{-1})\right).$$

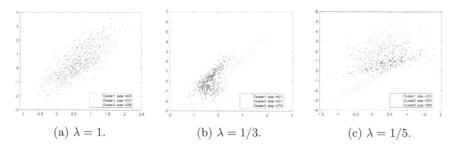

(a) $\lambda = 1$.   (b) $\lambda = 1/3$.   (c) $\lambda = 1/5$.

Figure 6.8: Data generated for various values of λ.

To simplify this expression, we use properties of determinants and traces of matrices. The product of determinants of two matrices is the determinant of the product of matrices. Therefore, $|\mathbf{\Sigma}|^{-N/2}||\mathbf{\Sigma}|^{-(\nu+d+1)/2} = |\mathbf{\Sigma}|^{-(N+\nu+d+1)/2}$. The expression $\mathbf{v}_n^T\mathbf{\Sigma}^{-1}\mathbf{v}_n$ is a scalar and the trace of a scalar is the scalar itself. Further, the trace is invariant to a cyclic permutation of the order of multiplication. Hence, $\mathbf{v}_n^T\mathbf{\Sigma}^{-1}\mathbf{v}_n = \mathrm{tr}(\mathbf{v}_n^T\mathbf{\Sigma}^{-1}\mathbf{v}_n) = \mathrm{tr}(\mathbf{v}_n\mathbf{v}_n^T\mathbf{\Sigma}^{-1})$. Using these identities, the posterior is proportional to

$$|\mathbf{\Sigma}|^{-(N+\nu+d+1)/2} \exp\left(-\frac{1}{2}\left[\sum_{n=1}^{N} \mathrm{tr}(\mathbf{v}_n\mathbf{v}_n^T\mathbf{\Sigma}^{-1}) + \mathrm{tr}(\mathbf{\Psi}\mathbf{\Sigma}^{-1})\right]\right) =$$

$$|\mathbf{\Sigma}|^{-(N+\nu+d+1)/2} \exp\left(-\frac{1}{2}\mathrm{tr}\left((\mathbf{\Psi} + \sum_{n=1}^{N} \mathbf{v}_n\mathbf{v}_n^T)\mathbf{\Sigma}^{-1}\right)\right),$$

since the sum of traces of matrices is the trace of the sum of matrices. Apart from the normalizing constant, this is exactly the probability density function of $\mathcal{W}^{-1}(\mathbf{\Sigma}|\mathbf{\Psi} + \sum_{n=1}^{N} \mathbf{v}_n\mathbf{v}_n^T, \nu + N)$. To summarize, after seeing the data, the inverse scale matrix is updated to $\mathbf{\Psi} + \sum_{n=1}^{N} \mathbf{v}_n\mathbf{v}_n^T$ and the degrees of freedom become $\nu+N$. Since this increases, the estimate for the distribution describing $\mathbf{\Sigma}$ becomes tighter, the more data samples are seen. The least informative choice for ν in the prior is $\nu = d$.

Having drawn $\mathbf{\Sigma}_k$, $\boldsymbol{\mu}_k$ is drawn from a normal distribution with mean $\mathbf{m}$ and covariance matrix $\frac{1}{\lambda}\mathbf{\Sigma}_k$. The parameter λ controls the spacing of the generated means $\boldsymbol{\mu}_k$. Figure 6.8 illustrates the effect of $\lambda = 1, 1/3, 1/5$ for $\alpha = 10K$, $\mathbf{m} = (0,0)^T$, the previous $\mathbf{\Psi}$ and $\nu = 6$. As λ decreases, the clusters separate. As λ increases, they overlap more and more. Figures 6.5 to 6.8 were generated by the function in Listing 6.4.

To summarize, the data is generated following the distributions:

$$\mathbf{v}_n|\mathbf{z}_n \sim \mathcal{N}(\boldsymbol{\mu}_k, \mathbf{\Sigma}_k)$$
$$p(z_{n,k} = 1) = \pi_k,$$
$$\boldsymbol{\pi} \sim \mathrm{Dir}(\boldsymbol{\alpha}),$$
$$\boldsymbol{\mu}_k, \mathbf{\Sigma}_k \sim G_0 = \mathrm{NIW}(\mathbf{m}, \lambda, \mathbf{\Psi}, \nu).$$

```
function [V,idx,MU,SIGMA] = BayesianMixture(N,K,alpha,mu0,...
    lambda,Psi,nu)
% Input:
% N:      number of data points to be generated,
% K:      number of clusters
% alpha:  concentration parameter,
% mu0:    location vector,
% lambda: mean fraction,
% Psi:    inverse scale matrix,
% nu:     degrees of freedom.
% Output:
% V:      data vector,
% idx:    index vector which process generated the sample,
% MU:     array of the means of all processes,
% SIGMA:  array of the covariance matrices of all processes.

% Generate vector of mixture coefficients from Dirichlet
% distribution given by alpha and K using the Gamma dsitribution.
a = alpha/K*ones(1,K);
PI = gamrnd(a,1);
PI = PI/sum(PI);

% Generate latent indicator variable for cluster mambership.
idx = randsample(K,N,true,PI);

% Generate distribution for each cluster.
SIGMA = zeros(2,2,K);
MU = zeros(2,K);
for k=1:K
    sigma = iwishrnd(Psi,nu);
    SIGMA(:,:,k) = sigma;
    MU(:,k) = mvnrnd(mu0,sigma/lambda);
end

% Generate data.
V = zeros(N,2);
for n=1:N
    v = mvnrnd(MU(:,idx(n)),SIGMA(:,:,idx(n)));
    V(n,:) = v;
end
```

Listing 6.4: Generating data from a distribution of distributions for a fixed number of clusters.

This can be viewed without involving the latent variables $\mathbf{z}_n$. Imagine a distribution G over the set of all possible pairs of $(\boldsymbol{\mu}, \boldsymbol{\Sigma})$ which can be generated from G_0. Since there is only a finite number of clusters, G is zero everywhere apart from the specific pairs $(\boldsymbol{\mu}_k, \boldsymbol{\Sigma}_k)$ where it has a point probability mass of π_k.

Using this distribution G, the generation of data can be described as:

$$\mathbf{v}_n \quad \sim \quad \mathcal{N}(\boldsymbol{\mu}_k, \boldsymbol{\Sigma}_k)$$
$$\boldsymbol{\pi} \quad \sim \quad \mathrm{Dir}(\boldsymbol{\alpha}),$$
$$\boldsymbol{\mu}_k, \boldsymbol{\Sigma}_k \quad \sim \quad G.$$

The base distribution G_0 is indirectly part of the distribution G, since the points where G is nonzero are drawn from G_0, while the probability masses at these points are drawn from $\mathrm{Dir}(\boldsymbol{\alpha})$.

Having explored the way the data is generated, we now look at inferring the cluster membership from the data. First, let $\mathcal{D} = \{\mathbf{v}_1, \ldots, \mathbf{v}_N\}$ be a set of data samples drawn from a multivariate normal distribution with mean $\boldsymbol{\mu}$ and covariance matrix $\boldsymbol{\Sigma}$, where $\boldsymbol{\mu}$ and $\boldsymbol{\Sigma}$ are themselves drawn from a normal inverse Wishart distribution. We assume a prior normal inverse Wishart distribution with parameters $\mathbf{m}, \lambda, \boldsymbol{\Psi}$ and ν. The posterior is also a normal inverse Wishart distribution. Let

$$\bar{\mathbf{v}} = \frac{1}{N} \sum_{n=1}^{N} \mathbf{v}_n$$

be the sample mean and

$$\mathbf{S} = \sum_{n=1}^{N} (\mathbf{v}_n - \bar{\mathbf{v}})(\mathbf{v}_n - \bar{\mathbf{v}})^T$$

the scaled sample covariance matrix. The posterior then has parameters

$$\mathbf{m}^{\mathrm{post}} = \frac{\lambda \mathbf{m} + N \bar{\mathbf{v}}}{\lambda + N},$$
$$\lambda^{\mathrm{post}} = \lambda + N,$$
$$\nu^{\mathrm{post}} = \nu + N,$$
$$\boldsymbol{\Psi}^{\mathrm{post}} = \boldsymbol{\Psi} + \mathbf{S} + \frac{\lambda N}{\lambda + N}(\bar{\mathbf{v}} - \mathbf{m})(\bar{\mathbf{v}} - \mathbf{m})^T.$$

In particular, if $N = 1$, we have one single data point $\mathbf{v}$ and $\bar{\mathbf{v}} = \mathbf{v}$ and $\mathbf{S} = 0$. With this,

$$\mathbf{m}^{\mathrm{post}} = \frac{\lambda \mathbf{m} + \mathbf{v}}{\lambda + 1},$$
$$\lambda^{\mathrm{post}} = \lambda + 1,$$
$$\nu^{\mathrm{post}} = \nu + 1,$$
$$\boldsymbol{\Psi}^{\mathrm{post}} = \boldsymbol{\Psi} + \frac{\lambda}{\lambda + 1}(\mathbf{v} - \mathbf{m})(\mathbf{v} - \mathbf{m})^T.$$

$$(6.11)$$

Thus the data samples can be considered one at a time and the posterior is calculated sequentially.

However, the above does not consider the probabilities of a sample belonging to a particular cluster which are governed by a Dirichlet prior distribution with parameter $\boldsymbol{\alpha} = (\alpha/K, \ldots, \alpha/K)^T$. We start with a random assignment of the indicator variables $\mathbf{z}_1, \ldots, \mathbf{z}_n$. For each individual cluster k, we let N_k be the number of samples in it with this particular assignment. We then calculate the parameters $\mathbf{m}_k, \lambda_k, \boldsymbol{\Psi}_k$ and ν_k of the posterior, normal, inverse Wishart distribution for each cluster and draw $\boldsymbol{\mu}_k$ and $\boldsymbol{\Sigma}_k$ from this distribution.

Gibbs sampling is used to reassign samples to clusters in a random order. To this end, a random sample $\mathbf{v}_n$ is selected. Let l be its current cluster assignment. Consider the data set without this sample, $\mathcal{D} \setminus \{\mathbf{v}_n\}$. It needs to be removed from cluster l by returning to the prior parameters of the normal, inverse Wishart distribution for the lth cluster. Using (6.11) backwards, we see that

$$
\begin{aligned}
\mathbf{m}_l^{\text{prior}} &= \frac{\lambda_l \mathbf{m}_l - \mathbf{v}_n}{\lambda_l - 1}, \\
\lambda_l^{\text{prior}} &= \lambda_l - 1, \\
\nu_l^{\text{prior}} &= \nu_l - 1, \\
\boldsymbol{\Psi}_l^{\text{prior}} &= \boldsymbol{\Psi}_l - \frac{\lambda_l}{\lambda_l - 1}(\mathbf{v}_n - \mathbf{m}_l)(\mathbf{v}_n - \mathbf{m}_l)^T.
\end{aligned}
$$

Note that this is independent of the order, in which samples were assigned clusters. A new mean $\boldsymbol{\mu}_l$ and covariance matrix $\boldsymbol{\Sigma}_l$ are drawn from this distribution, and N_l is replaced by $N_l - 1$.

We calculate for $k = 1, \ldots, K$ the probabilities that $\mathbf{v}_n$ belongs to cluster k given all the other data samples and their assignments,

$$
\begin{aligned}
p(z_{n,k} = 1 | \mathbf{v}_n, \mathcal{D} \setminus \{\mathbf{v}_n\}, \alpha, \mathbf{m}, \lambda, \boldsymbol{\Psi}, \nu) = \\
p(z_{n,k} = 1 | \mathcal{D} \setminus \{\mathbf{v}_n\}, \alpha) p(\mathbf{v}_n | \mathcal{D} \setminus \{\mathbf{v}_n\}, z_{n,k} = 1, \mathbf{m}, \lambda, \boldsymbol{\Psi}, \nu)
\end{aligned}
\tag{6.12}
$$

following the product rule. Here $\mathcal{D} \setminus \{\mathbf{v}_n\}$ means the set of samples as well as their assignments.

The first factor is the probability of $\mathbf{v}_n$ belonging to cluster k which is governed by the posterior Dirichlet distribution, given the prior Dirichlet distribution and all other cluster assignments. From the discussion about conjugate distributions leading to equation (2.23), this is

$$
\frac{N_k + \alpha/K}{N - 1 + \alpha}.
$$

The second factor in (6.12) is the likelihood of seeing sample $\mathbf{v}_n$ given $\mathcal{D} \setminus \{\mathbf{v}_n\}, z_{n,k} = 1$, and the parameters of the posterior, normal, inverse Wishart distribution for this cluster. This is approximated by the normal distribution with mean $\boldsymbol{\mu}_k$ and covariance matrix $\boldsymbol{\Sigma}_k$. Hence, $p(\mathbf{v}_n | \mathcal{D} \setminus \{\mathbf{v}_n\}, z_{n,k} = 1, \mathbf{m}, \lambda, \boldsymbol{\Psi}, \nu)$ is approximately

$$
\frac{1}{\sqrt{|2\pi\boldsymbol{\Sigma}_k|}} \exp\left(\frac{1}{2}(\mathbf{v}_n - \boldsymbol{\mu}_k)^T \boldsymbol{\Sigma}_k^{-1}(\mathbf{v}_n - \boldsymbol{\mu}_k)\right)
$$

```
function [idx,MU,SIGMA] = InferBayesianMixture(V, K,alpha,m, ...
    lambda,Psi,nu,iter)
% Input:
% V:      data,
% K:      number of clusters,
% alpha:  prior concentration parameter,
% m:      prior location vector,
% lambda: prior mean fraction,
% Psi:    prior inverse scale matrix,
% nu:     prior degrees of freedom,
% iter:   number of iterations
% Output:
% idx:    index vector of cluster assignments,
% MU:     array of the means sampled from the posterior normal
%         inverse Wishart distributions for each cluster,
% SIGMA:  array of the covariance matrices sampled from the posterior
%         normal inverse Wishart distributions for each cluster.

N = size(V,1);
% Start with random cluster assignments.
idx = randi(K,1,N);

% For each cluster store its size, posterior location vector,
% posterior mean fraction, posterior inverse scale matrix,
% posterior degrees of freedom.
nk = zeros(1,K);
M = zeros(2,K);
LAMBDA = zeros(1,K);
PSI = zeros(2,2,K);
NU = zeros(1,K);
% For each cluster store a draw from the posterior normal,
% inverse Wishart distribution.
SIGMA = zeros(2,2,K);
MU = zeros(2,K);
% For each cluster initialize these for the initial random cluster
% assignments and draw MU and SIGMA from that posterior normal,
% inverse Wishart distribution.
for k=1:K
    v = V(idx == k,:);
    nk(k) = size(v,1);
    sampleM = mean(v);
    sampleS = (nk(k)-1) * cov(v);
    M(:,k) = (lambda*m(:) + nk(k)*sampleM(:))/(lambda +nk(k));
    LAMBDA(k) = lambda + nk(k);
    PSIk = Psi(:,:) + sampleS + lambda*nk(k)*(sampleM(:) - m(:))*...
        (sampleM(:) - m(:))'/(lambda +nk(k));
    % Store Cholesky factorization to maintain positive definiteness.
    PSI(:,:,k) = chol(PSIk);
    NU(k) = nu + nk(k);
    SIGMA(:,:,k) = iwishrnd(PSI(:,:,k)'* PSI(:,:,k),NU(k));
    MU(:,k) = mvnrnd(M(:,k),SIGMA(:,:,k)/LAMBDA(k));
end
```

6.5a: Bayes' rule and Gibbs sampling determining the mixture model
for a fixed number of clusters - initialization.

```
for i = 1:iter
    % Consider the data in a random order.
    for n = randperm(N)
        v = V(n,:);
        l = idx(n);
        % Remove this sample from the data set and update the cluster
        % it was assigned to.
        priorlambda = LAMBDA(l)-1;
        % Rank one update on the Cholesky factorization to preserve
        % positive definiteness.
        update = sqrt(LAMBDA(l)/priorlambda) * (v' - M(:,l));
        PSI(:,:,l) = cholupdate(PSI(:,:,l),update,'-');
        M(:,l) = (LAMBDA(l)*M(:,l) - v(:))/priorlambda;
        NU(l) = NU(l)-1;
        LAMBDA(l) = priorlambda;
        SIGMA(:,:,l) = iwishrnd(PSI(:,:,l)'* PSI(:,:,l),NU(l));
        MU(:,l) = mvnrnd(M(:,l),SIGMA(:,:,l)/LAMBDA(l));
        nk(l) = nk(l)-1;
        % Calculate cluster assignment probabilities.
        p = zeros(1,K);
        for k=1:K
            p(k) = (nk(k) + alpha/K)/(N-1+alpha)*...
                mvnpdf(v',MU(:,k),SIGMA(:,:,k));
        end
        p = p/sum(p);
        % Sample new indicator variable.
        l = randsample(K,1,true,p);
        idx(n) = l;
        % Update the cluster the sample is now assigned to.
        postlambda = LAMBDA(l)+1;
        % Rank one update on the Cholesky factorization to preserve
        % positive definiteness.
        update = sqrt(LAMBDA(l)/postlambda) * (v' - M(:,l));
        PSI(:,:,l) = cholupdate(PSI(:,:,l),update,'+');
        M(:,l) = (LAMBDA(l)*M(:,l) + v(:))/postlambda;
        NU(l) = NU(l)+1;
        LAMBDA(l) = postlambda;
        SIGMA(:,:,l) = iwishrnd(PSI(:,:,l)'* PSI(:,:,l),NU(l));
        MU(:,l) = mvnrnd(M(:,l),SIGMA(:,:,l)/LAMBDA(l));
        nk(l) = nk(l)+1;
    end
end
```

6.5b: Bayes' rule and Gibbs sampling determining the mixture model for a fixed number of clusters - iterations.

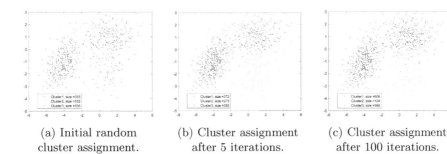

(a) Initial random cluster assignment.

(b) Cluster assignment after 5 iterations.

(c) Cluster assignment after 100 iterations.

Figure 6.9: Bayes' rule and Gibbs sampling determining the mixture model for a fixed number of clusters.

given $\mathcal{D} \setminus \{\mathbf{v}_n\}$, the current assignments and the prior.

With these results,

$$p(z_{n,k} = 1 | \mathbf{v}_n, \mathcal{D} \setminus \{\mathbf{v}_n\}, \alpha, \mathbf{m}, \lambda, \boldsymbol{\Psi}, \nu) \approx$$

$$\frac{N_k + \alpha/K}{N - 1 + \alpha} \frac{1}{\sqrt{|2\pi \boldsymbol{\Sigma}_k|}} \exp\left(\frac{1}{2}(\mathbf{v}_n - \boldsymbol{\mu}_k)^T \boldsymbol{\Sigma}_k^{-1}(\mathbf{v}_n - \boldsymbol{\mu}_k)\right).$$

Since these are approximations, we have to divide by the sum of probabilities to ensure they add to one. We use these probabilities to draw a new indicator variable $\mathbf{z}_n$ for $\mathbf{v}_n$.

Having assigned sample $\mathbf{v}_n$ to a new cluster m, all parameters for this cluster need to be updated following (6.11). A new mean $\boldsymbol{\mu}_m$ and covariance matrix $\boldsymbol{\Sigma}_m$ are drawn from this posterior distribution. N_m is increased by one. This completes the cluster re-assignment for $\mathbf{v}_n$.

An iteration is complete, if all samples have been considered in a random order. The process is repeated for a fixed number of iterations. Sometimes this form of Gibbs sampling is called *collapsed Gibbs sampling*, since the cluster assignments of the other samples are not explicitly used, when sampling $\mathbf{z}_n$ for $\mathbf{v}_n$. They are indirectly encoded in the parameters of the posterior, normal inverse, Wishart distribution for each cluster. This reduces the parameters influencing the sampling to a multiple of K.

Listings 6.5 is a function implementing this clustering algorithm. Some implementations do not draw from the posterior, normal, inverse Wishart distributions, but instead set $\boldsymbol{\mu}_k = \mathbf{m}_k$ and $\boldsymbol{\Sigma}_k = \boldsymbol{\Psi}_k$.

Figure 6.9 shows the results, when the method is applied to the same data as in Figures 6.3a and 6.3b. The algorithm was initialized with $\alpha = K$, $\lambda = 1$, $\nu = 2$. The parameter $\mathbf{m}$ was set to the sample mean of all data, while $\boldsymbol{\Psi}$ was set to the average squared distance of all samples to the mean divided by the number of dimensions times the identity matrix.

Figure 6.9a shows the initial random cluster assignments. The clusters start to separate after five iterations in 6.9b. 320 samples are given the wrong (up-to re-labeling) assignment. With more iterations this number reduces. After a hundred iterations 53 samples belong to the wrong cluster. A perfect result

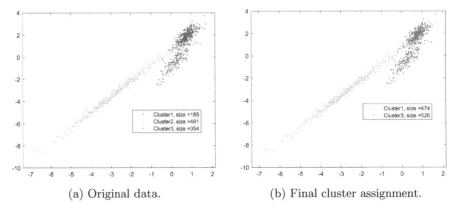

(a) Original data.　　　　　　　　(b) Final cluster assignment.

Figure 6.10: Incorrect cluster assignment.

is generally not possible, since for some samples the posterior probabilities $p(z_{n,k} = 1|\mathbf{v}_n, \mathcal{D} \setminus \{\mathbf{v}_n\}, \alpha, \mathbf{m}, \lambda, \mathbf{\Psi}, \nu)$ might be very similar for two clusters, with there being no clear maximal probability. In these cases, it is more useful to return the vector of posterior probabilities instead of a cluster assignment.

Even though the number of clusters is specified, this does not mean that this number of clusters is used. It is an upper bound. Figure 6.10 shows that the algorithm combined two clusters into one. Since all the generation of data as well as the inference of clusters is governed by probabilities, there is always the chance of the algorithm getting it wrong.

6.6　The Chinese Restaurant Process

In all the above, the number of clusters was fixed at the start of the algorithm. This is, however, undesirable, since in general it is not known beforehand, how many clusters there are. Ideally, the number of clusters should be determined by the data.

It is helpful to revisit the assumptions we made about how the data is generated. Previously, a new data sample was generated by choosing one of the K processes and then generating the sample according to the probability distribution for that process. We now take the view that there is a chance that the sample is generated by a completely new process, which is different from the processes which generated the previous samples.

It is more convenient to take a sequential view of the generation of samples. For the first sample, a process and its distribution are generated. The second sample can either be generated from the same distribution as the first sample or from a new distribution. We need to define a probability for these two cases. Let α be an additional parameter governing this probability. It is known as *dispersion, concentration, scaling* parameter or *strength*. The probability that the second sample is generated by the first process is $1/(1 + \alpha)$, while the

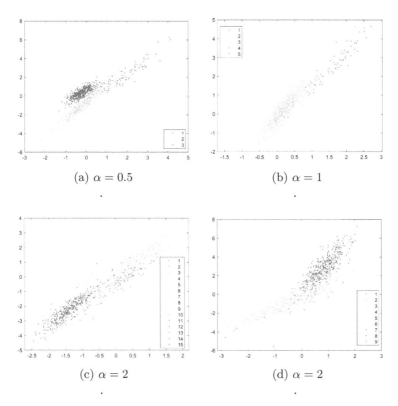

Figure 6.11: Data generated for various values of α.

probability that it is generated by a new process is $\alpha/(1+\alpha)$. If $\alpha = 1$, these two probabilities are the same and are both one half. If α is greater than 1, then a new process is favoured, while for $\alpha < 1$ the already existing process is more likely to be used to generate the second sample.

When generating the nth sample, it is generated by

$$\begin{cases} \text{process } k \text{ with probability} & \dfrac{n_k}{n-1+\alpha} \\[2ex] \text{a new process with probability} & \dfrac{\alpha}{n-1+\alpha} \end{cases},$$

where n_k denotes the number of samples generated by process k so far. When summing the samples generated by each process over all processes, the result is $n-1$. Therefore these probabilities sum to 1.

Note that as more and more samples are generated by a particular process k, it gets more likely that this process will generate further samples, since $n_k/(n-1+\alpha)$ increases relatively to the other probabilities. This is again the *rich-get-richer* phenomenon.

Listing 6.6 generates data in this fashion. The results for various values of α are shown in Figure 6.11. The distributions for each cluster were drawn

```
function [V,idx,MU,SIGMA] = CRP(N, alpha, mu0, lambda, Psi, nu)
% Input:
% N:      number of data points to be generated,
% alpha:  concentration parameter,
% mu0:    location vector,
% lambda: mean fraction,
% Psi:    inverse scale matrix,
% nu:     degrees of freedom.
% Output:
% V:      data vector,
% idx:    index vector which process generated the sample,
% MU:     array of the means of all processes,
% SIGMAS: array of the covariance matrices of all processes.

% Generate distribution of first process.
SIGMA = iwishrnd(Psi,nu);
MU =  mvnrnd(mu0,lambda * SIGMA);
% Generate first sample.
V = mvnrnd(MU,SIGMA);
idx = 1;
% The vector n_k tracks the number of samples generated by the k-th
% process, where k is the vector index.
n_k = 1;
for n =2:N
    % Calculate probabilities which process generates the next sample.
    p = n_k/(n-1+alpha);
    % Append the probability that a new process is generated.
    p = cat(1,p,alpha/(n-1+alpha));
    % Sample from which process the next data point is taken or
    % whether a new process is created.
    compidx = randsample(length(p),1,true,p/sum(p));
    if compidx < length(p)
        % New sample is generated from existing process indicated by
        % compidx. Generate new sample from this process.
        V = cat(1,V,mvnrnd(MU(compidx,:),SIGMA(:,:,compidx)));
        idx = cat(1,idx,compidx);
        % Update vector n_k of the number of samples generated by each
        % process.
        n_k(compidx) = n_k(compidx)+1;
    else
        % New process is generated.
        sigma = iwishrnd(Psi,nu);
        mu =  mvnrnd(mu0,lambda * sigma);
        % Generate new sample from this process.
        V = cat(1,V,mvnrnd(mu,sigma));
        idx = cat(1,idx,compidx);
        % Append mean and covariance matrix of new process to arrays
        % of means and covariances.
        MU = cat(1,MU,mu);
        SIGMA = cat(3,SIGMA,sigma);
        % Update vector n_k of the number of samples generated by each
        % process.
        n_k = cat(1,n_k,1);
    end
end
```

Listing 6.6: Generating data from a distribution of distributions where the number of clusters is variable.

from a normal, inverse, Wishart distribution with parameters $\mathbf{m} = (0,0)^T$, $\lambda = 3$, $\nu = 6$, and

$$\Psi = \begin{pmatrix} 1 & 3/2 \\ 3/2 & 1 \end{pmatrix}.$$

They illustrate the rich-get-richer property with some clusters dominating. As α increases, the number of clusters increases, since the probability that a new process is generated is larger. Different runs will give data sets which can look quite different as in Figures 6.11c and 6.11d. This last figure also shows that two processes could have very similar distributions. Without the knowledge of which process generated which, the data samples seem to be generated from the same process.

This way of generating data is known as the *Chinese Restaurant Process* (CRP). Many Chinese restaurants feature round tables with a centre that can be turned. The various dishes are placed there, and the diners can share by turning the centre. The data generation is likened to a hypothetical Chinese restaurant with an infinite availability of tables and each table can seat as many diners as necessary. The first customer comes, sits at the first table and orders a variety of food. This determines the distribution. They eat what they fancy which corresponds to a specific data sample. The second customer might choose to sit at the same table. They will eat from the same selection of food, but not exactly the same as the first customer. They sample a second distinct set of food. On the other hand, they might choose to sit at a new table and order a different set of dishes which constitutes the distribution of the second process. As new customers arrive, they either sit at an already open table or open a new one.

Since the final number of customers N is finite, so is the number of tables K. Once the final configuration is reached it is unique up to the relabeling of tables. This means there are $K!$ equivalent ways this configuration could have been reached. This in turn means that the order in which the customers arrive is of no importance. If in a different ordering the first customer sits at a different table which becomes the first table, we just relabel this table to 1 and change all subsequent labels.

6.7 Dirichlet Process

Having considered the process of how the data is generated, we need to change our viewpoint and deduce the number of clusters and which samples belong to which from the data. As before with Bayesian mixtures, we choose a *concentration* parameter α and prior probability distributions, and deduce posterior probabilities from the data using collapsed Gibbs sampling. However, now K is the *current* number of clusters and is variable. The normal, inverse, Wishart distribution with parameters $\mathbf{m}, \lambda, \Psi$ and ν is used as prior to illustrate.

At initialization stage the samples are considered in a random order. Let $\mathbf{v}_1$ be the first sample considered. It also constitutes the first cluster. The posterior normal, inverse Wishart distribution of this cluster is calculated from the prior as

$$
\begin{aligned}
\mathbf{m}_1 &= \frac{\lambda \mathbf{m} + \mathbf{v}_1}{\lambda + 1}, \\
\lambda_1 &= \lambda + 1, \\
\nu_1 &= \nu + 1, \\
\mathbf{\Psi}_1 &= \mathbf{\Psi} + \frac{\lambda}{\lambda + 1}(\mathbf{v}_1 - \mathbf{m})(\mathbf{v}_1 - \mathbf{m})^T.
\end{aligned}
$$

A covariance matrix $\mathbf{\Sigma}_1$ and mean $\boldsymbol{\mu}_1$ are drawn from this distribution for this cluster.

Let $\mathbf{v}_n$ be the nth sample considered and let $\mathcal{D}_n = \{\mathbf{v}_1, \ldots, \mathbf{v}_{n-1}\}$. That is $\mathcal{D}_n$ is the set of all samples considered so far and their cluster assignments.

For $k = 1, \ldots, K$, where K is the current number of clusters, we calculate the probabilities that $\mathbf{v}_n$ belongs to cluster k given the samples in $\mathcal{D}_n$ and their cluster assignments,

$$
p(z_{n,k} = 1 | \mathcal{D}_n, \alpha, \mathbf{m}, \lambda, \mathbf{\Psi}, \nu) =
$$

$$
p(z_{n,k} = 1 | \mathcal{D}_n, \alpha) p(\mathbf{v}_n | \mathcal{D}_n, z_{n,k} = 1, \mathbf{m}, \lambda, \mathbf{\Psi}, \nu)
$$

following the product rule. Now, following the Chinese Restaurant Process,

$$
p(z_{n,k} = 1 | \mathcal{D}_n, \alpha) = \frac{n_k}{n - 1 + \alpha},
$$

where n_k is the number of samples currently assigned to cluster k. On the other hand,

$$
p(\mathbf{v}_n | \mathcal{D}_n, z_{n,k} = 1, \mathbf{m}, \lambda, \mathbf{\Psi}, \nu) \approx
$$

$$
\frac{1}{\sqrt{|2\pi \mathbf{\Sigma}_k|}} \exp\left(\frac{1}{2}(\mathbf{v}_n - \boldsymbol{\mu}_k)^T \mathbf{\Sigma}_k^{-1}(\mathbf{v}_n - \boldsymbol{\mu}_k)\right). \tag{6.13}
$$

So the data $\mathcal{D}_n$ and the current cluster assignments are not used directly, but indirectly via the current $\boldsymbol{\mu}_k$ and $\mathbf{\Sigma}_k$.

We also calculate the probability that $\mathbf{v}_n$ belongs to a, so far unseen, empty cluster. It is

$$
p(z_{n,K+1} = 1 | \mathcal{D}_n, \alpha, \mathbf{m}, \lambda, \mathbf{\Psi}, \nu) =
$$

$$
p(z_{n,K+1} = 1 | \mathcal{D}_n, \alpha) p(\mathbf{v}_n | \mathcal{D}_n, z_{n,K+1} = 1, \mathbf{m}, \lambda, \mathbf{\Psi}, \nu).
$$

The first factor is

$$
p(z_{n,K+1} = 1 | \mathcal{D}_n, \alpha) = \frac{\alpha}{n - 1 + \alpha}
$$

according to the Chinese Restaurant Process. The data set $\mathcal{D}_n$ only influences this probability via its current size $n - 1$. The other factor is the probabil-

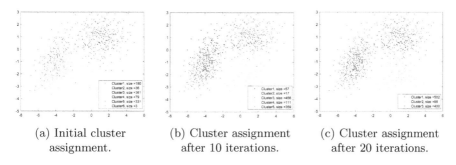

(a) Initial cluster assignment.

(b) Cluster assignment after 10 iterations.

(c) Cluster assignment after 20 iterations.

Figure 6.12: Dirichlet process method example.

ity of $\mathbf{v}_n$ belonging to an empty cluster. Since this cluster is still empty, its distribution is completely determined by the prior. Hence,

$$p(\mathbf{v}_n|\mathcal{D}_n, z_{n,K+1} = 1, \mathbf{m}, \lambda, \mathbf{\Psi}, \nu) =$$

$$\frac{1}{\sqrt{|2\pi\mathbf{\Psi}|}} \exp\left(\frac{1}{2}(\mathbf{v}_n - \mathbf{m})^T \mathbf{\Psi}^{-1}(\mathbf{v}_n - \mathbf{m})\right). \tag{6.14}$$

Having calculated these $K+1$ probabilities, they are used to sample which cluster $\mathbf{v}_n$ belongs to, either one of the existing clusters or the empty cluster. If the former, the posterior normal, inverse distribution of that cluster needs to be updated and a new covariance matrix and mean need to be drawn from it. If the latter, this cluster will no longer be empty and the parameters $\mathbf{m}_{K+1}, \lambda_{K+1}, \mathbf{\Psi}_{K+1}$ and ν_{K+1} of the posterior, normal, inverse Wishart distribution need to be calculated. A covariance matrix $\mathbf{\Sigma}_{K+1}$ and mean $\boldsymbol{\mu}_{K+1}$ are drawn from this distribution.

Once all samples have been considered, the initialization is complete. This phase can lead to very different results. In Figures 6.12a and 6.13a the same initialization procedure was applied to the data from Figures 6.3a and 6.3b. The results are very different; in Figure 6.12a three clusters dominate, while in 6.13a two dominate.

After initialization, several iterations follow and in each iteration all samples are considered in a random order. Let K be the current number of clusters, N_k the current number of samples in cluster k, $k = 1, \ldots, K$, and $\mathbf{v}_n$ be the sample under consideration. First, $\mathbf{v}_n$ is removed from the data set and the cluster it is currently assigned to following the same steps as described for Bayesian mixtures. If this removal results in an empty cluster, this cluster is removed completely, and K becomes $K - 1$.

The posterior probability that $\mathbf{v}_n$ belongs to cluster k given the samples in $\mathcal{D} \setminus \{\mathbf{v}_n\}$ and their cluster assignments is

$$p(z_{n,k} = 1|\mathcal{D} \setminus \{\mathbf{v}_n\}, \alpha, \mathbf{m}, \lambda, \mathbf{\Psi}, \nu) =$$

$$p(z_{n,k} = 1|\mathcal{D} \setminus \{\mathbf{v}_n\}, \alpha)p(\mathbf{v}_n|\mathcal{D} \setminus \{\mathbf{v}_n\}, z_{n,k} = 1, \mathbf{m}, \lambda, \mathbf{\Psi}, \nu).$$

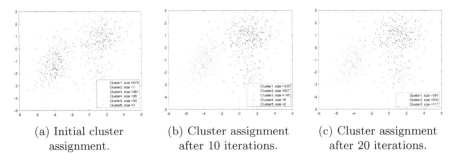

| (a) Initial cluster assignment. | (b) Cluster assignment after 10 iterations. | (c) Cluster assignment after 20 iterations. |

Figure 6.13: Dirichlet process method example.

In this situation, we have

$$p(z_{n,k} = 1 | \mathcal{D} \setminus \{\mathbf{v}_n\}, \alpha) = \frac{N_k}{N - 1 + \alpha},$$

while $p(\mathbf{v}_n | \mathcal{D} \setminus \{\mathbf{v}_n\}, z_{n,k} = 1, \mathbf{m}, \lambda, \mathbf{\Psi}, \nu)$ is as in Equation(6.13).

The possibility that $\mathbf{v}_n$ belongs to an empty cluster is

$$p(z_{n,K+1} = 1 | \mathcal{D} \setminus \{\mathbf{v}_n\}, \alpha, \mathbf{m}, \lambda, \mathbf{\Psi}, \nu) =$$
$$p(z_{n,K+1} = 1 | \mathcal{D} \setminus \{\mathbf{v}_n\}, \alpha) p(\mathbf{v}_n | \mathcal{D} \setminus \{\mathbf{v}_n\}, z_{n,K+1} = 1, \mathbf{m}, \lambda, \mathbf{\Psi}, \nu),$$

where

$$p(z_{n,K+1} = 1 | \mathcal{D} \setminus \{\mathbf{v}_n\}, \alpha) = \frac{\alpha}{N - 1 + \alpha}$$

and $p(\mathbf{v}_n | \mathcal{D} \setminus \{\mathbf{v}_n\}, z_{n,K+1} = 1, \mathbf{m}, \lambda, \mathbf{\Psi}, \nu)$ as in Equation(6.13).

The sample $\mathbf{v}_n$ is assigned to either one of the existing clusters or an empty cluster. In either case, the posterior, normal, inverse, Wishart distribution needs to be calculated and new samples for the covariance matrix and mean drawn for that cluster. Listing 6.7 implements this algorithm. Figures 6.12 and 6.13 are two runs on the same data.

Recall that, when K was fixed, the data generation was described as

$$\mathbf{v}_n \sim \mathcal{N}(\boldsymbol{\mu}_k, \mathbf{\Sigma}_k)$$
$$\boldsymbol{\pi} \sim \text{Dir}(\boldsymbol{\alpha}),$$
$$\boldsymbol{\mu}_k, \mathbf{\Sigma}_k \sim G.$$

where G was a distribution which is zero everywhere apart from K pairs $(\boldsymbol{\mu}_k, \mathbf{\Sigma}_k)$, which were generated from the base distribution $G_0 = \text{NIW}(\boldsymbol{m}, \lambda, \mathbf{\Sigma}, \nu)$ and the probability mass for these K pairs followed a Dirichlet distribution with parameter $\boldsymbol{\alpha}$.

```
function [idx,K,MU,SIGMA] = InferDP(V,alpha,m,lambda,Psi,nu,iter)
% Input:
% V:      data,
% alpha:  prior concentration parameter,
% m:      prior location vector,
% lambda: prior mean fraction,
% Psi:    prior inverse scale matrix,
% nu:     prior degrees of freedom,
% iter:   number of iterations
% Output:
% idx:    index vector of cluster assignments,
% K:      number of clusters
% MU:     array of the means sampled from the posterior normal
%         inverse Wishart distributions for each cluster,
% SIGMA: array of the covariance matrices sampled from the posterior
%         normal inverse Wishart distributions for each cluster.

N = size(V,1);   % Number of samples.
K = 0;           % Number of clusters.
idx =zeros(N,1); % Vector cluster assignments.
nk = [];         % Vector of cluster sizes.
% For each cluster:
M = [];          % Posterior location vector.
LAMBDA = [];     % Posterior mean fraction.
PSI = [];        % Cholesky factorization of posterior inverse
                 % scale matrix.
NU = [];         % Posterior degrees of freedom.
SIGMA = [];      % Covariance matrix draw.
MU = [];         % Mean draw.

% Use Cholesky factorization of Psi to ensure positive definiteness.
Psi = chol(Psi);

% Consider samples in a random order.
order = randperm(N);
% Initialize cluster assignments.
for n = 1:N
    v = V(order(n),:);
    % Calculate cluster assignment probabilities.
    p = zeros(1,K+1);
    for k=1:K
        p(k) = nk(k)/(n—1+alpha)*mvnpdf(v',MU(:,k),SIGMA(:,:,k));
    end
    p(K+1) = alpha/(n—1+alpha)*mvnpdf(v',m',Psi);
    p = p/sum(p);
    % Sample new indicator variable.
    l = randsample(K+1,1,true,p);
    idx(order(n)) = l;
```

6.7a: Dirichlet process method.

```
    if 1 ≤K
        % Update the cluster the sample is now assigned to.
        postlambda = LAMBDA(l)+1;
        % Rank one update on the Cholesky factorization to preserve
        % positive definiteness.
        update = sqrt(LAMBDA(l)/postlambda) * (v' - M(:,l));
        PSI(:,:,l) = cholupdate(PSI(:,:,l),update,'+');
        M(:,l) = (LAMBDA(l)*M(:,l) + v(:))/postlambda;
        NU(l) = NU(l)+1;
        LAMBDA(l) = postlambda;
        SIGMA(:,:,l) = iwishrnd(PSI(:,:,l)'* PSI(:,:,l),NU(l));
        MU(:,l) = mvnrnd(M(:,l),SIGMA(:,:,l)/LAMBDA(l));
        nk(l) = nk(l)+1;
    else
        % Create new cluster.
        LAMBDA = cat(2,LAMBDA,lambda+1);
        update = sqrt(lambda/(lambda+1)) * (v - m)';
        PSI = cat(3,PSI,cholupdate(Psi,update,'+'));
        M = cat(2,M,(lambda* m' +v' )/(lambda+1));
        NU = cat(2,NU,nu+1);
        SIGMA = cat(3,SIGMA,iwishrnd(PSI(:,:,l)'* PSI(:,:,l),NU(l)));
        MU = cat(2,MU,mvnrnd(M(:,l),SIGMA(:,:,l)/LAMBDA(l))');
        nk = cat(2,nk,1);
        K = K+1;
    end
end

for i = 1:iter
    % Consider the data in a random order.
    for n = randperm(N)
        v = V(n,:);
        l = idx(n);
        % Remove this sample from the data set and update the cluster
        % it was assigned to.
        nk(l) = nk(l)-1;
        if nk(l) == 0
            % Remove empty cluster.
            nk(l) = [];
            PSI(:,:,l) = [];
            M(:,l) = [];
            NU(l) = [];
            LAMBDA(l) = [];
            SIGMA(:,:,l) = [];
            MU(:,l) = [];
            % Adjust cluster numbering.
            temp = idx>l;
            idx(temp) = idx(temp)-1;
            K = K-1;
```

6.7b: Dirichlet process method.

```
        else
            priorlambda = LAMBDA(l)−1;
            % Rank one update on the Cholesky factorization to
            % preserve positive definiteness.
            update = sqrt(LAMBDA(l)/priorlambda) * (v' − M(:,l));
            PSI(:,:,l) = cholupdate(PSI(:,:,l),update,'−');
            M(:,l) = (LAMBDA(l)*M(:,l) − v(:))/priorlambda;
            NU(l) = NU(l)−1;
            LAMBDA(l) = priorlambda;
            SIGMA(:,:,l) = iwishrnd(PSI(:,:,l)'* PSI(:,:,l),NU(l));
            MU(:,l) = mvnrnd(M(:,l),SIGMA(:,:,l)/LAMBDA(l));
        end         % Calculate cluster assignment probabilities.
        p = zeros(1,K+1);
        for k=1:K
            p(k) = nk(k)/(N−1+alpha)*mvnpdf(v',MU(:,k),SIGMA(:,:,k));
        end
        p(K+1) = alpha/(N−1+alpha)*mvnpdf(v',m',Psi);
        p = p/sum(p);
        % Sample new indicator variable.
        l = randsample(K+1,1,true,p);
        idx(n) = l;
        if l ≤K
            % Update the cluster the sample is now assigned to.
            postlambda = LAMBDA(l)+1;
            % Rank one update on the Cholesky factorization to
            % preserve positive definiteness.
            update = sqrt(LAMBDA(l)/postlambda) * (v' − M(:,l));
            PSI(:,:,l) = cholupdate(PSI(:,:,l),update,'+');
            M(:,l) = (LAMBDA(l)*M(:,l) + v(:))/postlambda;
            NU(l) = NU(l)+1;
            LAMBDA(l) = postlambda;
            SIGMA(:,:,l) = iwishrnd(PSI(:,:,l)'* PSI(:,:,l),NU(l));
            MU(:,l) = mvnrnd(M(:,l),SIGMA(:,:,l)/LAMBDA(l));
            nk(l) = nk(l)+1;
        else
            % Create new cluster.
            LAMBDA = cat(2,LAMBDA,lambda+1);
            update = sqrt(lambda/(lambda+1)) * (v − m)';
            PSI = cat(3,PSI,cholupdate(Psi,update,'+'));
            M = cat(2,M,(lambda* m' +v' )/(lambda+1));
            NU = cat(2,NU,nu+1);
            SIGMA = cat(3,SIGMA,iwishrnd(PSI(:,:,l)'* ...
                PSI(:,:,l),NU(l)));
            MU = cat(2,MU,mvnrnd(M(:,l),SIGMA(:,:,l)/LAMBDA(l))');
            nk = cat(2,nk,1);
            K = K+1;
        end
    end
end
end
```

6.7c: Dirichlet process method.

A *Dirichlet process* extends the concept to an unknown variable number of clusters K. The notation is

$$\mathbf{v}_n \quad \sim \quad \mathcal{N}(\boldsymbol{\mu}_k, \boldsymbol{\Sigma}_k)$$

$$\boldsymbol{\mu}_k, \boldsymbol{\Sigma}_k \quad \sim \quad G,$$

$$G \quad \sim \quad \mathrm{DP}(\alpha, G_0).$$

The last line means that G is drawn from a Dirichlet process with parameters α and G_0. A Dirichlet process is a distribution of distributions. There are infinite many possibilities for pairs $(\boldsymbol{\mu}, \boldsymbol{\Sigma})$ and the vector $\boldsymbol{\pi}$ is often described as infinite giving each pair a probability mass. Conceptually, infinity is difficult. In practice, the number of clusters K is at most the number of samples N, when every cluster contains at least one sample. Therefore G gives a probability mass of $N_k/(N-1+\alpha)$ to at most N pairs $(\boldsymbol{\mu}, \boldsymbol{\Sigma})$ drawn from the base distribution G_0, and assigns the probability of $\alpha/((N-1+\alpha)$ to the set of all other possible pairs. G is a discrete distribution defined on a finite partition of the space of all pairs $(\boldsymbol{\mu}, \boldsymbol{\Sigma})$ which is denoted by S.

More formally, a distribution G is drawn from a Dirichlet process with parameters α and G_0, if for any finite, disjoint partition $S_1, \ldots, S_L$ of S, where L can be any finite number, the vector $(G(S_1), \ldots, G(S_L))^T$ follows a Dirichlet distribution with parameters $\alpha G_0(S_1), \ldots, \alpha G_0(S_L)$. In our example,

$$
\begin{aligned}
G_0(S_l) &= \int_{S_l} \mathrm{NIW}(\boldsymbol{\mu}, \boldsymbol{\Sigma} | \mathbf{m}, \lambda, \boldsymbol{\Psi}, \nu) \, d\boldsymbol{\mu} \, d\boldsymbol{\Sigma} \\
&= \int_{S_l} \mathcal{N}(\boldsymbol{\mu} | \mathbf{m}, \frac{1}{\lambda} \boldsymbol{\Sigma}) \mathcal{W}^{-1}(\boldsymbol{\Sigma} | \boldsymbol{\Psi}, \nu) \, d\boldsymbol{\mu} \, d\boldsymbol{\Sigma}.
\end{aligned}
$$

Similarly, $G(S_l)$ can be calculated from the values G takes on the partition, and we indeed have

$$(G(S_1), \ldots, G(S_L))^T \sim \mathrm{Dir}(\alpha G_0(S_1), \ldots, \alpha G_0(S_L)).$$

The Dirichlet process method infers G from the data by varying the number K of pairs $(\boldsymbol{\mu}_k, \boldsymbol{\Sigma}_k)$ and their positions. It belongs to a family of methods known as *Bayesian nonparametrics* which are Bayesian models operating on an infinite-dimensional parameter space. In our case, this is the space of all possible pairs $(\boldsymbol{\mu}, \boldsymbol{\Sigma})$. To delve further into Bayesian nonparametrics consult [17].

CHAPTER 7

Dimensionality Reduction

While samples can be described by many features, the ones which essentially define the sample are often few, the many features being expressions of the few in different ways. The chapter explores techniques to find these starting with Principal Component Analysis introducing it as geometric concept. It then takes a probabilistic viewpoint maximizing the likelihood of the data arriving at the same conclusion. The Expectation-Maximization algorithm is shown to be an alternative method to maximize the data likelihood. The method is generalized to factor analysis. Lastly, kernel principal component analysis tackles data samples lying on a non-linear manifold.

In the previous chapters, the features were taken as given. The kernel trick was introduced through a mapping of the feature space to a higher dimensional space, where the data samples are more easily separable. This higher dimensional space can be viewed as creating new features from the given, by combining them in some way. Often, however, the data samples lie in a lower dimensional set of the feature space.

For example in an 8-bit gray scale image, each pixel has a value between 0 (black) and 255 (white). A standard sized image of 1280 by 720 pixels therefore is encoded in $921,600$ bytes or approaching one megabyte. The human visual system is, however, not too fussed over exact details. Or in other words, humans are quite well adapted to interpret missing information in an image. The Joint Photographic Experts group (JPEG) used this fact in 1992 to create a standard for lossy image compression.

The image is first divided into 8×8 blocks. For the example size above, this results into $14,400$ blocks. Instead of storing 64 pixel values for each block, each block is built by overlaying blocks from Figure 7.1 with appropriate weights. The weights are only allowed to have integer values. Many weights will be zero, because few blocks in the image will have high variations; most

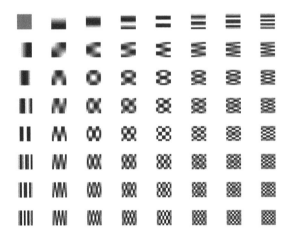

Figure 7.1: Discrete Cosine Transform basis of the JPEG standard.

will have a moderate variation between pixels, some will be mostly one smooth shade of gray. By changing the focus away from the individual pixel to groups of pixels, i.e. the blocks, fewer bytes are needed to encode the image.

In mathematical terms, this represents a change of basis. When storing the values for each pixel in a block, this is equivalent to using a basis where all values are zero but one, and the position of this non-zero pixel transverses the block. The weights for this basis are the individual pixel values, and most will be nonzero. The other basis is shown in Figure 7.1. When representing the block in this basis, only a few non-zero weights are necessary.

In this chapter, we introduce techniques to find representations of the data samples with different, but fewer features.

7.1 Principal Component Analysis

Principal Component Analysis seeks a subspace of a given dimension K of the feature space such that projections of the data samples onto that subspace are as spread out as possible. If there is space between groups of data samples, this has advantages when clustering or classifying. This subspace is known as the *principal subspace*.

Let $\mathcal{D} = \{\mathbf{v}_1, \ldots, \mathbf{v}_N\}$ be a set of data samples. We first consider the projection onto a one-dimensional subspace defined by the vector $\mathbf{w}$. Just as in the case of linear classification, only the direction of the vector is important, not its length. The length is chosen to be $\|\mathbf{w}\| = 1$, so that the projection of the sample $\mathbf{v}_n$ is calculated as $\mathbf{w}^T \mathbf{v}_n$.

The sample mean is given by

$$\boldsymbol{\mu} = \frac{1}{N} \sum_{n=1}^{N} \mathbf{v}_n,$$

while the sample covariance matrix is

$$\Sigma = \frac{1}{N} \sum_{n=1}^{N} (\mathbf{v}_n - \boldsymbol{\mu})(\mathbf{v}_n - \boldsymbol{\mu})^T, \tag{7.1}$$

which is symmetric.

The variance of the projected samples is

$$\frac{1}{N} \sum_{n=1}^{N} (\mathbf{w}^T \mathbf{v}_n - \mathbf{w}^T \boldsymbol{\mu})^2 = \frac{1}{N} \sum_{n=1}^{N} \mathbf{w}^T \mathbf{v}_n \mathbf{v}_n^T \mathbf{w} - 2\mathbf{w}^T \mathbf{v}_n \boldsymbol{\mu}^T \mathbf{w} + \mathbf{w}^T \boldsymbol{\mu}\boldsymbol{\mu}^T \mathbf{w}$$

$$= \mathbf{w}^T \left[\frac{1}{N} \sum_{n=1}^{N} (\mathbf{v}_n - \boldsymbol{\mu})(\mathbf{v}_n - \boldsymbol{\mu})^T \right] \mathbf{w} = \mathbf{w}^T \Sigma \mathbf{w}.$$

The objective is to maximize this, subject to the constraint $\|\mathbf{w}\|^2 = \mathbf{w}^T\mathbf{w} = 1$. The Lagrangian function is given by

$$L(\mathbf{w}, \lambda) = \mathbf{w}^T \Sigma \mathbf{w} - \lambda(\mathbf{w}^T\mathbf{w} - 1).$$

A stationary point of $L(\mathbf{w}, \lambda)$ is a maximum of the constraint optimization. Using Appendices A.2.3 and A.2.2, the derivative of $L(\mathbf{w}, \lambda)$ with respect to $\mathbf{w}$ is

$$\frac{d}{d\mathbf{w}} L(\mathbf{w}, \lambda) = \Sigma \mathbf{w} - \lambda \mathbf{w}.$$

Setting this to zero, gives

$$\Sigma \mathbf{w} = \lambda \mathbf{w},$$

and thus $\mathbf{w}$ is an eigenvector of Σ.

Using $\mathbf{w}^T\mathbf{w} = 1$, the eigenvalue is

$$\lambda = \lambda \mathbf{w}^T \mathbf{w} = \mathbf{w}^T \Sigma \mathbf{w},$$

which is the variance of the projected data. Thus, the projected data will be as much spread out as possible, if $\mathbf{w}$ is chosen to be the eigenvector of Σ with the largest eigenvalue.

Being a covariance matrix, Σ has non-negative eigenvalues. Since Σ is symmetric, its eigenvectors are orthogonal to each other. Therefore the principal space of dimension K, where the projected data has the largest variance, is the subspace spanned by the eigenvectors of the K largest eigenvalues. These are known as *principal components*.

We illustrate this on the MNIST data set of handwritten digits [26]. This data set contains $60,000$ images of handwritten digits of size 28×28 pixels. We first concentrate on distinguishing between the digits zero and one. Figure 7.2a shows some examples of these digits from the data set, while Figures 7.2b and 7.2c show the representations of the same images using only two or three principal components. In 7.2b the original digits can be vaguely discerned in

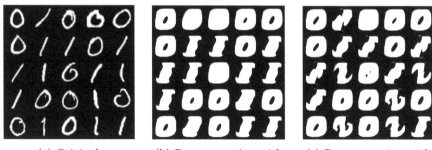

(a) Original.
(b) Reconstruction with two principal components.
(c) Reconstruction with three principal components.

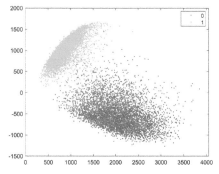
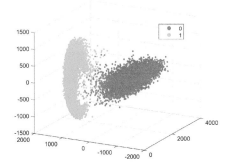

(d) Sample locations in subspace spanned by two principal components.

(e) Sample locations in subspace spanned by three principal components.

Figure 7.2: PCA for two digits.

most cases, but in 7.2c there seems to be little resemblance to the original digits. Figures 7.2d and 7.2e, however, display the location of samples in the space spanned by two or three principal components respectively. Note that the sets of zeros and ones are separated. The objective was not the reconstruction of the images, but the separation of digits. To reconstruct the images perfectly in all cases $28 * 28 = 784$ pixels or principal components would be necessary. Even a reasonably good reconstruction would require many more principal components than just two or three.

Figures 7.3 and 7.4 do the same for three or four digits respectively. Three digits could possibly be separated using two principal components as Figure 7.3d shows, but there is some overlap of the data sets. The separation is better with three principal components in Figure 7.3e. Separating four digits is impossible with two principal components as Figure 7.4d illustrates. Going to three dimensions with three principal components is an improvement in Figure 7.4e, but using a further principal component is advisable. Note, that

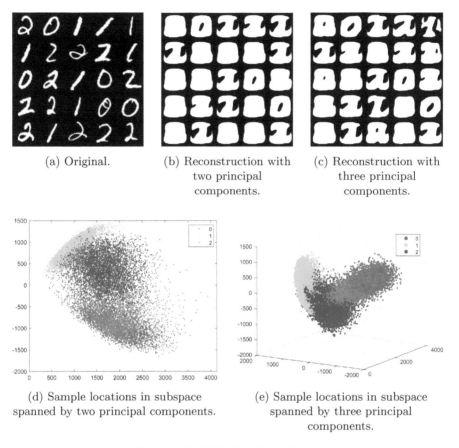

(a) Original.

(b) Reconstruction with two principal components.

(c) Reconstruction with three principal components.

(d) Sample locations in subspace spanned by two principal components.

(e) Sample locations in subspace spanned by three principal components.

Figure 7.3: PCA for three digits.

for each of the figures 7.2, 7.3 and 7.4 the principal components are different, since including data samples of a further digit changes the covariance matrix Σ and thus the eigenvectors are different for different data sets.

The original dimension of the feature space is $D = 784$, the number of pixels. Let $\mathbf{w}_1, \ldots, \mathbf{w}_D$ be the complete set of orthonormal eigenvectors of Σ with corresponding eigenvalues $\lambda_1, \ldots, \lambda_D$ sorted by decreasing size of eigenvalue. This is also a basis of the feature space. Each sample $\mathbf{v}_n$ can be expressed in this basis as

$$\mathbf{v}_n = \sum_{d=1}^{D} (\mathbf{v}_n^T \mathbf{w}_d) \mathbf{w}_d,$$

while its projection on the principal subspace is given by

$$\sum_{d=1}^{K} (\mathbf{v}_n^T \mathbf{w}_d) \mathbf{w}_d.$$

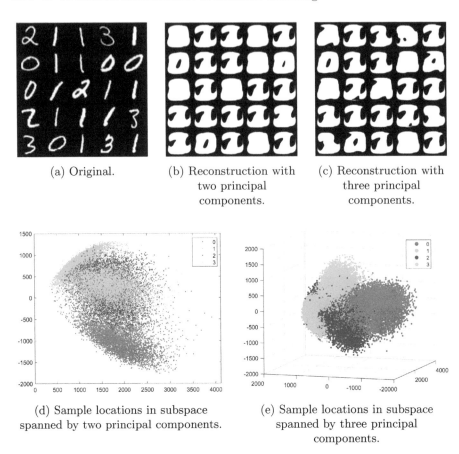

(a) Original.

(b) Reconstruction with two principal components.

(c) Reconstruction with three principal components.

(d) Sample locations in subspace spanned by two principal components.

(e) Sample locations in subspace spanned by three principal components.

Figure 7.4: PCA for four digits.

Thus, the squared distance between a sample and its projection is

$$\| \sum_{d=K+1}^{D} (\mathbf{v}_n^T \mathbf{w}_d)\mathbf{w}_d \|^2 = \left(\sum_{d=K+1}^{D} (\mathbf{v}_n^T \mathbf{w}_d)\mathbf{w}_d \right)^T \left(\sum_{d=K+1}^{D} (\mathbf{v}_n^T \mathbf{w}_d)\mathbf{w}_d \right)$$
$$= \sum_{d=K+1}^{D} (\mathbf{v}_n^T \mathbf{w}_d)^2$$

due to $\mathbf{w}_1, \ldots, \mathbf{w}_D$ being orthonormal.

Summing this over all samples and dividing by N gives the average squared distance by which a sample is moved, when restricting the representation to the principal subspace,

$$\frac{1}{N}\sum_{n=1}^{N}\sum_{d=K+1}^{D}(\mathbf{v}_n^T\mathbf{w}_d)^2 \quad = \quad \sum_{d=K+1}^{D}\mathbf{w}_d^T\left(\frac{1}{N}\sum_{n=1}^{N}\mathbf{v}_n\mathbf{v}_n^T\right)\mathbf{w}_d$$

$$= \quad \sum_{d=K+1}^{D}\mathbf{w}_d^T\left(\mathbf{\Sigma}+\boldsymbol{\mu}\boldsymbol{\mu}^T\right)\mathbf{w}_d$$

$$= \quad \sum_{d=K+1}^{D}\lambda_d+(\boldsymbol{\mu}^T\mathbf{w}_d)^2$$

where we used Equation (2.10) as an alternative calculation of the covariance matrix. The first part is the sum over the $D - K$ smallest eigenvalues of $\mathbf{\Sigma}$, while the second part is the squared distance between the mean and its projection. As a consequence as more eigenvectors are used as principal components, the average squared distance between samples and their projection becomes smaller. This means the reconstructions become more accurate. We will revisit this when covering principal component regression in Section 8.8.

The eigenvalues and eigenvectors of $\mathbf{\Sigma}$ can be used to perform a technique known as *whitening* or *sphering* the data. This transforms the data to a different feature space where the samples have zero mean and the covariance is the identity matrix. If measurements of different features differ in magnitude or the variability of different features is disparate, this may become necessary.

However, after such a transformation all eigenvalues of the covariance matrix are one. Principal component analysis and any other technique which relies on distinct eigenvalues of the sample covariance matrix are no longer applicable. For $\mathbf{v}_n$ the transformed value is

$$\hat{\mathbf{v}}_n = \mathbf{\Lambda}^{-1/2}\mathbf{W}^T(\mathbf{v}_n - \boldsymbol{\mu}),$$

where $\mathbf{W}$ is the orthogonal matrix formed from the eigenvectors $\mathbf{w}_d$ as columns and $\mathbf{\Lambda}^{-1/2}$ is the diagonal matrix with $\lambda_d^{-1/2}$ on the diagonal. This is well defined, if $\mathbf{\Sigma}$ has no zero eigenvalues, which is the case, if the data samples are distinct and $N > D$. Because of the subtraction of $\boldsymbol{\mu}$, the mean of $\hat{\mathbf{v}}_1, \dots, \hat{\mathbf{v}}_N$ is zero. On the other hand, the covariance matrix of the transformed data is

$$\frac{1}{N}\sum_{n=1}^{N}\hat{\mathbf{v}}_n\hat{\mathbf{v}}_n^T \quad = \quad \frac{1}{N}\sum_{n=1}^{N}\mathbf{\Lambda}^{-1/2}\mathbf{W}^T(\mathbf{v}_n - \boldsymbol{\mu})(\mathbf{v}_n - \boldsymbol{\mu})^T\mathbf{W}\mathbf{\Lambda}^{-1/2}$$

$$= \quad \mathbf{\Lambda}^{-1/2}\mathbf{W}^T\mathbf{\Sigma}\mathbf{W}\mathbf{\Lambda}^{-1/2} = \mathbf{\Lambda}^{-1/2}\mathbf{\Lambda}\mathbf{\Lambda}^{-1/2} = \mathbf{I}.$$

All this relies on stable and efficient methods to find eigenvalues and eigenvectors. For some algorithms see for example [12]. It is important to keep the dimensionality and size of the data set in mind. To illustrate this, let $\mathbf{V}$ be the matrix where the nth row is $\mathbf{v}_n - \boldsymbol{\mu}$, i.e. the data sample shifted by the sample mean. Then

$$\mathbf{\Sigma} = \frac{1}{N}\mathbf{V}^T\mathbf{V},$$

and the eigenvector equation is

$$\frac{1}{N}\mathbf{V}^T\mathbf{V}\mathbf{w}_d = \lambda_d\mathbf{w}_d.$$

Multiplying through with $\mathbf{V}$ it becomes

$$\frac{1}{N}(\mathbf{V}\mathbf{V}^T)\mathbf{V}\mathbf{w}_d = \lambda_d\mathbf{V}\mathbf{w}_d.$$

So $\mathbf{V}\mathbf{w}_d$ is an eigenvector of $\mathbf{V}\mathbf{V}^T/N$. If $D > N$, instead of finding the K largest eigenvalues and eigenvectors of $\boldsymbol{\Sigma}$, it is easier to find the K largest eigenvalues and eigenvectors of $\mathbf{V}\mathbf{V}^T/N$ and then transform these eigenvectors back and normalize them. This makes sense, if the data lies in a subspace of much lower dimension which needs to be found.

7.2   Probabilistic View

It is helpful to think about the process which generates the data. The assumption is that $\mathbf{v} \in \mathbb{R}^D$ is related to $\mathbf{u} \in \mathbb{R}^K$ via a linear mapping. More formally we are trying to find a $D \times K$ matrix $\mathbf{W}$, which has full rank, and $\mathbf{m} \in \mathbb{R}^D$, such that

$$\mathbf{v} = \mathbf{W}\mathbf{u} + \mathbf{m} + \boldsymbol{\epsilon}. \tag{7.2}$$

We can assume that the columns of $\mathbf{W}$ are orthogonal to each other. If not, orthogonal columns can be achieved by using for example the Gram–Schmidt algorithm in [12] and a suitable basis transformation in $\mathbb{R}^K$. $\boldsymbol{\epsilon}$ is normally distributed noise with zero mean and covariance matrix $\sigma^2\mathbf{I}$. This is known as an *isotropic* or *spherical covariance matrix*. The noise explains why the data samples do not exactly lie in a lower dimensional subspace. Further, we assume that $\mathbf{u}$ follows a normal distribution with mean $\hat{\mathbf{m}}$ and covariance matrix $\mathbf{S}$.

If a more general distribution for $\mathbf{u}$ is assumed, then the technique arising from this is known as *Independent Component Analysis (ICA)*. Such a more general distribution could be one which factorizes as

$$p(\mathbf{u}) = \prod_{k=1}^{K} p(u_k),$$

where each u_k follows the distribution given by

$$p(u_k) = \frac{2}{\pi(\exp(u_k) + \exp(-u_k))}.$$

This distribution has a larger kurtosis than the normal distribution, often displayed in real world applications. A thorough treatment of Independent Component Analysis is given in [23]. In this text, however, we continue to assume that $\mathbf{u}$ has a normal distribution.

The linear transformation of a normally distributed variable is also normally distributed with mean $\mathbf{m} + \mathbf{W}\hat{\mathbf{m}}$ and covariance matrix $\mathbf{W}\mathbf{S}\mathbf{W}^T$. Since S is symmetric, it can be diagonalized by an orthogonal matrix $\mathbf{Q}$ such that $\mathbf{S} = \mathbf{Q}\mathbf{D}\mathbf{Q}^T$, where $\mathbf{D}$ is a diagonal matrix with positive entries, since $\mathbf{S}$

as covariance matrix is positive definite. The covariance matrix can then be written as $(\mathbf{WQD}^{1/2})(\mathbf{WQD}^{1/2})^T$. Thus we can assume without loss of generality that $\mathbf{S} = \mathbf{I}$, since this assumption only results in a different $\mathbf{W}$ to be found. Since Q is an orthogonal matrix, we can continue to assume that the columns of $\mathbf{W}$ are orthogonal. Equally, we can assume zero mean for $\mathbf{u}$, since any non-zero mean can be absorbed in $\mathbf{m}$. With this in mind, we have

$$\mathbf{v} \sim \mathcal{N}(\mathbf{m}, \mathbf{WW}^T + \sigma^2 \mathbf{I}).$$

Let $\mathbf{C} = \mathbf{WW}^T + \sigma^2 \mathbf{I}$. The likelihood of the data set $\mathcal{D} = \{\mathbf{v}_1, \ldots, \mathbf{v}_N\}$ is given by

$$\prod_{n=1}^{N} \frac{1}{\sqrt{|2\pi\mathbf{C}|}} \exp\left(-\frac{1}{2}(\mathbf{v}_n - \mathbf{m})^T \mathbf{C}^{-1}(\mathbf{v}_n - \mathbf{m})\right).$$

Instead of maximizing the likelihood, it is more convenient to maximize its logarithm, known as the *log likelihood*,

$$\mathcal{L} = -\frac{ND}{2}\log(2\pi) - \frac{N}{2}\log|\mathbf{C}| - \frac{1}{2}\sum_{n=1}^{N}(\mathbf{v}_n - \mathbf{m})^T \mathbf{C}^{-1}(\mathbf{v}_n - \mathbf{m}).$$

Using Appendix A.2.3, the derivative with respect to $\mathbf{m}$ is

$$\frac{\partial}{\partial \mathbf{m}}\mathcal{L} = \sum_{n=1}^{N} \mathbf{C}^{-1}(\mathbf{v}_n - \mathbf{m}) = \mathbf{C}^{-1}\left[\sum_{n=1}^{N}(\mathbf{v}_n - \mathbf{m})\right].$$

This can only be zero, if

$$\mathbf{m} = \frac{1}{N}\sum_{n=1}^{N}\mathbf{v}_n = \boldsymbol{\mu}.$$

Hence, $\boldsymbol{\mu}$ is the maximum likelihood estimate for $\mathbf{m}$ and we insert this back into the expression for $\mathcal{L}$.

Now $(\mathbf{v}_n - \boldsymbol{\mu})^T \mathbf{C}^{-1}(\mathbf{v}_n - \boldsymbol{\mu})$ is scalar and therefore equal to its trace. A trace of a three term product is invariant to cyclic permutations. Therefore,

$$\begin{aligned}
\mathcal{L} &= -\frac{ND}{2}\log(2\pi) - \frac{N}{2}\log|\mathbf{C}| - \frac{1}{2}\sum_{n=1}^{N}\mathrm{tr}(\mathbf{C}^{-1}(\mathbf{v}_n - \boldsymbol{\mu})(\mathbf{v}_n - \boldsymbol{\mu})^T) \\
&= -\frac{ND}{2}\log(2\pi) - \frac{N}{2}\log|\mathbf{C}| - \frac{1}{2}\mathrm{tr}\left(\mathbf{C}^{-1}\sum_{n=1}^{N}(\mathbf{v}_n - \boldsymbol{\mu})(\mathbf{v}_n - \boldsymbol{\mu})^T\right),
\end{aligned}$$

since the sum of traces is the trace of the sum of matrices. Using the definition (7.1) for the sample covariance matrix, this becomes

$$\mathcal{L} = -\frac{N}{2}\left[D\log(2\pi) + \log|\mathbf{C}| + \mathrm{tr}\left(\mathbf{C}^{-1}\boldsymbol{\Sigma}\right)\right].$$

This needs to be maximized with respect to $\mathbf{W}$.

Let $\mathbf{w}_1, \ldots, \mathbf{w}^K$ be the columns of $\mathbf{W}$, then

$$\mathbf{C} = \sigma^2 \mathbf{I} + \sum_{k=1}^{K} \mathbf{w}_k \mathbf{w}_k^T.$$

The inverse of $\mathbf{C}$ is

$$\mathbf{C}^{-1} = \sigma^{-2} \left(I - \sum_{k=1}^{K} \frac{\mathbf{w}_k \mathbf{w}_k^T}{\sigma^2 + \mathbf{w}_k^T \mathbf{w}_k} \right). \tag{7.3}$$

This can easily be checked by multiplying the two expressions and using the fact that the columns of $\mathbf{W}$ are mutually orthogonal.

We consider the derivative with respect to the jth column $\mathbf{w}_j$. To this end let

$$\mathbf{C}_{-j} = \sigma^2 \mathbf{I} + \sum_{\substack{k=1 \\ k \neq j}}^{K} \mathbf{w}_k \mathbf{w}_k^T.$$

Then $\mathbf{C} = \mathbf{C}_{-j} + \mathbf{w}_j \mathbf{w}_j^T$. Using both the *Sherman–Morrison formula* in Appendix A.1.4 and the *matrix determinant lemma* in Appendix A.1.5, the inverse and the determinant of $\mathbf{C}$ are

$$\mathbf{C}^{-1} = \mathbf{C}_{-j}^{-1} - \frac{1}{1 + \mathbf{w}_j^T \mathbf{C}_{-j}^{-1} \mathbf{w}_j} \mathbf{C}_{-j}^{-1} \mathbf{w}_j \mathbf{w}_j^T \mathbf{C}_{-j}^{-1},$$

$$|\mathbf{C}| = |\mathbf{C}_{-j}|(1 + \mathbf{w}_j^T \mathbf{C}_{-j}^{-1} \mathbf{w}_j).$$

The inverse of $\mathbf{C}_{-j}$ is of the same form as the inverse of $\mathbf{C}$ in (7.3) with the jth term missing in the sum. Therefore, $\mathbf{C}_{-j}^{-1} \mathbf{w}_j = \sigma^{-2} \mathbf{w}_j$, because $\mathbf{w}_j$ is orthogonal to the other column vectors. The above expressions become

$$\mathbf{C}^{-1} = \mathbf{C}_{-j}^{-1} - \frac{1}{1 + \sigma^{-2} \mathbf{w}_j^T \mathbf{w}_j} \sigma^{-4} \mathbf{w}_j \mathbf{w}_j^T = \mathbf{C}_{-j}^{-1} - \frac{\sigma^{-2}}{\sigma^2 + \mathbf{w}_j^T \mathbf{w}_j} \mathbf{w}_j \mathbf{w}_j^T,$$

$$|\mathbf{C}| = |\mathbf{C}_{-j}|(1 + \sigma^{-2} \mathbf{w}_j^T \mathbf{w}_j).$$

With this we can simplify the terms in $\mathcal{L}$. First,

$$\log |\mathbf{C}| = \log |\mathbf{C}_{-j}| + \log(1 + \sigma^{-2} \mathbf{w}_j^T \mathbf{w}_j). \tag{7.4}$$

Differentiating this with respect to $\mathbf{w}_j$ using Appendix A.2.1 gives

$$\frac{\partial}{\partial \mathbf{w}_j} \log |\mathbf{C}| = \frac{1}{1 + \sigma^{-2} \mathbf{w}_j^T \mathbf{w}_j} 2\sigma^{-2} \mathbf{w}_j = \frac{2}{\sigma^2 + \mathbf{w}_j^T \mathbf{w}_j} \mathbf{w}_j.$$

On the other hand,

$$\text{tr} \left(\mathbf{C}^{-1} \mathbf{\Sigma} \right) = \text{tr}(\mathbf{C}_{-j}^{-1} \mathbf{\Sigma}) - \frac{\sigma^{-2}}{\sigma^2 + \mathbf{w}_j^T \mathbf{w}_j} \text{tr} \left(\mathbf{w}_j \mathbf{w}_j^T \mathbf{\Sigma} \right)$$

$$= \text{tr}(\mathbf{C}_{-j}^{-1} \mathbf{\Sigma}) - \frac{\sigma^{-2} \mathbf{w}_j^T \mathbf{\Sigma} \mathbf{w}_j}{\sigma^2 + \mathbf{w}_j^T \mathbf{w}_j},$$

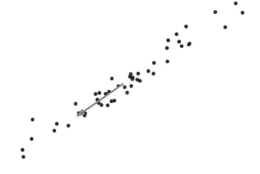

Figure 7.5: Probabilistic Principal Component Analysis for two-dimensional data.

where we again used the cyclic permutation property of the trace. Using the product rule, Appendices A.2.1 and A.2.3, the derivative with respect to $\mathbf{w}_j$ is

$$\frac{\partial}{\partial \mathbf{w}_j} \operatorname{tr}\left(\mathbf{C}^{-1}\mathbf{\Sigma}\right) = -\frac{\sigma^{-2}}{(\sigma^2 + \mathbf{w}_j^T\mathbf{w}_j)^2}\left[(\sigma^2 + \mathbf{w}_j^T\mathbf{w}_j)2\mathbf{\Sigma}\mathbf{w}_j - (\mathbf{w}_j^T\mathbf{\Sigma}\mathbf{w}_j)2\mathbf{w}_j\right].$$

Therefore, the derivative of the log likelihood with respect to $\mathbf{w}_j$ is

$$\frac{\partial}{\partial \mathbf{w}_j}\mathcal{L} = -\frac{\sigma^{-2}N}{\sigma^2 + \mathbf{w}_j^T\mathbf{w}_j}\left[\left(\sigma^2 + \frac{\mathbf{w}_j^T\mathbf{\Sigma}\mathbf{w}_j}{\sigma^2 + \mathbf{w}_j^T\mathbf{w}_j}\right)\mathbf{w}_j - \mathbf{\Sigma}\mathbf{w}_j\right].$$

This vanishes, if $\mathbf{w}_j$ is an eigenvector of $\mathbf{\Sigma}$ with eigenvalue λ_j and the length of $\mathbf{w}_j$ is such that

$$\sigma^2 + \frac{\mathbf{w}_j^T\mathbf{\Sigma}\mathbf{w}_j}{\sigma^2 + \mathbf{w}_j^T\mathbf{w}_j} = \lambda_j.$$

Using $\mathbf{\Sigma}\mathbf{w}_j = \lambda_j\mathbf{w}_j$, we arrive at $\mathbf{w}_j^T\mathbf{w}_j = \lambda_j - \sigma^2$, or equivalently $\sigma^2 + \mathbf{w}_j^T\mathbf{w}_j = \lambda_j$. Since $\mathbf{\Sigma}$ is symmetric, it has a set of mutually orthogonal eigenvectors and hence this agrees with our assumption.

Figure 7.5 shows two dimensional data and the principal component $\mathbf{w}_1$ is drawn as the red arrow with length $\sqrt{\lambda_1 - \sigma^2}$ starting at the sample mean, where λ_1 is the larger eigenvalue of the sample covariance matrix. In the following we will see that choosing the largest eigenvalues is the right choice to maximize the log likelihood.

Letting all $\mathbf{w}_k$, $k = 1,\ldots,K$, be some eigenvectors of $\mathbf{\Sigma}$ with squared length $\mathbf{w}_k^T\mathbf{w}_k = \lambda_k - \sigma^2$ and inserting this into (7.4), we get

$$|\mathbf{C}| = |\mathbf{C}_{-j}|(1 + \sigma^{-2}(\lambda_j - \sigma^2)) = \sigma^{-2}\lambda_j|\mathbf{C}_{-j}| = \ldots$$

$$= (\sigma^{-2})^K \prod_{k=1}^{K} \lambda_k |\sigma^2 \mathbf{I}| = (\sigma^2)^{D-K} \prod_{k=1}^{K} \lambda_k.$$

Equally using this in (7.3), we arrive at

$$\mathbf{C}^{-1} = \sigma^2 \left(\mathbf{I} - \sum_{k=1}^{K} \frac{\mathbf{w}_k \mathbf{w}_k^T}{\lambda_k} \right),$$

and

$$\mathrm{tr}(\mathbf{C}^{-1}\boldsymbol{\Sigma}) = \sigma^{-2} \left(\mathrm{tr}(\boldsymbol{\Sigma}) - \sum_{k=1}^{K} \frac{\mathrm{tr}(\mathbf{w}_k \mathbf{w}_k^T \boldsymbol{\Sigma})}{\lambda_k} \right)$$

$$= \sigma^{-2} \left(\mathrm{tr}(\boldsymbol{\Sigma}) - \sum_{k=1}^{K} \mathbf{w}_k^T \mathbf{w}_k \right) = \sigma^{-2} \left(\mathrm{tr}(\boldsymbol{\Sigma}) - \sum_{k=1}^{K} \lambda_k \right) + K,$$

where we used the cyclic permutation property of the trace and the fact that $\mathbf{w}_k$ is an eigenvector of $\boldsymbol{\Sigma}$.

With these two results, the expression for the log likelihood is

$$\mathcal{L} = -\frac{N}{2} \left[D \log(2\pi) + (D - K) \log \sigma^2 + \sum_{k=1}^{K} \log \lambda_k + \sigma^{-2} \left(\sum_{k=K+1}^{D} \lambda_k \right) + K \right],$$

since the trace of a matrix is the sum of its eigenvalues. We differentiate this with respect to σ^2 in order to optimize with respect to the noise variance.

$$\frac{\partial}{\partial \sigma^2} \mathcal{L} = -\frac{N}{2} \left((D - K)(\sigma^2)^{-1} - (\sigma^2)^{-2} \sum_{k=K+1}^{D} \lambda_k \right).$$

This will be zero for

$$\sigma^2 = \frac{1}{D - K} \sum_{k=K+1}^{D} \lambda_k.$$

In other words, σ^2 is the average of all the other eigenvalues apart from $\lambda_1, \ldots, \lambda_K$. As mentioned before, σ^2 explains the data protruding beyond the K-dimensional subspace of the feature space. This needs to be as small as possible. Therefore $\lambda_1, \ldots, \lambda_K$ are chosen to be the K largest eigenvalues of $\boldsymbol{\Sigma}$. The value of K itself is chosen such that σ^2 is acceptably small. The log likelihood takes the value

$$\mathcal{L} = -\frac{N}{2} \left[D \log(2\pi) + (D - K) \log \left(\frac{1}{D - K} \sum_{k=K+1}^{D} \lambda_k \right) + \sum_{k=1}^{K} \log \lambda_k + D \right].$$

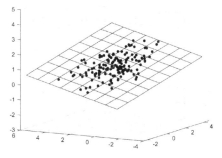
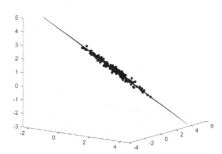

(a) Viewing the plane spanned by two principal components .

(b) Viewing the data along one of the principal components.

Figure 7.6: Probabilistic Principal Component Analysis for three-dimensional data.

Figure 7.6 illustrates the concepts. The data is viewed from two angles, firstly looking onto the plane spanned by the two principal components and secondly looking along one of the principal components. The latter angle shows that not all data points lie in the plane of the two principal components. The noise variance σ^2 explains this deviation and is equal to the smallest eigenvalue of the sample covariance matrix of this data.

7.3 Expectation-Maximization

The previous section was based on the assumption that the data samples $\mathbf{v}_1, \ldots, \mathbf{v}_N$ are related to vectors in a lower dimensional space via a linear mapping given by (7.2) with additive Gaussian noise with zero mean and co-variance matrix $\sigma^2 \mathbf{I}$. Therefore, each $\mathbf{v}_n$ follows a conditional normal distribution with mean $\mathbf{W}\mathbf{u}_n + \mathbf{m}$ and covariance matrix $\sigma^2 \mathbf{I}$, where $\mathbf{u}_n$, $n = 1, \ldots, N$, are hidden latent variables following the standard multivariate distribution. The solution in the previous section effectively calculated the marginal distribution from which the $\mathbf{v}_n$ are drawn by marginalizing the latent variables. The likelihood of the data set was then maximized to find $\mathbf{m}$, $\mathbf{W}$ and σ^2. The likelihood takes a maximum with respect to $\mathbf{m}$, when $\mathbf{m} = \boldsymbol{\mu}$, the sample mean which is easily obtained. The maximization with respect to $\mathbf{W}$ and σ^2, however, involves finding the K largest eigenvalues and eigenvectors of the sample covariance matrix which is in itself a computationally expensive problem. Firstly, calculating the covariance matrix is of complexity $O(ND^2)$. The computational complexity of then finding the K largest eigenvalues is $O(KD^2)$ (see for example [12]).

In the following, we will use the result $\mathbf{m} = \boldsymbol{\mu}$ and we employ the Expectation-Maximization technique introduced in Section 6.4 to find $\mathbf{W}$ and σ^2, making use of the joint distribution of $\mathbf{v}$ and $\mathbf{u}$ and the distribution of $\mathbf{u}$

conditioned on $\mathbf{v}$. Note that while in Section 6.4 the latent variables are discrete, here they are continuous. The principles remain the same though.

By the product rule, the joint probability is given by

$$
\begin{aligned}
p(\mathbf{v}, \mathbf{u}) &= p(\mathbf{u})p(\mathbf{v}|\mathbf{u}) \\
&= (2\pi)^{-K/2} \exp\left(-\frac{1}{2}\mathbf{u}^T\mathbf{u}\right) \times \\
&\quad (2\pi\sigma^2)^{-D/2} \exp\left(-\frac{1}{2}\sigma^{-2}(\mathbf{v} - \mathbf{W}\mathbf{u} - \boldsymbol{\mu})^T(\mathbf{v} - \mathbf{W}\mathbf{u} - \boldsymbol{\mu})\right).
\end{aligned}
$$

Gathering the terms in the exponential gives

$$
-\tfrac{1}{2}\sigma^{-2}\Big[(\mathbf{v} - \boldsymbol{\mu})^T(\mathbf{v} - \boldsymbol{\mu}) - (\mathbf{v} - \boldsymbol{\mu})^T\mathbf{W}\mathbf{u} -
$$
$$
\mathbf{u}^T\mathbf{W}^T(\mathbf{v} - \boldsymbol{\mu}) + \mathbf{u}^T(\mathbf{W}^T\mathbf{W} + \sigma^2\mathbf{I})\mathbf{u}\Big]. \tag{7.5}
$$

This can be written as

$$
-\frac{1}{2}\begin{pmatrix} \mathbf{v} - \boldsymbol{\mu} \\ \mathbf{u} \end{pmatrix}^T \sigma^{-2}\begin{pmatrix} \mathbf{I} & -\mathbf{W} \\ -\mathbf{W}^T & \mathbf{W}^T\mathbf{W} + \sigma^2\mathbf{I} \end{pmatrix}\begin{pmatrix} \mathbf{v} - \boldsymbol{\mu} \\ \mathbf{u} \end{pmatrix}.
$$

Hence the mean of the joint normal distribution is $(\ \boldsymbol{\mu}\ \ 0\ )^T$, while the matrix with the scalar factor σ^{-2} is the *precision matrix*, i.e. the inverse of the covariance matrix. Therefore, using the inversion formula for a block matrix in Appendix A.1.1, the covariance matrix of the joint distribution is

$$
\begin{pmatrix} \sigma^2\mathbf{I} + \mathbf{W}\mathbf{W}^T & \mathbf{W} \\ \mathbf{W}^T & \mathbf{I} \end{pmatrix}.
$$

Using the formula for the calculation of the determinant of a block matrix in Appendix A.1.2, the determinant of the joint covariance matrix is $(\sigma^2)^D$ which agrees with the factor of the joint probability distribution.

Also the conditional distribution $p(\mathbf{u}|\mathbf{v})$ can be deduced from (7.5) by treating $\mathbf{v}$ as a constant. The quadratic term in $\mathbf{u}$ is

$$
-\frac{1}{2}\sigma^{-2}\mathbf{u}^T(\mathbf{W}^T\mathbf{W} + \sigma^2\mathbf{I})\mathbf{u},
$$

and thus the covariance matrix of $p(\mathbf{u}|\mathbf{v})$ is

$$
\boldsymbol{\Sigma}_{\mathbf{u}|\mathbf{v}} = \sigma^2(\mathbf{W}^T\mathbf{W} + \sigma^2\mathbf{I})^{-1}, \tag{7.6}
$$

which is independent of $\mathbf{v}$.

The linear term in $\mathbf{u}$ is

$$
-\frac{1}{2}\sigma^{-2}\left(-2\mathbf{u}^T\mathbf{W}^T(\mathbf{v} - \boldsymbol{\mu})\right) = \mathbf{u}^T\sigma^{-2}\mathbf{W}^T(\mathbf{v} - \boldsymbol{\mu}).
$$

This needs to equal $\mathbf{u}^T\boldsymbol{\Sigma}_{\mathbf{u}|\mathbf{v}}^{-1}\boldsymbol{\mu}_{\mathbf{u}|\mathbf{v}}$. Therefore, the mean of the conditional distribution is

$$
\boldsymbol{\mu}_{\mathbf{u}|\mathbf{v}} = \boldsymbol{\Sigma}_{\mathbf{u}|\mathbf{v}}\sigma^{-2}\mathbf{W}^T(\mathbf{v} - \boldsymbol{\mu}) = (\mathbf{W}^T\mathbf{W} + \sigma^2\mathbf{I})^{-1}\mathbf{W}^T(\mathbf{v} - \boldsymbol{\mu}), \tag{7.7}
$$

which depends on $\mathbf{v}$.

The Expectation-Maximization algorithm calculates the expectation of the logarithm of the joint data likelihood with respect to the conditional probability distribution of the latent variables $\mathbf{u}_1, \ldots, \mathbf{u}_N$. The logarithm of the joint data likelihood is

$$\mathcal{L} = \log \prod_{n=1}^{N} p(\mathbf{v}_n, \mathbf{u}_n) = -\frac{1}{2} \sum_{n=1}^{N} \Big[(K+D) \log(2\pi) + D \log(\sigma^2) + \mathbf{u}_n^T \mathbf{u}_n + \\ \sigma^{-2}(\mathbf{v}_n - \boldsymbol{\mu})^T (\mathbf{v}_n - \boldsymbol{\mu}) - 2\sigma^{-2} \mathbf{u}_n^T \mathbf{W}^T (\mathbf{v}_n - \boldsymbol{\mu}) + \\ \sigma^{-2} \mathbf{u}_n^T \mathbf{W}^T \mathbf{W} \mathbf{u}_n \Big].$$

When taking the expectation with respect to the latent variables, we have

$$\mathbb{E}[\mathbf{u}_n] = \boldsymbol{\mu}_{\mathbf{u}|\mathbf{v}_n}. \tag{7.8}$$

To find the expectation of the inner product of $\mathbf{u}_n$ with itself, we use the cyclic permutation property of the trace and the fact that the operations of calculating the trace and expectation commute,

$$\mathbb{E}[\mathbf{u}_n^T \mathbf{u}_n] = \mathbb{E}[\mathrm{tr}(\mathbf{u}_n^T \mathbf{u}_n)] = \mathbb{E}[\mathrm{tr}(\mathbf{u}_n \mathbf{u}_n^T)] = \mathrm{tr}(\mathbb{E}[\mathbf{u}_n \mathbf{u}_n^T]).$$

Similarly,

$$\mathbb{E}[\mathbf{u}_n^T \mathbf{W}^T \mathbf{W} \mathbf{u}_n] = \mathbb{E}[\mathrm{tr}(\mathbf{u}_n^T \mathbf{W}^T \mathbf{W} \mathbf{u}_n)] = \mathbb{E}[\mathrm{tr}(\mathbf{W}^T \mathbf{W} \mathbf{u}_n \mathbf{u}_n^T)] \\ = \mathrm{tr}(\mathbf{W}^T \mathbf{W} \mathbb{E}[\mathbf{u}_n \mathbf{u}_n^T]).$$

We can then use (2.10) to calculate

$$\mathbb{E}[\mathbf{u}_n \mathbf{u}_n^T] = \boldsymbol{\Sigma}_{\mathbf{u}|\mathbf{v}} + \mathbb{E}[\mathbf{u}_n]\mathbb{E}[\mathbf{u}_n]^T = \boldsymbol{\Sigma}_{\mathbf{u}|\mathbf{v}} + \boldsymbol{\mu}_{\mathbf{u}|\mathbf{v}_n} \boldsymbol{\mu}_{\mathbf{u}|\mathbf{v}_n}^T. \tag{7.9}$$

The expectation of the logarithm of the joint data likelihood is therefore

$$\mathbb{E}[\mathcal{L}] = -\frac{1}{2} \sum_{n=1}^{N} \Big[(K+D) \log(2\pi) + D \log(\sigma^2) + \mathrm{tr}(\mathbb{E}[\mathbf{u}_n \mathbf{u}_n^T]) + \\ \sigma^{-2} \|\mathbf{v}_n - \boldsymbol{\mu}\|^2 - 2\sigma^{-2} \mathbb{E}[\mathbf{u}_n]^T \mathbf{W}^T (\mathbf{v}_n - \boldsymbol{\mu}) + \\ \sigma^{-2} \mathrm{tr}(\mathbf{W}^T \mathbf{W} \mathbb{E}[\mathbf{u}_n \mathbf{u}_n^T]) \Big].$$

Now, employing the cyclic permutation property of the trace, the formulae given in Appendices A.2.6 and A.2.9 and the symmetry of $\mathbb{E}[\mathbf{u}_n \mathbf{u}_n^T]$, the derivative with respect to $\mathbf{W}$ is

$$\frac{\partial}{\partial \mathbf{W}} \mathbb{E}[\mathcal{L}] = -\frac{1}{2} \sum_{n=1}^{N} \Big[-2\sigma^{-2}(\mathbf{v}_n - \boldsymbol{\mu})\mathbb{E}[\mathbf{u}_n]^T + 2\sigma^{-2} \mathbf{W} \mathbb{E}[\mathbf{u}_n \mathbf{u}_n^T] \Big].$$

Setting this to zero and solving for $\mathbf{W}$ results in an update formula for it,

$$\mathbf{W}_{\text{new}} = \left[\sum_{n=1}^{N} (\mathbf{v}_n - \boldsymbol{\mu})\mathbb{E}[\mathbf{u}_n]^T\right]\left[\sum_{n=1}^{N} \mathbb{E}[\mathbf{u}_n\mathbf{u}_n^T]\right]^{-1}. \qquad (7.10)$$

Note that this is truly an update formula, since it indirectly depends on $\mathbf{W}$ in the conditional mean and covariance matrix given in (7.6) and (7.7) via the formulae given in (7.9) and (7.8).

On the other hand, differentiating with respect to σ^2 results in

$$\frac{\partial}{\partial \sigma^2}\mathbb{E}[\mathcal{L}] = -\frac{1}{2}\sum_{n=1}^{N}\left[\frac{D}{\sigma^2} - \frac{1}{(\sigma^2)^2}\left(\|\mathbf{v}_n - \boldsymbol{\mu}\|^2\right.\right.$$
$$\left.\left. -2\mathbb{E}[\mathbf{u}_n]^T\mathbf{W}^T(\mathbf{v}_n - \boldsymbol{\mu}) + \text{tr}(\mathbf{W}^T\mathbf{W}\mathbb{E}[\mathbf{u}_n\mathbf{u}_n^T]))\right].$$

The update formula for σ^2 follows as

$$\sigma^2_{\text{new}} = \frac{1}{ND}\sum_{n=1}^{N}\left[\|\mathbf{v}_n - \boldsymbol{\mu}\|^2 - 2\mathbb{E}[\mathbf{u}_n]^T\mathbf{W}^T(\mathbf{v}_n - \boldsymbol{\mu}) + \text{tr}(\mathbf{W}^T\mathbf{W}\mathbb{E}[\mathbf{u}_n\mathbf{u}_n^T])\right].$$
$$(7.11)$$

It is common to use the already updated value $\mathbf{W}_{\text{new}}$, when updating σ^2. Using (7.9), the update can be written as

$$\sigma^2_{\text{new}} = \frac{1}{ND}\sum_{n=1}^{N}\|\mathbf{v}_n - \boldsymbol{\mu} - \mathbf{W}_{\text{new}}\mathbb{E}[\mathbf{u}_n])\|^2 + \frac{1}{D}\text{tr}(\mathbf{W}_{\text{new}}^T\mathbf{W}_{\text{new}}\boldsymbol{\Sigma}_{\mathbf{u}|\mathbf{v}}).$$

This can be interpreted as measuring how close on average $\mathbb{E}[\mathbf{u}_n]$ is to the sample $\mathbf{v}_n - \boldsymbol{\mu}$ under the mapping given by $\mathbf{W}_{\text{new}}$ and how close $\mathbf{W}_{\text{new}}^T\mathbf{W}_{\text{new}}$ is to the inverse of the conditional covariance matrix; that is how close it is to the precision matrix.

Another interpretation of Equation (7.11) is as follows. Since the sum of traces is the trace of the sum, we can sum inside the trace. Using (7.10) for $\mathbf{W}_{\text{new}}^T$, the sum cancels with its inverse. Further, using the cyclic permutation property of the trace and that the trace of a scalar is a scalar, the update formula simplifies to

$$\sigma^2_{\text{new}} = \frac{1}{ND}\sum_{n=1}^{N}\left[\|\mathbf{v}_n - \boldsymbol{\mu}\|^2 - \mathbb{E}[\mathbf{u}_n]^T\mathbf{W}_{\text{new}}^T(\mathbf{v}_n - \boldsymbol{\mu})\right]$$
$$= \frac{1}{ND}\sum_{n=1}^{N}(\mathbf{v}_n - \boldsymbol{\mu} - \mathbf{W}_{\text{new}}\mathbb{E}[\mathbf{u}_n])^T(\mathbf{v}_n - \boldsymbol{\mu}). \qquad (7.12)$$

Thus the noise variance is the average inner product of $\mathbf{v}_n - \boldsymbol{\mu}$ with the difference of $\mathbf{v}_n - \boldsymbol{\mu}$ and the image of $\mathbb{E}[\mathbf{u}_n]$ under $\mathbf{W}_{\text{new}}$ scaled by the number of dimensions D. Each term in the sum is small if either $\mathbf{W}_{\text{new}}\mathbb{E}[\mathbf{u}_n]$ is close to $\mathbf{v}_n - \boldsymbol{\mu}$ or the difference is nearly orthogonal to $\mathbf{v}_n - \boldsymbol{\mu}$. It measures, how close on average $\mathbf{W}_{\text{new}}\mathbb{E}[\mathbf{u}_n]$ is to the orthogonal projection of $\mathbf{v}_n - \boldsymbol{\mu}$ onto the subspace spanned by the columns of $\mathbf{W}_{\text{new}}$.

These update formulae for $\mathbf{W}_{\text{new}}$ and σ^2_{new} form the M-step of the Expectation-Maximization algorithm. They are sums of length N, where each element in the sum can be calculated in $O(KD)$ operations. So the overall complexity of the M-step is $O(KDN)$.

The E-step calculates $\mathbb{E}[\mathbf{u}_n]$ via (7.7) and (7.8) and $\mathbb{E}[\mathbf{u}_n\mathbf{u}_n^T]$ using (7.6) and (7.9). The E-step needs to calculate the inverse of a matrix, but this matrix is of size $K \times K$ so this is $O(K^3)$. Calculating this matrix itself is of order $O(K^2D)$. Once the inverse is found, calculating $\mathbb{E}[\mathbf{u}_n]$ is $O(KD)$ and $\mathbb{E}[\mathbf{u}_n\mathbf{u}_n^T]$ is $O(K^2) = O(KD)$, since $K \leq D$. These need to be calculated for $n = 1, \ldots, N$ and the complexity for the E-step is $O(KDN)$. However, it can be done in parallel for each data point, since $\mathbb{E}[\mathbf{u}_n]$ and $\mathbb{E}[\mathbf{u}_n\mathbf{u}_n^T]$ can be calculated independently of each other.

While the Expectation-Maximization algorithm is an iterative procedure, the number of operations per iteration is significantly less if $K \ll D$. This can offset the extra time due to the iterations against the direct maximization of the log likelihood which has complexity $O(ND^2)$.

So far in this section, K was chosen and not determined by the data. For a Bayesian treatment of Principal Component Analysis which determines K see [30].

7.4 Factor Analysis

So far the noise was modeled by $\sigma^2\mathbf{I}$ implying that all directions are treated equally. A more suitable model would be to let the noise variance be a diagonal matrix $\mathbf{D}$ such that each direction can be treated separately. The diagonal elements of $\mathbf{D}$ are known as *uniquenesses*, while the columns of $\mathbf{W}$ are known as *factor loadings*.

Following a similar calculation as in the previous section, the joint distribution $p(\mathbf{v}, \mathbf{u})$ is given by

$$(2\pi)^{-(K+D)/2}|\mathbf{D}|^{-1/2}$$

$$\exp\left(-\frac{1}{2}\begin{pmatrix} \mathbf{v} - \boldsymbol{\mu} \\ \mathbf{u} \end{pmatrix}^T \begin{pmatrix} \mathbf{D}^{-1} & -\mathbf{D}^{-1}\mathbf{W} \\ -\mathbf{W}^T\mathbf{D}^{-1} & \mathbf{W}^T\mathbf{D}^{-1}\mathbf{W}+\mathbf{I} \end{pmatrix} \begin{pmatrix} \mathbf{v} - \boldsymbol{\mu} \\ \mathbf{u} \end{pmatrix}\right).$$

As before, the mean of the joint normal distribution is $(\boldsymbol{\mu} \quad 0)^T$, where $\boldsymbol{\mu}$ is the sample mean. The inversion formula for block matrices gives the joint covariance matrix as

$$\begin{pmatrix} \mathbf{D} + \mathbf{W}^T\mathbf{W} & \mathbf{W} \\ \mathbf{W}^T & \mathbf{I} \end{pmatrix}.$$

The formula in Appendix A.1.2 for calculating the determinant of block matrices shows that the determinant of the joint covariance matrix is $|\mathbf{D}|$.

The logarithm of the joint data likelihood is

$$\mathcal{L} = \log \prod_{n=1}^{N} p(\mathbf{v}_n, \mathbf{u}_n) = -\frac{1}{2} \sum_{n=1}^{N} \Big[(K+D)\log(2\pi) + \log(|\mathbf{D}|) + \mathbf{u}_n^T \mathbf{u}_n +$$
$$(\mathbf{v}_n - \boldsymbol{\mu})^T \mathbf{D}^{-1}(\mathbf{v}_n - \boldsymbol{\mu}) - 2\mathbf{u}_n^T \mathbf{W}^T \mathbf{D}^{-1}(\mathbf{v}_n - \boldsymbol{\mu}) +$$
$$\mathbf{u}_n^T \mathbf{W}^T \mathbf{D}^{-1} \mathbf{W} \mathbf{u}_n \Big].$$

As before, we take the expectation with respect to the latent variables $\mathbf{u}_n$ which results in

$$\mathbb{E}[\mathcal{L}] \;=\; -\frac{1}{2} \sum_{n=1}^{N} \Big[(K+D)\log(2\pi) + \log(|\mathbf{D}|) + \mathrm{tr}(\mathbb{E}[\mathbf{u}_n \mathbf{u}_n^T]) +$$
$$(\mathbf{v}_n - \boldsymbol{\mu})^T \mathbf{D}^{-1}(\mathbf{v}_n - \boldsymbol{\mu}) - 2\mathbb{E}[\mathbf{u}_n]^T \mathbf{W}^T \mathbf{D}^{-1}(\mathbf{v}_n - \boldsymbol{\mu}) +$$
$$\mathrm{tr}(\mathbf{W} \mathbb{E}[\mathbf{u}_n \mathbf{u}_n^T] \mathbf{W}^T \mathbf{D}^{-1}) \Big],$$

where the cyclic permutation property of the trace was used.

Differentiating this with respect to $\mathbf{W}$ using Appendices A.2.6 and A.2.9, gives

$$\frac{\partial}{\partial \mathbf{W}} \mathbb{E}[\mathcal{L}] = -\frac{1}{2} \sum_{n=1}^{N} \Big[-2\mathbf{D}^{-1}(\mathbf{v}_n - \boldsymbol{\mu})\mathbb{E}[\mathbf{u}_n]^T + 2\mathbf{D}^{-1}\mathbf{W}\mathbb{E}[\mathbf{u}_n \mathbf{u}_n^T] \Big].$$

Setting this to zero, multiplying through with $\mathbf{D}$ from the left and solving for $\mathbf{W}$ results in the update formula

$$\mathbf{W}_{\text{new}} = \left[\sum_{n=1}^{N} (\mathbf{v}_n - \boldsymbol{\mu})\mathbb{E}[\mathbf{u}_n]^T \right] \left[\sum_{n=1}^{N} \mathbb{E}[\mathbf{u}_n \mathbf{u}_n^T] \right]^{-1}, \qquad (7.13)$$

which is exactly the same as before.

To differentiate with respect to $\mathbf{D}$, we first rewrite the expression for the expectation of the joint log likelihood using the cyclic permutation property of the trace,

$$\mathbb{E}[\mathcal{L}] = -\frac{1}{2} \sum_{n=1}^{N} \Big[(K+D)\log(2\pi) + \log(|\mathbf{D}|) + \mathrm{tr}(\mathbb{E}[\mathbf{u}_n \mathbf{u}_n^T]) +$$
$$\mathrm{tr}\Big(\big[(\mathbf{v}_n - \boldsymbol{\mu})(\mathbf{v}_n - \boldsymbol{\mu})^T - 2(\mathbf{v}_n - \boldsymbol{\mu})\mathbb{E}[\mathbf{u}_n]^T \mathbf{W}^T + \mathbf{W}\mathbb{E}[\mathbf{u}_n \mathbf{u}_n^T]\mathbf{W}^T \big] \mathbf{D}^{-1} \Big) \Big].$$

This way we can use the derivative formulae in Appendices A.2.4, A.2.7 and A.2.10. The derivative with respect to $\mathbf{D}$ is therefore

$$
\frac{\partial}{\partial \mathbf{D}} \mathbb{E}[\mathcal{L}] = -\frac{1}{2} \sum_{n=1}^{N} \Big[\frac{1}{|\mathbf{D}|} |\mathbf{D}| \mathbf{D}^{-1} - \mathbf{D}^{-1} \mathrm{diag}\Big((\mathbf{v}_n - \boldsymbol{\mu})(\mathbf{v}_n - \boldsymbol{\mu})^T - 2(\mathbf{v}_n - \boldsymbol{\mu})\mathbb{E}[\mathbf{u}_n]^T \mathbf{W}^T + \mathbf{W}\mathbb{E}[\mathbf{u}_n \mathbf{u}_n^T] \mathbf{W}^T \Big) \mathbf{D}^{-1} \Big].
$$

Setting this to zero, multiplying through by $\mathbf{D}$ from both left and right and solving for $\mathbf{D}$, gives the following update formula

$$
\mathbf{D}_{\text{new}} = \mathrm{diag}\Bigg(\frac{1}{N} \sum_{n=1}^{N} (\mathbf{v}_n - \boldsymbol{\mu})(\mathbf{v}_n - \boldsymbol{\mu})^T
$$
$$
-2\frac{1}{N} \sum_{n=1}^{N} (\mathbf{v}_n - \boldsymbol{\mu})\mathbb{E}[\mathbf{u}_n]^T \mathbf{W}^T
$$
$$
+\frac{1}{N} \mathbf{W} \sum_{n=1}^{N} \mathbb{E}[\mathbf{u}_n \mathbf{u}_n^T] \mathbf{W}^T \Bigg).
$$

It is common to use the already updated $\mathbf{W}_{\text{new}}$, when updating $\mathbf{D}_{\text{new}}$. When inserting formula (7.13) into the left of the last line, the sum over $\mathbb{E}[\mathbf{u}_n \mathbf{u}_n^T]$, $n = 1, \ldots, N$, cancels with its inverse and the last line combines with the one above. Hence, the update formula is

$$
\mathbf{D}_{\text{new}} = \frac{1}{N} \sum_{n=1}^{N} \mathrm{diag}\left((\mathbf{v}_n - \boldsymbol{\mu})(\mathbf{v}_n - \boldsymbol{\mu} - \mathbf{W}_{\text{new}}\mathbb{E}[\mathbf{u}_n])^T \right).
$$

This can be viewed as the D-dimensional version of (7.12).

These updates depend on $\mathbf{D}$ and the previous $\mathbf{W}$ via the expectations, which are calculated in the E-step. To this end, the conditional probability distribution $p(\mathbf{u}|\mathbf{v})$ of the latent variable $\mathbf{u}$ given $\mathbf{v}$ has variance

$$
\boldsymbol{\Sigma}_{\mathbf{u}|\mathbf{v}} = (\mathbf{W}^T \mathbf{D}^{-1} \mathbf{W} + \mathbf{I})^{-1},
$$

which is independent of $\mathbf{v}$, and mean

$$
\boldsymbol{\mu}_{\mathbf{u}|\mathbf{v}} = \boldsymbol{\Sigma}_{\mathbf{u}|\mathbf{v}} \mathbf{W}^T \mathbf{D}^{-1}(\mathbf{v} - \boldsymbol{\mu}),
$$

which depends on $\mathbf{v}$. Therefore, the two calculations forming the E-step are

$$
\mathbb{E}[\mathbf{u}_n] = \boldsymbol{\Sigma}_{\mathbf{u}|\mathbf{v}} \mathbf{W}^T \mathbf{D}^{-1}(\mathbf{v}_n - \boldsymbol{\mu})
$$

and using (2.10)

$$
\mathbb{E}[\mathbf{u}_n \mathbf{u}_n^T] = \boldsymbol{\Sigma}_{\mathbf{u}|\mathbf{v}} + \mathbb{E}[\mathbf{u}_n]\mathbb{E}[\mathbf{u}_n]^T.
$$

7.5   Kernel Principal Component Analysis

The previous three sections relied on the assumption that the data samples are related to latent variables in a lower dimensional space via a *linear* mapping as defined in (7.2). Geometrically, this means that the data lies close to a line, plane or hyperplane as illustrated in Figures 7.5 and 7.6. This assumption is, however, unrealistic in many practical applications.

The data samples may lie close to a K-dimensional *manifold*. A manifold is a set of points which locally resembles Euclidean space. An example for a two-dimensional manifold is the sphere. Flat-earthers succumb to the impression that locally the Earth does look flat. The sphere resembles a plane locally. Globally, however, it does not, because traveling continuously in one direction will return to the starting point. This is not possible on a plane. Figure 7.7a gives an example of data being close to the unit sphere. Lighter shaded data points lie behind the surface of the sphere.

It is possible to choose a mapping $\phi : \mathbb{R}^D \to \mathbb{R}^{\hat{D}}$ into a higher dimensional space such that the data then is linearly related to latent variables in a K-dimensional subspace. In Figure 7.7b, the data was first mapped to a four-dimensional space, by letting the fourth component be the sum of squares of the other three. Since the data lies approximately on the unit sphere, the fourth component will be close to one for all data samples. Thus the data lies on a hyperplane. After finding the first two principal components, the data is projected onto the plane given by these two principal components. This is shown in Figure 7.7b. That the two data sets are grouped towards the top or bottom, means that they can be separated along one principal component.

However, like in the kernel trick in Section 5.2, performing the mapping explicitly is avoided. Instead, all steps in the calculation use the *kernel function* or *kernel* for short $k : \mathbb{R}^D \times \mathbb{R}^D \to \mathbb{R}$ defined as

$$k(\mathbf{x}, \mathbf{y}) = \phi(\mathbf{x})^T \phi(\mathbf{y}).$$

Let $\mathbf{E}$ be the $N \times N$ matrix with each entry being equal to one. This matrix can be used to *mean centre* a vector $\mathbf{a}^T = (a_1, \ldots, a_N)$ by

$$\mathbf{a}^T (\mathbf{I} - \frac{1}{N} \mathbf{E}) = \left(a_1 - \frac{1}{N} \sum_{n=1}^{N} a_n, \quad \cdots, \quad a_N - \frac{1}{N} \sum_{n=1}^{N} a_n \right).$$

It can also be used to shift data samples so that their mean is zero. Let $\phi(\mathbf{v}_1), \ldots, \phi(\mathbf{v}_N) \in \mathbb{R}^{\hat{D}}$ be the images under the mapping ϕ of the data samples $\mathbf{v}_1, \ldots, \mathbf{v}_N \in \mathbb{R}^D$, which are known as the *pre-images*. Further let, $\mathbf{\Phi}$ be the $\hat{D} \times N$ matrix whose columns are the vectors $\phi(\mathbf{v}_1), \ldots, \phi(\mathbf{v}_N)$, then

$$\mathbf{\Phi}(\mathbf{I} - \frac{1}{N} \mathbf{E}) = \left(\phi(\mathbf{v}_1) - \frac{1}{N} \sum_{n=1}^{N} \phi(\mathbf{v}_n), \quad \cdots, \quad \phi(\mathbf{v}_N) - \frac{1}{N} \sum_{n=1}^{N} \phi(\mathbf{v}_n) \right),$$

since, for every element in each row of $\mathbf{\Phi}$, we subtract the mean of that row of $\mathbf{\Phi}$.

Recall that Principal Component Analysis calculates the covariance matrix $\mathbf{\Sigma}$ of the data samples. In this context, it means the covariance matrix of the

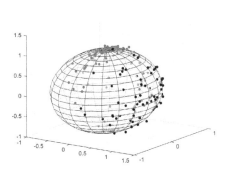

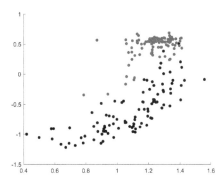

(a) Data samples lying approximately on the unit sphere in two groups.

(b) Projection onto two principal components after a transformation to a four-dimensional space.

Figure 7.7: Principal Component Analysis on a manifold.

shifted images, which is the average of the outer products of the nth column of $\mathbf{\Phi}(\mathbf{I} - \frac{1}{N}\mathbf{E})$ with its transpose. This can be written as

$$\mathbf{\Sigma} = \frac{1}{N}\sum_{n=1}^{N}\mathbf{\Phi}(\mathbf{I} - \frac{1}{N}\mathbf{E})\mathbf{e}_n\mathbf{e}_n^T(\mathbf{I} - \frac{1}{N}\mathbf{E})\mathbf{\Phi}^T,$$

where $\mathbf{e}_n$ is the nth standard unit basis vector, with all elements being equal to zero, apart from the nth being one.

Now, $\mathbf{\Phi}(\mathbf{I} - \frac{1}{N}\mathbf{E})$ and its transpose are independent of n in the summation and can be taken outside the sum. Summing $\mathbf{e}_n\mathbf{e}_n^T$ over all $n = 1, \ldots, N$ results in the identity matrix. Further, the matrix $\mathbf{I} - \frac{1}{N}\mathbf{E}$ is *idempotent*, which means any power of it is the matrix itself, since $\mathbf{E}^2 = N\mathbf{E}$ and

$$(\mathbf{I} - \frac{1}{N}\mathbf{E})(\mathbf{I} - \frac{1}{N}\mathbf{E}) = I - \frac{2}{N}\mathbf{E} + \frac{1}{N^2}\mathbf{E}^2 = \mathbf{I} - \frac{1}{N}\mathbf{E}.$$

Therefore, the covariance matrix can be expressed as

$$\mathbf{\Sigma} = \frac{1}{N}\mathbf{\Phi}(\mathbf{I} - \frac{1}{N}\mathbf{E})\mathbf{\Phi}^T.$$

We seek the largest eigenvalues λ_k of $\mathbf{\Sigma}$ with corresponding eigenvectors $\mathbf{w}_k$, $k = 1, \ldots, K$. Using the above expression for $\mathbf{\Sigma}$, these vectors satisfy

$$\mathbf{\Phi}(\mathbf{I} - \frac{1}{N}\mathbf{E})\mathbf{\Phi}^T\mathbf{w}_k = N\lambda_k\mathbf{w}_k.$$

Multiplying through with $(\mathbf{I} - \frac{1}{N}\mathbf{E})\mathbf{\Phi}^T$ from the left and using the idempotence of $\mathbf{I} - \frac{1}{N}\mathbf{E}$, this becomes

$$(\mathbf{I} - \frac{1}{N}\mathbf{E})\mathbf{\Phi}^T\mathbf{\Phi}(\mathbf{I} - \frac{1}{N}\mathbf{E})(\mathbf{I} - \frac{1}{N}\mathbf{E})\mathbf{\Phi}^T\mathbf{w}_k = N\lambda_k(\mathbf{I} - \frac{1}{N}\mathbf{E})\mathbf{\Phi}^T\mathbf{w}_k.$$

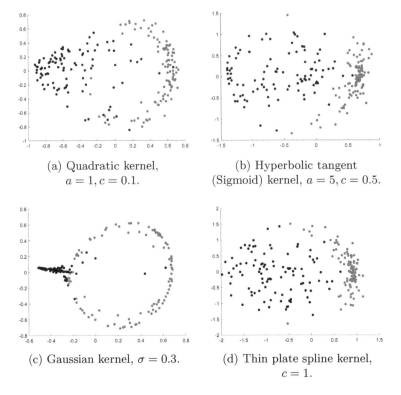

(a) Quadratic kernel,
$a = 1, c = 0.1.$

(b) Hyperbolic tangent
(Sigmoid) kernel, $a = 5, c = 0.5.$

(c) Gaussian kernel, $\sigma = 0.3.$

(d) Thin plate spline kernel,
$c = 1.$

Figure 7.8: Kernel Principal Component Analysis.

Hence the vector

$$\hat{\mathbf{w}}_k = (\mathbf{I} - \frac{1}{N}\mathbf{E})\mathbf{\Phi}^T\mathbf{w}_k \qquad (7.14)$$

is an eigenvector of the $N \times N$, symmetric matrix $(\mathbf{I} - \frac{1}{N}\mathbf{E})\mathbf{\Phi}^T\mathbf{\Phi}(\mathbf{I} - \frac{1}{N}\mathbf{E})$ with eigenvalue $\hat{\lambda}_k = N\lambda_k$. Note that due to $\mathbf{I} - \frac{1}{N}\mathbf{E}$ being idempotent, we have $(\mathbf{I} - \frac{1}{N}\mathbf{E})\hat{\mathbf{w}}_k = \hat{\mathbf{w}}_k$. Thus, the components of $\hat{\mathbf{w}}_k$ sum to zero.

The (i, j) entry of $\mathbf{\Phi}^T\mathbf{\Phi}$ is the inner product

$$\phi(\mathbf{v}_i)^T\phi(\mathbf{v}_j) = k(\mathbf{v}_i, \mathbf{v}_j).$$

$\mathbf{K} = \mathbf{\Phi}^T\mathbf{\Phi}$ is known as the *kernel matrix*. Kernel Principal Component Analysis calculates the eigenvalues $N\lambda_k$ and eigenvectors $\hat{\mathbf{w}}_k$ of $(\mathbf{I} - \frac{1}{N}\mathbf{E})\mathbf{K}(\mathbf{I} - \frac{1}{N}\mathbf{E})$. Multiplying Equation (7.14) from the left with $\mathbf{\Phi}/N$, gives

$$\frac{1}{N}\mathbf{\Phi}\hat{\mathbf{w}}_k = \frac{1}{N}\mathbf{\Phi}(\mathbf{I} - \frac{1}{N}\mathbf{E})\mathbf{\Phi}^T\mathbf{w}_k = \Sigma\mathbf{w}_k = \lambda_k\mathbf{w}_k.$$

Hence, $\mathbf{w}_k$ is a linear combination of the images $\phi(\mathbf{v}_1), \ldots, \phi(\mathbf{v}_N)$, since

$$\mathbf{w}_k = \frac{1}{N\lambda_k}\sum_{n=1}^{N}\hat{w}_{kn}\phi(\mathbf{v}_n),$$

where $\hat{w}_{kn}$, $n = 1, \ldots, N$, are the components of $\hat{\mathbf{w}}_k$. In practice, however, this calculation is never done, since we are only interested in expressing any data point $\mathbf{v}$ in the basis formed by the principal components, i.e. the value of the projections

$$
\begin{aligned}
\mathbf{w}_k^T \phi(\mathbf{v})^T &= \frac{1}{N\lambda_k} \hat{\mathbf{w}}_k^T \mathbf{\Phi}^T \phi(\mathbf{v}) = \frac{1}{N\lambda_k} \hat{\mathbf{w}}_k^T \begin{pmatrix} \phi(\mathbf{v})^T \phi(\mathbf{v}_1) \\ \vdots \\ \phi(\mathbf{v})^T \phi(\mathbf{v}_N) \end{pmatrix} \\
&= \frac{1}{N\lambda_k} \hat{\mathbf{w}}_k^T \begin{pmatrix} k(\mathbf{v}, \mathbf{v}_1) \\ \vdots \\ k(\mathbf{v}, \mathbf{v}_N) \end{pmatrix}.
\end{aligned}
$$

It remains to specify the length of $\hat{\mathbf{w}}_k$. The principal components have length one. This translates into a condition on the length of $\hat{\mathbf{w}}_k$,

$$
\begin{aligned}
1 &= \mathbf{w}_k^T \mathbf{w}_k = \frac{1}{(N\lambda_k)^2} \hat{\mathbf{w}}_k^T \mathbf{\Phi}^T \mathbf{\Phi} \hat{\mathbf{w}}_k \\
&= \frac{1}{(N\lambda_k)^2} \hat{\mathbf{w}}_k^T (\mathbf{I} - \frac{1}{N}\mathbf{E}) \mathbf{K} (\mathbf{I} - \frac{1}{N}\mathbf{E}) \hat{\mathbf{w}}_k = \frac{1}{N\lambda_k} \hat{\mathbf{w}}_k^T \hat{\mathbf{w}}_k,
\end{aligned}
$$

since $\hat{\mathbf{w}}_k = (\mathbf{I} - \frac{1}{N}\mathbf{E}) \hat{\mathbf{w}}_k$ is an eigenvector with eigenvalue $N\lambda_k$. Thus, $\hat{\mathbf{w}}_k$ has to have length $\sqrt{N\lambda_k}$.

The kernel function does not need to be defined via a mapping ϕ and an inner product in a higher dimensional space. Section 5.2 gives several examples of kernel functions. Figure 7.8 shows the results of kernel Principal Component Analysis applied to the data in Figure 7.7a for various choices of kernels. Visually the results are quite different, but what they have in common is that the two data sets are grouped towards the left or right. This means that they can be separated along one principal component.

CHAPTER 8

Regression

Regression developed from interpolation and is introduced as building a model of the process generating the data where the building blocks are chosen. The first choice of building blocks are polynomials. Ordinary Least Squares are introduced as the simplest method arising from assumed normally distributed noise. The concepts of over- and under-fitting are explained using the degree of polynomials as a descriptor for the complexity of the model space. Bias and variance are two contributors to the expected error of the model. Cross-validation is a method to decide the goodness of the model. If the building blocks are too similar to each other, multicollinearity becomes a problem. Principal component regression ensures small correlation between building blocks, while Partial Least Squares emphasizes strong correlation with the data at the same time. Regularization controls model complexity. The chapter concludes with taking a probabilistic viewpoint in form of Bayesian regression leading to Gaussian processes.

In regression, we try find the relationship between variables. Regression is related to curve fitting, interpolation, and data prediction. As an example, we consider colour vision. Spectral colours, as seen in Figure 8.1, are evoked by a single wavelength. Colours change continuously as the wavelength changes. A physical colour is a combination of pure spectral colours. Hence there are infinitely many possibilities.

Figure 8.1: Spectral colours.

Around 1854, James Clerk Maxwell [8] showed that all colours can be mixed by different combinations of the three primary colours, red, green and blue. He did so by using a colour-mixing top which can be seen in Figure 8.2

consisting of three coloured paper discs overlapping by different amounts. A smaller central disc contained a wedge of a colour sample. The brightness could be adjusted by the width of the wedge. When the top spun, the outer colours would mix and one could adjust how much each colour contributed until it matched the colour in the centre.

Figure 8.2: Maxwell's colour-mixing top.
Photo courtesy of Professor John Mollon, Cambridge University.

There is a biological explanation for this. The human eye has only three types of measuring devices to perceive colours – the cones. Figure 8.3 shows to which wavelengths different type of cones are sensitive. The blue peak is well separated from the red and green peaks. It is believed that the separation between red and green is evolutionarily fairly new and that the curves will move further apart with time. Colour blindness is in most cases due to one type of cone missing or malfunctioning. With three measurements, the human eye can distinguish ten million colours. This is an example of regression, and shows how powerful our brain is. From only three measurements our brain is able to infer millions of colours.

These are not just spectral colours, but also non-spectral ones. Magenta is an extraordinary colour. It is a mixture of the primary colours red and blue which lie far apart in the colour spectrum. It occurs in nature when the green component of white light is absorbed, since it is the complementary colour to green. In fact, magenta did not exist as a dye until 1859/60, when it was invented by François-Emmanuel Verguin (Lyon), Chambers Nicolson and George Maule (London). It is an anilin dye, and was named after the battle of Magenta. It caused a fashion craze with those who could afford it, showing off in magenta clothes some of which can be seen in the Victoria and Albert Museum in London, England.

Different animals and insects have different numbers of cone types. Refer to Table 8.1 gives an overview.

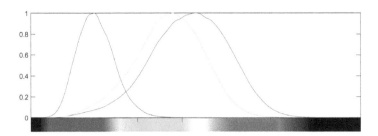

Figure 8.3: Cone sensitivity.

Table 8.1: Chromacy

Chromacy	Types of cone cells	Rough number of colours perceived	Examples
Monochromacy	1	100	Marine mammals
Dichromacy	2	10,000	Most terrestrial mammals
Trichromacy	3	10 million	Humans, great apes, some insects
Tetrachromacy	4	100 million	Most reptiles, amphibians, birds and insects, rarely humans
Pentachromacy	5	10 billion	Some insects (butterflies), some birds (pigeons).

We cannot imagine how a butterfly sees the world. Our brain has never experienced the extra information and thus has not learned how to interpret it.

There are three primary colours: red, green, and blue. All other colours are mixed from these. In computer imaging, they are represented in an array of bytes taking values between 0 and 255: $[r, g, b]$, where red $= (255, 0, 0)$, green $= (0, 255, 0)$, and blue $= (0, 0, 255)$. Linearisation means multiplications by a number and additions are allowed. For example

- red + green $= (255, 255, 0) =$ yellow,

- red + 0.5∗ green $= (255, 128, 0) =$ orange,

- red + 0.5∗ blue $= (255, 0, 128) =$ pink,

- 0.6∗ red + 0.8∗ blue $= (153, 0, 204) =$ purple.

In regression we estimate the relationship between variables. Or in other words, given the intensity of red, green, and blue for some samples, what is the relationship between intensity values and the colour? Can we predict the colour given the intensity values, as our eyes and brain do?

Regression has its applications in

- Computer Vision: e.g. image compression and restoration,

- Engineering: e.g. machine degradation,

- Medicine: e.g. epidemiology, mammography,

- Finance: e.g. volatility prediction, pricing models,

- Econometrics: e.g. cost and benefit optimization,

- Hydrology: e.g. flow prediction, ground water level forecasting,

- Seismology: e.g. soil liquefaction, seismic surveys.

8.1   Problem Description

Given scalar measurements $t_1, \ldots, t_N$, each measurement depends on parameters we know, $\mathbf{x}_1, \ldots, \mathbf{x}_N$, the intensities of red, green, blue in the previous example. These are quantities which can be measured with more or less effort, for example by photocells, geophones, hydrophones, PET and MRI neuroimaging, EEG technology, etc. They are also known as *targets* the machine learning technique has to achieve in its task to create a model. In the following it is assumed that the measurements are *mean centred*, i.e. $1/N(t_1 + \cdots + t_N) = 0$. This eliminates the need for a bias in the model. The *bias* is a constant which when added to all measurements makes them mean centred.

The measurements also depend on parameters we do not know. Any real world application depends on factors which cannot be measured. Or these measurements would be disproportionally difficult, costly or invasive.

We assume that the measurements are the result of underlying processes following some laws. In some applications, especially physical ones, these laws are known, and for example described by partial differential equations, but the specific parameters are not known. Sometimes nothing is known and we have to try to find this out. For applications which incorporate human interactions such as the financial markets, it is inherently difficult to infer underlying processes.

For example, the physics of waves are well understood. However, they depend on the mixture of media the wave travels in, the materials and their properties, as well as their interfaces. These are the unknown parameters of the process.

If we had a solution to the underlying process, we could predict the measurement from a function $t(\mathbf{x})$ as

$$t_n = t(\mathbf{x}_n),$$

where parameters of the function depend on the process. On the other hand, if we had a set of candidate functions $d_1(\mathbf{x}), \ldots, d_M(\mathbf{x})$, we could try which fits the measurements, and thus infer the underlying process. We say the functions $d_1(\mathbf{x}), \ldots, d_M(\mathbf{x})$ form a *dictionary* and let

$$f(\mathbf{x}) = \sum_{m=1}^{M} c_m d_m(\mathbf{x}),$$

be an approximation to $t(\mathbf{x})$, where $c_1, \ldots, c_M$ are *coefficients*, which need to be determined. The dictionary defines a model space in which ideally the true solution $t(\mathbf{x})$ lies. That is the underlying process can be described entirely by the dictionary. In a less ideal world, the approximation $f(\mathbf{x})$ obtained in the model space shall be close enough to the true solution. We will refer to $f(\mathbf{x})$ as our *model* describing the data.

8.2 Linear Regression

Often the term linear regression refers to fitting a straight line to the data. More generally, however, the term linear means that the model is a sum of building blocks with suitable coefficients, a *linear combination*. This gives us flexibility in the choice of building blocks. They can be non-linear functions. It also gives us flexibility in the choice of noise model and in the method to determine the coefficients.

Linear regression developed in several fields. Therefore the terminology is varied. First there is the scalar function value t, also known as *dependent*, *endogenous, response, measured, criterion* variable, or *regressand* or *target*. In the previous example this is the colour.

The $d_1(\mathbf{x}), \ldots, d_M(\mathbf{x})$ are known as *independent, exogenous, input, explanatory, predictor variables* or *regressors*. Remember d_m is not a coordinate of a vector. It is a basis function.

The coefficients $c_1, \ldots, c_M$ are also referred to as *parameter vector* or *vector of weights*, also known as *effects, regression coefficients*. They can also be viewed as *latent* variables. Regression determines these by various techniques.

The relationship to the measurements is

$$t_n = f(\mathbf{x}_n) + \epsilon_n,$$

where ϵ_n is *noise* intrinsic to the measurement process and assumed to be *independent and identically distributed* (i.i.d.) following the normal distribution $\mathcal{N}(0, \sigma^2)$. This is known as *homoscedasticity*. Note that the assumption of the same constant noise variance for each measurement might be wrong for two reasons. Firstly, there might be different sources of the noise with different effects. In this case the noise should be modeled by a mixture of probability distributions. Secondly, restrictions on the data might make different noise distributions necessary. For example if it is known that the data is always positive, the noise variance for smaller values has to be smaller to ensure positive predicted values.

Now

$$t_n = f(\mathbf{x}_n) + \epsilon_n = \sum_{m=1}^{M} c_m d_m(\mathbf{x}_n) + \epsilon_n,$$

Let $\mathbf{D}$ be the matrix with entries

$$D_{n,m} = d_m(\mathbf{x}_n)$$

and let $\mathbf{t}^T = (t_1, \ldots, t_N)$, $\mathbf{c}^T = (c_1, \ldots, c_M)$ and $\boldsymbol{\epsilon}^T = (\epsilon_1, \ldots, \epsilon_N)$, then

$$\mathbf{t} = \mathbf{D}\mathbf{c} + \boldsymbol{\epsilon}.$$

The regression problem has become a system of linear equations. Solutions of these have been studied extensively; see for example [12]. However, it differs from a standard system of linear equations in that it incorporates a noise model.

The $N \times M$ matrix $\mathbf{D}$ is called the *design matrix*. The challenge is to find the dictionary of basis functions and the coefficients. Once these are found, predictions for new unseen data $\mathbf{x}$ can be made by

$$t = \sum_{m=1}^{M} c_m d_m(\mathbf{x}).$$

Or, by defining $\mathbf{d}(\mathbf{x})^T = (d_1(\mathbf{x}), \ldots, d_M(\mathbf{x}))$,

$$t = \mathbf{d}(\mathbf{x})^T \mathbf{c}.$$

In the following we assume that the dictionary of basis functions and thus the design matrix are fixed.

8.3 Polynomial Regression

Polynomial regression dates back to Lagrange (1805) and Gauss (1809). Fitting a line to the data is a special case of polynomial regression, where the polynomial is a linear one. Here t is modeled as a polynomial of degree $M - 1$. The dictionary is $d_1(x) = 1, d_2(x) = x, \ldots, d_{M-1}(x) = x^{M-2}, d_M(x) = x^{M-1}$ which form a basis of polynomials of degree $M - 1$. The design matrix is

$$\mathbf{D} = \begin{pmatrix} 1 & x_1 & \cdots & x_1^{M-2} & x_1^{M-1} \\ \vdots & \vdots & \ddots & \cdots & \vdots \\ 1 & x_N & \cdots & x_N^{M-2} & x_N^{M-1} \end{pmatrix}$$

Assuming no noise, the solution is unique if $N = M$.

Figure 8.4 shows polynomial regression for $y = \sin(x)$ for various degrees of polynomials and number of measurements which are equally spaced in $[0, 2\pi]$. In Figure 8.4a the regression solution is a horizontal line through zero. This is because the locations of measurements are ill chosen. For $N = 1$ the measurement point is at π, the middle of the interval where $\sin(\pi) = 0$ while for $N = 2$ the measurement points are at the ends of the interval and again $\sin(0) = \sin(2\pi) = 0$. For $N = 3$, the end points and the middle of the interval are used. With more measurement points and thus higher degrees of polynomials the reconstruction improves. It seems that a fifth degree polynomial describes the sine adequately, even though this is an infinite sum of odd

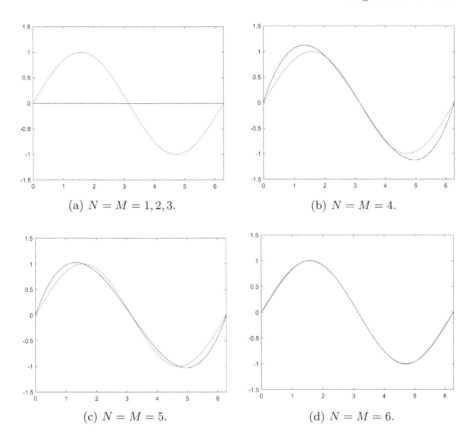

(a) $N = M = 1, 2, 3$.

(b) $N = M = 4$.

(c) $N = M = 5$.

(d) $N = M = 6$.

Figure 8.4: Polynomial regression for $y = \sin(x)$ (red). The black curve is the regression solution.

powers and thus lies in an infinite model space:

$$\sin(x) = \sum_{n=0}^{\infty} \frac{(-1)^n}{(2n+1)!} x^{2n+1}$$

However, this is only the case because we are considering the interval $[0, 2\pi]$, where the convergence properties of the infinite sum are very good, because the terms in the sum become small quickly due to the factorial in the denominator. Already after three terms the approximation to the true value $\sin(x)$ is adequate. If, however, when $x = 20$ for example, the powers first grow much quicker than the factorial. The 26th term in the sum is the first one with an absolute value less than one. The solution here is to use the identity $\sin(x \pm 2\pi) = \sin(x)$ to shift the argument into the area of good convergence. However, this uses knowledge about the underlying process, which in many real world applications is not available. Sometimes it exists in the form of professional experience, which is difficult to quantify.

This shows that the kind and location of measurements are an important consideration in regression.

Could we have been more clever in the design phase by using knowledge about the problem? Firstly, $\sin(0) = 0$. This means that the constant 1 is not necessary as regressor. Secondly, $\sin(x)$ is an odd function, that is $\sin(-x) = -\sin(x)$. Therefore the regressors should also be odd polynomials. Hence the polynomials with even exponent are not required. A better choice might have been

$$\mathbf{D} = \begin{pmatrix} x_1 & x_1^3 & \cdots & x_1^{2M-3} & x_1^{2M-1} \\ \vdots & \vdots & \ddots & \cdots & \vdots \\ x_N & x_N^3 & \cdots & x_N^{2M-3} & x_N^{2M-1} \end{pmatrix}$$

Still M regressors, but possibly higher accuracy.

Therefore the choice of regressors depends on the problem.

8.4 Ordinary Least Squares

Recall that we assume the dictionary of basis functions is fixed. How best to solve $\mathbf{t} = \mathbf{Dc} + \boldsymbol{\epsilon}$ generally? One approach is to consider the probabilistic nature of the noise. Remember, the ϵ_i are independent and identical normally distributed random variables with zero mean and variance σ^2. The *likelihood* of observing $\mathbf{t}$ given the model specified by $\mathbf{D}$, $\mathbf{c}$ and σ^2 is

$$\mathcal{L}(\mathbf{t}|\mathbf{D}, \mathbf{c}, \sigma^2) = (2\pi\sigma^2)^{-N/2} \exp\left(-\frac{1}{2\sigma^2}(\mathbf{t} - \mathbf{Dc})^T(\mathbf{t} - \mathbf{Dc})\right).$$

The likelihood quantifies whether the difference between the predictions $\mathbf{Dc}$ and the targets $\mathbf{t}$ can be explained by the noise.

Since the formula involves the exponential function, it makes sense to consider the logarithm of the likelihood known as the *log likelihood*:

$$\log \mathcal{L}(\mathbf{t}|\mathbf{D}, \mathbf{c}, \sigma^2) = -\frac{N}{2}\log 2\pi\sigma^2 - \frac{1}{2\sigma^2}(\mathbf{t} - \mathbf{Dc})^T(\mathbf{t} - \mathbf{Dc}) \quad (8.1)$$

A suitable choice for $\mathbf{c}$ is the one which maximizes the log likelihood. Since the first term is independent of $\mathbf{c}$, this is equivalent to minimizing

$$(\mathbf{t} - \mathbf{Dc})^T(\mathbf{t} - \mathbf{Dc}) = \|\mathbf{t} - \mathbf{Dc}\|^2 = \sum_{n=1}^N r_n^2 = \sum_{n=1}^N (t_n - \mathbf{d}(\mathbf{x}_n)^T\mathbf{c})^2,$$

where $\|\cdot\|$ denotes the *Euclidean norm* or L_2 *norm* and where $\mathbf{d}(\mathbf{x}_n)^T = (d_1(\mathbf{x}_n), \ldots, d_M(\mathbf{x}_n))^T$. The expression $r_n = t_n - \mathbf{d}(\mathbf{x}_n)^T\mathbf{c}$ is called the nth *residual*. It is the difference between the nth target and the prediction for it. Note that $\mathbf{d}(\mathbf{x}_n)^T$ is the nth row of $\mathbf{D}$. The sum is known as *sum of squared residual (SSR)*, *error sum of squares (ESS)*, or *residual sum of squares (RSS)*. The technique is known as *Ordinary Least Squares (OLS)*, since a sum of squares is minimized.

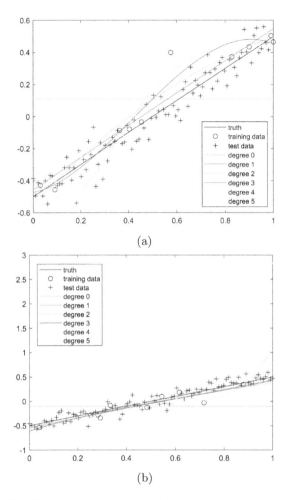

(a)

(b)

Figure 8.5: Examples of polynomial regression for $y = x - 0.5$.

Writing

$$(\mathbf{t} - \mathbf{Dc})^T(\mathbf{t} - \mathbf{Dc}) = \mathbf{c}^T\mathbf{D}^T\mathbf{Dc} - 2\mathbf{c}^T\mathbf{D}^T\mathbf{t} + \mathbf{t}^T\mathbf{t},$$

we can differentiate with regards to $\mathbf{c}$ using Appendix A.2.3 and set to zero to find the extremum,

$$\mathbf{D}^T\mathbf{Dc} = \mathbf{D}^T\mathbf{t}. \tag{8.2}$$

Since $\mathbf{D}$ is a $N \times M$ matrix, $\mathbf{D}^T\mathbf{D}$ is an $M \times M$ matrix. If the inverse $(\mathbf{D}^T\mathbf{D})^{-1}$ exists, then the choice for $\mathbf{c}$ is

$$\mathbf{c} = (\mathbf{D}^T\mathbf{D})^{-1}\mathbf{D}^T\mathbf{t}.$$

However, even if the inverse exists, the inversion might be ill-conditioned leading to large numerical errors. Numerical analysis provides many methods to solve equation (8.2) safely; see for example [12].

We now have a model of the data specified by $\mathbf{D}$ and $\mathbf{c}$. For $\mathbf{x}_1, \ldots, \mathbf{x}_N$ the predicted values $\mathbf{f} = (f_1, \ldots, f_N)^T$ are given by

$$\mathbf{f} = \mathbf{D}\mathbf{c} = \mathbf{D}(\mathbf{D}^T\mathbf{D})^{-1}\mathbf{D}^T\mathbf{t}.$$

Geometrically, this is the projection of $\mathbf{t}$ onto the space spanned by the columns of $\mathbf{D}$. If $\mathbf{t}$ already lies in that space, then $\mathbf{f} = \mathbf{t}$. In this case, the regressors are perfectly suited to predict the data. For the minimal value of $\mathbf{c}$ the error sum of squares takes the value

$$
\begin{aligned}
\sum_{n=1}^{N} r_n^2 &= (\mathbf{t} - \mathbf{D}(\mathbf{D}^T\mathbf{D})^{-1}\mathbf{D}^T\mathbf{t})^T(\mathbf{t} - \mathbf{D}(\mathbf{D}^T\mathbf{D})^{-1}\mathbf{D}^T\mathbf{t}) \\
&= \mathbf{t}^T\mathbf{t} - \mathbf{t}^T\mathbf{D}(\mathbf{D}^T\mathbf{D})^{-1}\mathbf{D}^T\mathbf{t} = \mathbf{t}^T(\mathbf{t} - \mathbf{D}\mathbf{c}).
\end{aligned}
$$

It is the inner product between $\mathbf{t}$ and the error the model makes when predicting the data.

The regression problem can be viewed as finding an approximate solution in a finite dimensional space spanned by the basis functions, which is in some sense closest to to the true solution in a possibly infinite dimensional space. A projection gives the closet solution with respect to the *Euclidean norm* $\| \cdot \|$, also known as the L_2 *norm*.

8.5   Over- and Under-fitting

Figure 8.5 shows polynomial reconstructions of various degrees using data generated from the simple line $y = x - 0.5$ with added noise. The noise is modeled as a mixture of two probabilities, in order to generate outliers. The MATLAB code in Listing 8.1 gives details regarding how the data was generated.

The technique employed to fit the polynomials was ordinary least squares. The training data are marked by ∘. A small training set of size 10 was deliberately chosen to illustrate various effects. Also 90 test data were generated and are marked by +. One can see that some reconstructions are preferable over others.

Under-fitting occurs if there are not enough explanatory variables to explain the data. This can be seen in Figures 8.5a, and 8.5b, with the polynomial of degree 0. Note that the horizontal line is put at the mean of the training data, which is close to 0, since this is the mean of the true solution. Since the data comes from a linear degree polynomial, a polynomial of degree at least 1 is necessary to describe it. If this is not the case, the predictors fail to capture the complexity of the underlying process. The solution is to amend the dictionary of basis functions.

```
function [x,y] = data(f,x1,x2,N,s,sigma0,sigma1)
% This function creates N noisy data pairs (x,y) from a function f
% in the interval [x1,x2] with added noise where the noise is a
% mixture of zero mean normal distributions with variances sigma0
% and sigma1 with probability s and 1−s.
% Input arguments:
%    f,        function handle
%    x1,x2,    end points of interval
%    N,        number of data pairs to be generated
%    s,        mixing probability of error distributions
%    sigma0, standard deviation of first error distribution
%    sigma1, standard deviation of second error distribution
x = linspace(x1,x2,N);
y = zeros(1,N);
for i = 1:N
    r = rand;
    if (r≤s)
       y(i) = f(x(i)) + randn*sigma1;
    else
       y(i) = f(x(i)) + randn*sigma0;
    end
end
```

Listing 8.1: Generating noisy data.

The question is, how complex should the dictionary we choose be? Or in other words, what should be the dimension of the model space? With too much complexity, other effects come into play. One known as *over-fitting* is modeling noise as well as the process. In Figure 8.5, the higher order polynomials try to follow the training data more closely, and to achieve this they sometimes have to bend more, taking the curve away from the true solution. This is especially noticeable with the polynomial of degree 5 in Figure 8.5b.

A further problem occurs if the regressors are not suitably chosen. For example, high degree polynomials are unsuitable, since the solutions diverge from the true solution, especially at the end of intervals, as Figure 8.5a shows at the right hand side. Every training data point can be viewed as a carrier of information. If the approximation in the middle of the interval is evaluated, it benefits from the fact that information flowed from training data points from both sides. At the end of the interval information can only arrive from one side. Higher degree polynomials fan away from the true solution at the end of the interval. The effect is more dominant, the further away from training data one is. The approximations of degree 4 or below in Figure 8.5b benefit from the fact that a training data point is close to each end of the interval.

In this context, the *law of parsimony* states that if two models explain the data equally well, the one with less complexity should be chosen. This is also known as *Occam's razor*. William Ockham (also Occam) writes in [34]

```
% This script generates noisy data from a polynomial and splits this
% into test and training data. It fits polynomials of various degrees
% to it using least squares regression. The line, training and test
% data are plotted along with the fitted polynomials. The absolute
% training and test errors together with their variances are also
% plotted.

pt = [1 -0.5];    % target polynomial
t = poly2sym(pt);
t = matlabFunction(t);
x1 = 0;
x2 = 1;
N =100;    % number of data pairs
sig = 0.2;   % mixing probability of error distributions
sigma0 = 0.1;   % standard deviation of first error distribution
sigma1 = 0.05;   % standard deviation of second error distribution
% generate data
[x,y] = data(t,x1,x2,N,sig,sigma0,sigma1);
% choose training set
K = 10;     % size of training set
trainindex = randi([1 N],1,K);
trainx = x(trainindex);
trainy = y(trainindex);
% let other data be test set
testx = x;
testx(trainindex) = [];
testy = y;
testy(trainindex) = [];

figure;
plot(x, t(x), 'k', 'DisplayName','truth');
hold on;
plot(trainx, trainy, 'ko', 'DisplayName','training data');
plot(testx, testy, 'k+', 'DisplayName','test data');

D = 6;  % D-1 highest degree of fitted polynomial
trainmean = zeros(1,D);
trainvar = zeros(1,D);
testmean = zeros(1,D);
testvar = zeros(1,D);
for d = 1:D
    p = polyfit(trainx,trainy,d-1);% returns the coefficients for a
                                   % polynomial p of degree i-1 that
                                   % is a best fit (in a least-squares
                                   % sense) for the data
    trainerror = (trainy - polyval(p, trainx)).^2; % squared error
    trainmean(d) = mean(trainerror);
    trainvar(d) = var(trainerror);
    testerror = (testy - polyval(p, testx)).^2;
    testmean(d) = mean(testerror);
    testvar(d) = var(testerror);
    y1 = polyval(p,x);
    plot(x,y1,'DisplayName',['degree ' num2str(d-1)]);
end
legend('show', 'Location', 'southeast');
figure;
errorbar(trainmean,trainvar, 'Displayname', 'training error');
hold on;
errorbar(testmean,testvar, 'Displayname', 'test error');
legend('show');
xticks(1:1:D);
xticklabels(0:1:D-1);
xlim([0 D+1]);
```

Listing 8.2: Polynomial regression.

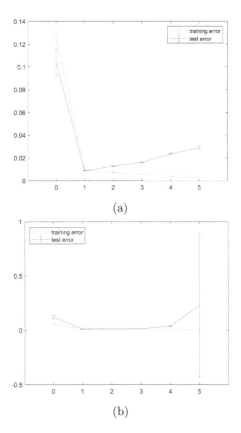

(a)

(b)

Figure 8.6: Mean squared test errors and their standard deviations for the examples in Figure 8.5.

Numquam ponenda est pluralitas sine necessitate.

(Plurality is never to be posited without necessity.)

The razor refers to 'shaving away' unnecessary complexity. To decide what is necessary and what is unnecessary, or perhaps even detrimental to a good reconstruction, the test error is employed. The MATLAB code in Listing 8.2 generates the data, performs polynomial OLS regression for different degrees of polynomials, and plots both mean squared training and test error together with their variances.

Figure 8.6 shows the mean squared training and test errors and their variances of the examples given in Figure 8.5 against the degree of polynomials which serves as a descriptor of the complexity. More generally, the dimension of the model space describes the complexity. Figure 8.6a shows clearly that introducing any further complexity beyond a linear polynomial generalizes

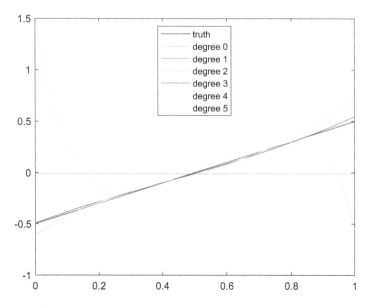

Figure 8.7: Average polynomial approximations.

badly to unseen test data, while in Figure 8.6b it is not so clear whether polynomials of degree 1, 2, 3, or 4 are best chosen.

8.6 Bias and Variance

Both example data sets in the previous section were generated from the same underlying process. They are two different snapshots of the process. The reconstructions are inconclusive regarding which model space should be chosen. Taking many snapshots, the picture becomes clearer. For example, running the experiment generating noisy data and fitting polynomials of various degrees 100 times results in the following average mean squared training and test errors and variances:

degree of polynomial	mean squared training error	variance of squared training error	mean squared test error	variance of squared test error
0	0.82	$8.6 * 10^{-3}$	1.0	$13 * 10^{-3}$
1	0.076	$0.14 * 10^{-3}$	0.1	$0.24 * 10^{-3}$
2	0.066	$0.11 * 10^{-3}$	0.15	$1.3 * 10^{-3}$
3	0.057	$0.083 * 10^{-3}$	0.62	0.27
4	0.045	$0.064 * 10^{-3}$	7.5	100
5	0.038	$0.053 * 10^{-3}$	24	$156 * 10^{3}$

We can see that the model space of linear polynomials generalizes best to unseen data, since the mean squared test error is minimal there. We can also see that the variance in the mean squared test error increases significantly with increasing complexity that is the degree of polynomial.

Since we have averaged over 100 experiments, we can look at the average polynomial fitted. They are shown in Figure 8.7. It can clearly be seen that the higher degree average polynomials fan out at the edges of the interval. The coefficients of the average polynomials are given in the following table:

degree	x^5	x^4	x^3	x^2	x	1
0	0	0	0	0	0	−0.011
1	0	0	0	0	1.0	−0.501
2	0	0	0	0.079	0.92	−0.49
3	0	0	0.50	−0.61	1.1	−0.49
4	0	−1.5	4.8	−4.2	2.3	−0.61
5	−63	163	−164	81	−19	1.5

The coefficients of the linear average polynomial are nearly the correct solution $x - 0.5$, and even the quadratic polynomial gets close to the true solution with a relatively small coefficient of x^2. It is worth looking at the variance of the coefficients:

degree	x^5	x^4	x^3	x^2	x	1
0	0	0	0	0	0	0.0092
1	0	0	0	0	0.011	0.0035
2	0	0	0	0.42	0.51	0.033
3	0	0	0.50	31.9	9.03	0.3
4	0	1227	5030	2869	324.1	5.36
5	338000	2016000	1829000	395848	20873	182.1

With increasing complexity of the model space, the models for given data sets become very different, showing that the models are susceptible to small changes in the training data. Two competing principles are at work, fitting a model well and avoiding too much complexity which leads to over-fitting and a large variance in the models. These can be separated formally.

We need to generalize to all possible data sets. This means looking at the expected error that a particular model space gives rise to. Given a model space, let $f(\mathbf{x})$ be the approximation to the underlying process generating the data which is described by $t(\mathbf{x})$. It is common to look at the expected squared error,

$$\mathbb{E}\left[(t - f(\mathbf{x}))^2\right],$$

where $t = t(\mathbf{x}) + \epsilon$. In the following, this is decomposed into different error components.

Now $t(\mathbf{x})$ is a deterministic process, and thus $\mathbb{E}[t(\mathbf{x})] = t(\mathbf{x})$, which implies

$\mathbb{E}[t] = \mathbb{E}[t(\mathbf{x}) + \epsilon] = t(\mathbf{x})$, since the noise has zero mean. Similarly,

$$
\begin{aligned}
\mathrm{var}[t] &= \mathbb{E}\left[(t - \mathbb{E}[t])^2\right] = \mathbb{E}\left[(t - t(\mathbf{x}))^2\right] = \mathbb{E}\left[(t(\mathbf{x}) + \epsilon - t(\mathbf{x}))^2\right] \\
&= \mathbb{E}\left[\epsilon^2\right] = \mathrm{var}\left[\epsilon\right] + \mathbb{E}\left[\epsilon\right]^2 = \sigma^2.
\end{aligned}
$$

Contrary to $t(\mathbf{x})$, $f(\mathbf{x})$ is not deterministic, because it depends on the data set which was used to calculate the model. However, it is independent of the noise ϵ, which leads to

$$
\begin{aligned}
\mathbb{E}\left[(t - f(\mathbf{x}))^2\right] &= \mathbb{E}\left[t^2 + f(\mathbf{x})^2 - 2tf(\mathbf{x})\right] \\
&= \mathbb{E}\left[t^2\right] + \mathbb{E}\left[f(\mathbf{x})^2\right] - 2\mathbb{E}\left[tf(\mathbf{x})\right] \\
&= \mathrm{var}\left[t\right] + \mathbb{E}\left[t\right]^2 + \mathrm{var}\left[f(\mathbf{x})\right] + \mathbb{E}\left[f(\mathbf{x})\right]^2 - 2\mathbb{E}\left[tf(\mathbf{x})\right] \\
&= \sigma^2 + t(\mathbf{x})^2 + \mathrm{var}\left[f(\mathbf{x})\right] + \mathbb{E}\left[f(\mathbf{x})\right]^2 - 2t(\mathbf{x})\mathbb{E}\left[f(\mathbf{x})\right] \\
&= \sigma^2 + \mathrm{var}\left[f(\mathbf{x})\right] + \left(\mathbb{E}\left[f(\mathbf{x})\right] - t(\mathbf{x})\right)^2.
\end{aligned}
$$

The first term is the irreducible error due to the noise in the data. It is also the minimum achievable error. The second term is the *variance*. It gives the extent to which different solutions arising from different snapshots of data vary around their mean. Or in other words, how sensitive the model is to a particular data set. The last term is the square of the *bias*. The bias represents the extent to which the average prediction over all snapshots differs from the true process. In other words a small bias indicates that the generated models fit the process well, while small variance indicates that the models are robust to perturbations in the data. In general, making one of them smaller increases the other. This is called the *bias-variance trade-off*

Note that the analysis so far was only concerned with one particular unseen data point $\mathbf{x}$. To make a general statement about the goodness of the model, we need to take the expectation with regards to $\mathbf{x}$.

$$
\int \mathbb{E}\left[(t - f(\mathbf{x}))^2\right] p(\mathbf{x})d\mathbf{x} = \int \left(\sigma^2 + \mathrm{var}\left[f(\mathbf{x})\right] + \left(\mathbb{E}\left[f(\mathbf{x})\right] - t(\mathbf{x})\right)^2\right) p(\mathbf{x})d\mathbf{x}.
$$

This simplifies greatly, if $p(\mathbf{x})$ is a uniform distribution. That is measurements are equally likely taken everywhere.

In general, it is not possible to calculate $\mathbb{E}\left[f(\mathbf{x})\right]$ and $\mathrm{var}\left[f(\mathbf{x})\right]$. But it can be estimated, if for example multiple runs of the experiment are possible. For the 100 runs above, the approximate integrated squared bias, integrated variance and their sums are given in the following table:

degree	integrated squared bias	integrated variance	sum
0	$83.5 * 10^3$	$0.047 * 10^{-9}$	$83.5 * 10^3$
1	$2.7 * 10^{-6}$	$7.4 * 10^{-9}$	$2.75 * 10^{-6}$
2	$35.8 * 10^{-6}$	$0.78 * 10^{-6}$	$36.6 * 10^{-6}$
3	$0.23 * 10^{-3}$	$0.14 * 10^{-3}$	$0.37 * 10^{-3}$
4	$3.4 * 10^{-3}$	0.11	0.11
5	0.23	309	309

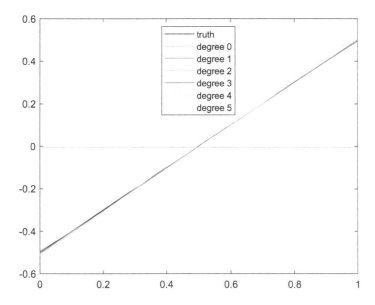

Figure 8.8: Average polynomial approximations.

These results also show that linear polynomials are the right model space to choose, since the sum of integrated squared bias and variance is smallest there.

To summarize, high bias and low variance indicate that the model space cannot capture the underlying process and is too restrictive. Low bias and low variance show the model space is adequately describing the underlying process, and does so for differing data sets. Low bias and high variance mean that with a good distribution of training data, the predictions are good, but other training data might result in poorer predictions. High bias and high variance are altogether undesirable.

As mentioned before, the training data set was chosen deliberately small for illustration purposes. Doubling the training data set size to 20 in our experiments causes the average polynomial approximation of degree 1 or more to be closer to $x - 0.5$ when plotted (Figure 8.8), especially in the middle of the interval. Polynomial approximation deteriorates towards the ends of the interval, since locally less and less information flows from one side and most of the information flows from the other, skewing the prediction. The following table gives the coefficients:

degree	x^5	x^4	x^3	x^2	x	1
0	0	0	0	0	0	-0.006
1	0	0	0	0	1.007	-0.506
2	0	0	0	-0.011	1.017	-0.507
3	0	0	-0.118	-0.189	0.918	-0.494
4	0	0.1143	-0.103	-0.019	1.046	-0.514
5	-3.325	7.935	-6.383	1.870	0.952	-0.539

8.7 Cross-validation

The analysis of the previous section relied on the fact that experiments are repeatable, and thus data can easily be generated. However, often the acquisition of data is associated with substantial costs, and it is essential to use data well. In *cross-validation*, the available data are partitioned into complementary subsets, one for training and one *validation set*. Multiple rounds of cross-validation are performed, and the results are averaged over the rounds. This way the goodness of the model can be assessed. There are two types of cross-validation: *exhaustive* and *non-exhaustive*. The former examines all possible ways to divide the original data set into a training and a validation set.

Leave-p-out cross-validation sets aside p samples for validation as the names suggests. All possibilities to choose p validation samples from n samples are considered. This means the regression algorithm has to be run $\frac{n!}{p!(n-p)!}$ times which even for moderate n and p can become computationally prohibitive. *Leave-one-out cross-validation* is a special case of leave-p-out cross-validation with $p = 1$. In this case the algorithm is run n times.

In *k-fold cross-validation* the data set is randomly subdivided into k equal sized subsets. One of these is used as the validation set while the other $k - 1$ form the training data set. The model generation is repeated k times (the *folds*) with each of the subsets being the validation set exactly once. The k results are then averaged for assessment. All data samples are used for both training and validation, and each sample is used exactly once for validation. When $k = n$, k-fold cross-validation is the same as leave-one-out cross-validation.

8.8 Multicollinearity and Principal Component Regression

Multicollinearity is another problem which arises when one or more predictor variables are highly correlated. In this case, one of the basis functions can be modeled by the others. One dimension of the model space is redundant. A high degree of correlation increases the variance in the coefficients, since different models are equivalent. They all model the measurements, but differ in the coefficients. As a consequence, small changes in the input data can lead to large changes in the model. *Perfect multicollinearity* means the regressors are linearly dependent, that is one can be exactly expressed by the others. In this case $(\mathbf{D}^T\mathbf{D})^{-1}$ does not exist.

In a good regression model, the regressors correlate minimally with each other, but are each highly correlated with the regressand. The aim is to find a linear combination of few regressors which summarize and explain the data without too much loss of information. This section and the following show how to obtain new regressors from the dictionary.

Principal component regression uses the principal components of $\mathbf{D}$ as regressors instead of the columns of $\mathbf{D}$. In particular, let $\mathbf{d}_1, \ldots, \mathbf{d}_M$ denote the columns of $\mathbf{D}$. That is $\mathbf{d}_i$ is the ith basis function evaluated at $\mathbf{x}_1, \ldots, \mathbf{x}_n$. In the following, we assume that the columns are standardized; that is they have mean 0 and length 1. The first assumption is valid, since the measurements are mean centred, and the second assumption is valid, since regressors are invariant to scaling. Any scaling will be absorbed in the multiplicative coefficient.

The correlation between the ith and j^{th} regressors evaluated at the N measuring points is then

$$\mathbf{d}_i^T \mathbf{d}_j.$$

The correlation matrix of all regressors is $\mathbf{D}^T \mathbf{D}$.

A new set of regressors is generated as linear combinations of basis functions from the dictionary, say

$$v_1 d_1(\mathbf{x}) + \ldots + v_M d_M(\mathbf{x}).$$

Evaluating this at the N sampling points, we arrive at a linear combination of the columns of $\mathbf{D}$,

$$v_1 \mathbf{d}_1 + \ldots + v_M \mathbf{d}_M = \mathbf{D}\mathbf{v},$$

where $\mathbf{v} = (v_1, \ldots, v_M)^T$ is chosen such that $\mathbf{D}\mathbf{v} \neq 0$. The zero vector would make a very poor regressor.

The correlation between two such linear combinations is

$$\frac{\mathbf{v}_1^T \mathbf{D}^T \mathbf{D} \mathbf{v}_2}{\|\mathbf{D}\mathbf{v}_1\| \|\mathbf{D}\mathbf{v}_2\|},$$

where $\mathbf{v}_1$ and $\mathbf{v}_2$ are such that $\mathbf{D}\mathbf{v}_1 \neq 0$ and $\mathbf{D}\mathbf{v}_2 \neq 0$. Here we need to divide by the lengths, since $\mathbf{D}\mathbf{v}_1$ and $\mathbf{D}\mathbf{v}_2$ no longer will have length 1.

Now, the matrix $\mathbf{D}^T \mathbf{D}$ is symmetric and positive semi-definite. Hence it has M non-negative eigenvalues and corresponding orthonormal eigenvectors. Thus choosing $\mathbf{v}_1$ and $\mathbf{v}_2$ to be eigenvectors of $\mathbf{D}^T \mathbf{D}$, the new regressors are uncorrelated, since

$$\mathbf{v}_1^T \mathbf{D}^T \mathbf{D} \mathbf{v}_2 = \lambda_2 \mathbf{v}_1^T \mathbf{v}_2 = 0.$$

The eigenvectors corresponding to K nonzero eigenvalues are chosen. The new design matrix $\hat{\mathbf{D}}$ has columns $\mathbf{D}\mathbf{v}_1, \ldots, \mathbf{D}\mathbf{v}_K$. The corresponding functions span a K-dimensional subspace of the original model space, and are uncorrelated.

Let $\mathbf{V}$ be the matrix with columns $\mathbf{v}_1, \ldots, \mathbf{v}_K$. Then $\hat{\mathbf{D}} = \mathbf{D}\mathbf{V}$. The predictions are

$$\mathbf{f} = \hat{\mathbf{D}}(\hat{\mathbf{D}}^T \hat{\mathbf{D}})^{-1} \hat{\mathbf{D}}^T \mathbf{t} = \mathbf{D}\mathbf{V}(\mathbf{V}^T \mathbf{D}^T \mathbf{D} \mathbf{V})^{-1} \mathbf{V}^T \mathbf{D}^T \mathbf{t}.$$

Now $\mathbf{D}^T\mathbf{D}\mathbf{V}$ is the matrix with columns $\lambda_1\mathbf{v}_1,\ldots,\lambda_K\mathbf{v}_K$, since $\mathbf{v}_1,\ldots,\mathbf{v}_K$ are eigenvectors of $\mathbf{D}^T\mathbf{D}$. It then follows from the orthonormality of $\mathbf{v}_1,\ldots,\mathbf{v}_K$ that

$$
(\mathbf{V}^T\mathbf{D}^T\mathbf{D}\mathbf{V})^{-1} = \begin{pmatrix} \frac{1}{\lambda_1} & 0 & \cdots & 0 \\ 0 & \ddots & & \vdots \\ \vdots & & \ddots & 0 \\ 0 & \cdots & 0 & \frac{1}{\lambda_K} \end{pmatrix}.
$$

With this result, we obtain

$$
\begin{aligned}
\mathbf{f} &= \begin{pmatrix} \mathbf{D}\mathbf{v}_1 & \cdots & \mathbf{D}\mathbf{v}_k \end{pmatrix} \begin{pmatrix} \frac{1}{\lambda_1} & 0 & \cdots & 0 \\ 0 & \ddots & & \vdots \\ \vdots & & \ddots & 0 \\ 0 & \cdots & 0 & \frac{1}{\lambda_K} \end{pmatrix} \begin{pmatrix} (\mathbf{D}\mathbf{v}_1)^T \\ \vdots \\ (\mathbf{D}\mathbf{v}_K)^T \end{pmatrix} \mathbf{t} \\
&= \sum_{k=1}^{K} \frac{(\mathbf{D}\mathbf{v}_k)^T\mathbf{t}}{\lambda_k}\mathbf{D}\mathbf{v}_k.
\end{aligned}
$$

Geometrically, this is the projection of the regressand $\mathbf{t}$ onto the space spanned by the columns of the new design matrix $\hat{\mathbf{D}}$. The coefficient for the new regressor $\mathbf{D}\mathbf{v}_i$ is $c_i = (\mathbf{D}\mathbf{v}_i)^T\mathbf{t}/\lambda_i$ in this model.

To analyze further, we decompose the regressand $\mathbf{t}$ into one portion lying in the subspace spanned by $\mathbf{D}\mathbf{v}_1,\ldots,\mathbf{D}\mathbf{v}_K$ and a remainder,

$$
\mathbf{t} = \sum_{k=1}^{K} a_k\mathbf{D}\mathbf{v}_k + \mathbf{a},
$$

where $\mathbf{a}$ is orthogonal to $\mathbf{D}\mathbf{v}_1,\ldots,\mathbf{D}\mathbf{v}_K$. We then have

$$
c_i = \frac{1}{\lambda_i}(\mathbf{D}\mathbf{v}_i)^T\left(\sum_{k=1}^{K} a_k\mathbf{D}\mathbf{v}_k + \mathbf{a}\right) = \frac{1}{\lambda_i}\sum_{k=1}^{K} a_k\mathbf{v}_i^T\mathbf{D}^T\mathbf{D}\mathbf{v}_k = a_i,
$$

as expected.

We can calculate the distance between $\mathbf{t}$ and $\mathbf{f}$,

$$
\begin{aligned}
\|\mathbf{t} - \mathbf{f}\|^2 &= \mathbf{t}^T\mathbf{t} - 2\left(\sum_{i=1}^{K} \frac{(\mathbf{D}\mathbf{v}_i)^T\mathbf{t}}{\lambda_i}\mathbf{D}\mathbf{v}_i\right)^T \mathbf{t} \\
&\quad + \sum_{i,j=1}^{K} \frac{(\mathbf{D}\mathbf{v}_i)^T\mathbf{t}}{\lambda_i}\frac{(\mathbf{D}\mathbf{v}_j)^T\mathbf{t}}{\lambda_j}(\mathbf{D}\mathbf{v}_i)^T\mathbf{D}\mathbf{v}_j \\
&= \mathbf{t}^T\mathbf{t} - \sum_{i=1}^{K} \frac{((\mathbf{D}\mathbf{v}_i)^T\mathbf{t})^2}{\lambda_i} = \|\mathbf{a}\|^2,
\end{aligned}
$$

where we used the fact that $(\mathbf{D}\mathbf{v}_i)^T\mathbf{D}\mathbf{v}_j = \lambda_j$, if $i = j$ and zero otherwise. Thus the predictions are closest, if all eigenvectors with non-zero eigenvalues are used. Using the decomposition of $\mathbf{t}$, we have

$$(\mathbf{D}\mathbf{v}_i)^T\mathbf{t} = \sum_{k=1}^{K} a_k \mathbf{v}_i^T \mathbf{D}^T \mathbf{D}\mathbf{v}_k + \mathbf{v}_i^T \mathbf{D}^T \mathbf{a} = a_i \lambda_i,$$

where we again made use of the orthogonality. Inserting this into the distance between $\mathbf{t}$ and $\mathbf{f}$, we obtain

$$\|\mathbf{t} - \mathbf{f}\|^2 = \mathbf{t}^T\mathbf{t} - \sum_{i=1}^{K} a_i^2 \lambda_i.$$

If sparsity is required and not all eigenvectors can be used, those for which $a_i^2\lambda_i$ is largest should be chosen. However, this implicitly depends on the data $\mathbf{t}$ via the coefficient a_i, and can cause the model to not generalize well to unseen data. To avoid this, it is customary to choose the eigenvectors with the largest eigenvalues. It should be noted here that determining the eigenvectors and eigenvalues of a matrix is not a trivial matter in itself, especially if M is large.

Note that generating new regressors is equivalent to generating new features. These new features are linear combinations of the original features.

Principal component regression does not address the correlation with the regressand. The method presented in the next section does so.

8.9 Partial Least Squares

Partial Least Squares (PLS) aims to maximize the correlation between regressors and regressand. This time a new set of regressors is generated iteratively as linear combinations of the original regressors; in the first iteration say

$$z_1\mathbf{d}_1 + \ldots + z_M\mathbf{d}_M = \mathbf{D}\mathbf{z}.$$

Without loss of generality, we can assume $\|\mathbf{D}\mathbf{z}\| = 1$, since regressors are invariant to scaling.

The matrix $\mathbf{D}^T\mathbf{t}\mathbf{t}^T\mathbf{D}$ is a symmetric, positive semi-definite $M \times M$ matrix, since

$$(\mathbf{D}^T\mathbf{t}\mathbf{t}^T\mathbf{D})^T = \mathbf{D}^T\mathbf{t}\mathbf{t}^T\mathbf{D} \text{ and } \mathbf{v}^T\mathbf{D}^T\mathbf{t}\mathbf{t}^T\mathbf{D}\mathbf{v} = (\mathbf{v}^T\mathbf{D}^T\mathbf{t})^2 \geq 0.$$

It has M non-negative eigenvalues and corresponding orthonormal eigenvectors $\mathbf{v}_1, \ldots, \mathbf{v}_M$, where the eigenvectors are ordered with regards to the corresponding eigenvalues from largest to smallest.

Now $\mathbf{z} \in \mathbb{R}^M$, and thus can be expressed as a linear combination of these eigenvectors:

$$\mathbf{z} = \hat{z}_1\mathbf{v}_1 + \ldots + \hat{z}_M\mathbf{v}_M.$$

The square of the correlation between $\mathbf{t}$ and the new regressor $\mathbf{Dz}$ is

$$\left(\frac{\mathbf{t}^T \mathbf{Dz}}{\|\mathbf{t}\|\|\mathbf{Dz}\|}\right)^2 = \frac{1}{\|\mathbf{t}\|^2} \mathbf{z}^T \mathbf{D}^T \mathbf{t}\mathbf{t}^T \mathbf{Dz} = \frac{1}{\|\mathbf{t}\|^2}\left(\lambda_1 \hat{z}_1^2 + \ldots \lambda_M \hat{z}_M^2\right).$$

This is maximal for $\hat{z}_2 = \ldots = \hat{z}_M = 0$, since the length of $\mathbf{z}$ is fixed by the condition $\|\mathbf{Dz}\| = 1$. Thus the first new regressor is $\mathbf{t}_1 = \mathbf{Dv}_1/\|\mathbf{Dv}_1\|$. We know that $\mathbf{Dv}_1 \neq 0$, since otherwise $\mathbf{D}^T \mathbf{t}\mathbf{t}^T \mathbf{D}$ would only have zero eigenvalues, since λ_1 is the largest of the non-negative eigenvalues.

Having generated $\mathbf{t}_1$, we calculate

$$\mathbf{D}_1 = \left(\mathbf{I} - \mathbf{t}_1 \mathbf{t}_1^T\right) \mathbf{D}.$$

Note

$$\mathbf{D}_1 \mathbf{v}_1 = \left(\mathbf{I} - \mathbf{t}_1 \mathbf{t}_1^T\right) \mathbf{Dv}_1 = \left(\mathbf{I} - \mathbf{t}_1 \mathbf{t}_1^T\right) \|\mathbf{Dv}_1\|\mathbf{t}_1 = \|\mathbf{Dv}_1\|(1 - \|\mathbf{t}_1\|^2)\mathbf{t}_1 = 0.$$

This implies that the rank of $\mathbf{D}_1$ is less than the rank of $\mathbf{D}$, because a vector $(\mathbf{v}_1)$ which previously was not mapped to zero, now is mapped to zero.

Now, let $\mathbf{v}_2$ with $\|\mathbf{v}_2\| = 1$ be the eigenvector corresponding to the largest eigenvalue of $\mathbf{D}_1^T \mathbf{t}\mathbf{t}^T \mathbf{D}_1$. The second new regressor is $\mathbf{t}_2 = \mathbf{D}_1 \mathbf{v}_2$ normalized such that $\|\mathbf{t}_2\| = 1$. Again, the correlation is maximal.

Next

$$\mathbf{D}_2 = \left(\mathbf{I} - \mathbf{t}_2 \mathbf{t}_2^T\right) \mathbf{D}_1,$$

and the process continues until $\mathbf{D}_r$ is a null matrix, i.e. its rank is zero. This will be achieved, since $\mathbf{D}_j$ maps $\mathbf{v}_j$ to zero, but $\mathbf{D}_{j-1}\mathbf{v}_j$ is non-zero, and hence $\text{rank}\mathbf{D}_j \leq \text{rank}\mathbf{D}_{j-1} - 1$.

PLS can be used for multivariate regression; that is the regressand has multiple components and measuring at $\mathbf{x}_1, \ldots, \mathbf{x}_N$ gives rise to a matrix $\mathbf{t}$ with N rows. However, this is beyond this text. More information can be found in [20].

8.10   Regularization

Recall, we are trying to find suitable coefficients $\mathbf{c}$ minimizing $\|\mathbf{t} - \mathbf{Dc}\|^2$ while avoiding unnecessary complexity which leads to over-fitting. This is achieved by introducing a *penalty term* $\Omega(\mathbf{c})$ and minimizing

$$\frac{1}{2}\|\mathbf{t} - \mathbf{Dc}\|^2 + \lambda\Omega(\mathbf{c}),$$

where $\Omega(\mathbf{c}) \geq 0$ for all $\mathbf{c}$ and $\Omega(\mathbf{0}) = 0$.

The penalty $\Omega(\mathbf{c})$ is also known as the *entropy measure*. The parameter λ controls the trade-off between fitting the data and reducing complexity and has to be chosen well. Often, a model is obtained for different choices of λ, and the goodness of the model is assessed on a *validation set*. Note that if

$\lambda = 0$, we obtain the ordinary least squares solution, while if $\lambda = \infty$, $\mathbf{c} = 0$. The zero vector of coefficients is the least complex one. For λ in between these two extremes, we are balancing between fitting a linear model and keeping the model complexity small. Note that we still explore the same model space as before, but the inclusion of $\Omega(\mathbf{c})$ gives preference to certain areas of the model space. The parameter λ controls how strong this preference is.

In the following, norms which are different to the Euclidean norm (L_2 norm) are distinguished by subscripts. The choices for $\Omega(\mathbf{c})$ are numerous:

$\Omega(\mathbf{c})$	Regression method
$\|\mathbf{c}\|_0$ = number of nonzero elements in $\mathbf{c}$	L_0 regularization
$\|\mathbf{c}\|_1$ = $\displaystyle\sum_{m=1}^{M} |c_m|$	L_1 regularization Least Absolute Shrinkage and Selection Operator (LASSO)
$\dfrac{1}{2}\|\mathbf{c}\|^2$ = $\dfrac{1}{2}\displaystyle\sum_{m=1}^{M} c_m^2$	L_2 regularization Ridge regression
$\lambda\|\mathbf{c}\|_1 + \dfrac{1-\lambda}{2}\|\mathbf{c}\|^2$	Elastic net regularization

L_0 regularization is an *NP hard* problem. This means there is no known algorithm to solve this in polynomial time. The aim of L_0 regularization is to arrive at a *sparse* vector of coefficients, i.e. a vector where many entries are zero. That is only a few components of $\mathbf{c}$ are non-zero. If it is known that $\mathbf{c}$ is 1-*sparse*, i.e. only one component is non-zero, we have $\binom{M}{1}$ possibilities and each needs to be checked whether it is the minimum. Similarly, if we know $\mathbf{c}$ is k-*sparse*, $\binom{M}{k}$ possibilities need to be checked. However, we do not know beforehand how many components are non-zero. Hence, all M possible values for k need to be checked. The resulting complexity is

$$\sum_{k=1}^{M} \binom{M}{k} = 2^M - 1.$$

Since the size of the problem, M, appears in the exponent, the complexity is exponential. The time to find the solution increases exponentially with the size of the input. We encounter a different method to get a sparse solution when looking at Bayesian Learning in Section 8.13.

The L_0 regularization is related to the *Bayesian Information Criterion* (BIC) and the *Akaike Information Criterion* (AIC). These criteria are used to select the model structure, k being the model parameter to be selected. For more information see [9].

The L_1 regularization has the disadvantage that in the case of perfect multicollinearity there is not a unique minimum, but a continuum of minima. For example, if two columns $\mathbf{d}_i$ and $\mathbf{d}_j$ are the same and c_i and c_j are nonzero coefficients in the minimal solution, then for any $a \in [0,1]$ replacing c_i by $a(\mathbf{c}_i + \mathbf{c}_j)$ and replacing $\mathbf{c}_j$ by $(1-a)(\mathbf{c}_i + \mathbf{c}_j)$ is also minimal.

If the columns of $\mathbf{D}$ are orthogonal, then they are orthonormal, since we assume they are standardized. This means $\mathbf{D}^T\mathbf{D}$ is the identity matrix. In this case, the solution to L_1 regularization can be found by *soft thresholding*. We minimize

$$\frac{1}{2}\|\mathbf{t} - \mathbf{Dc}\|^2 + \lambda\|\mathbf{c}\|_1 = \frac{1}{2}\mathbf{t}^T\mathbf{t} - \mathbf{c}^T\mathbf{D}^T\mathbf{t} + \frac{1}{2}\mathbf{c}^T\mathbf{c} + \lambda\|\mathbf{c}\|_1.$$

Since $\mathbf{t}^T\mathbf{t}$ is independent of the variables of interest, the first term can be dropped. Furthermore, let $\mathbf{c}^{\mathrm{OLS}}$ be the ordinary least squares solution

$$\mathbf{c}^{\mathrm{OLS}} = \left(\mathbf{D}^T\mathbf{D}\right)^{-1}\mathbf{D}^T\mathbf{t} = \mathbf{D}^T\mathbf{t}.$$

Hence the minimization is equivalent to minimizing

$$-\mathbf{c}^T\mathbf{c}^{\mathrm{OLS}} + \frac{1}{2}\mathbf{c}^T\mathbf{c} + \lambda\|\mathbf{c}\|_1 = \sum_{m=1}^{M} -c_m c_m^{\mathrm{OLS}} + \frac{1}{2}c_m^2 + \lambda|c_m|,$$

where each summand can be minimized individually. That is, for each m, we minimize

$$-c_m c_m^{\mathrm{OLS}} + \frac{1}{2}c_m^2 + \lambda|c_m| = -c_m c_m^{\mathrm{OLS}} + \frac{1}{2}c_m^2 + \lambda\,\mathrm{sgn}(c_m)c_m,$$

where

$$\mathrm{sgn}(c_m) = \begin{cases} +1 & \text{if} \quad c_m > 0, \\ 0 & \text{if} \quad c_m = 0, \\ -1 & \text{if} \quad c_m < 0. \end{cases}$$

If $c_m^{\mathrm{OLS}} = 0$, then $c_m = 0$ is minimal. Otherwise, the minimal c_m has to have the same sign as c_m^{OLS}, since otherwise the sign could be flipped and a smaller value could be achieved. Thus, it is equivalent to minimize

$$-c_m c_m^{\mathrm{OLS}} + \frac{1}{2}c_m^2 + \lambda\,\mathrm{sgn}(c_m^{\mathrm{OLS}})c_m.$$

Differentiating with respect to c_m and setting to zero gives

$$c_m = c_m^{\mathrm{OLS}} - \lambda\,\mathrm{sgn}(c_m^{\mathrm{OLS}}) = \mathrm{sgn}(c_m^{\mathrm{OLS}})(|c_m^{\mathrm{OLS}}| - \lambda).$$

To ensure that c_m has the same sign as c_m^{OLS}, we set $c_m = 0$, if $|c_m^{\mathrm{OLS}}| < \lambda$. This effectively prunes the corresponding basis function. For $\lambda = 0$, we recover the ordinary least squares solution. If λ is chosen too large, all coefficients are set to zero.

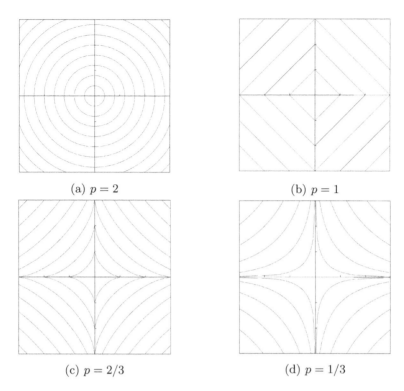

(a) $p = 2$ (b) $p = 1$

(c) $p = 2/3$ (d) $p = 1/3$

Figure 8.9: Contour lines of various L_p norms.

For L_2 regularization we minimize

$$\frac{1}{2}\|\mathbf{t} - \mathbf{D}\mathbf{c}\|^2 + \frac{1}{2}\lambda\|\mathbf{c}\|^2 = \frac{1}{2}\mathbf{t}^T\mathbf{t} - \mathbf{c}^T\mathbf{D}^T\mathbf{t} + \frac{1}{2}\mathbf{c}^T(\mathbf{D}^T\mathbf{D} + \lambda\mathbf{I})\mathbf{c}.$$

The gradient of this function is

$$-\mathbf{D}^T\mathbf{t} + (\mathbf{D}^T\mathbf{D} + \lambda\mathbf{I})\mathbf{c}.$$

Setting the gradient to zero, we have a minimum for

$$\mathbf{c} = (\mathbf{D}^T\mathbf{D} + \lambda\mathbf{I})^{-1}\mathbf{D}^T\mathbf{t}, \tag{8.3}$$

since the matrix $\mathbf{D}^T\mathbf{D} + \lambda\mathbf{I}$ is positive definite. If $\mathbf{D}^T\mathbf{D}$ is close to being singular, adding λ to each diagonal term will move it away from singularity. Thus ridge regression stabilizes ordinary least squares.

It is interesting to consider the case of orthonormal basis functions $(\mathbf{D}^T\mathbf{D} = \mathbf{I})$. Then L_2 regularization reduces each component of the ordinary least squares solution by a factor of $(1 + \lambda)^{-1}$. L_1 regularization on the other hand moves each component towards zero by λ. If the component is closer to zero than λ, it is set to zero. Elastic net regularization combines the two effects.

Ridge regression is implemented in MATLAB as `ridge`, while LASSO is implemented as `lasso`. The latter also incorporates elastic net regularization, since the elastic net becomes LASSO for $\lambda = 1$.

More generally, the L_p *norm* is defined as

$$\|\mathbf{c}\|_p = \left(\sum_{m=1}^{M} |c_m|^p \right)^{1/p}.$$

Figure 8.9 shows the contour lines for various choices of p. The L_2 regularization does not favour any axes, while for decreasing $p < 2$, the L_p regularization favours points closer to the axes. In the limit the L_0 norm snaps the solutions to the axis. This makes the importance of the choice of axes more apparent. Ideally axes aligned with the principal components should be chosen. This also makes the algorithms more stable.

8.11   Bayesian Regression

We have seen that finding suitable coefficients $\mathbf{c}$ to satisfy $\mathbf{t} = \mathbf{Dc} + \boldsymbol{\epsilon}$ by minimizing $\|\mathbf{t} - \mathbf{Dc}\|^2 = \|\boldsymbol{\epsilon}\|^2$ leads to the noise being modeled and overfitting. This can be mitigated by minimizing $\|\mathbf{t} - \mathbf{Dc}\|^2 + \Omega(\mathbf{c})$ for some penalty function $\Omega(\mathbf{c})$. This "tweaks" the model to favour less complex models. The question we are asking now is whether the "tweaking" can be formalized.

To this end, we utilize *Bayes' Rule* which we recall as

$$P(A|B) = \frac{P(B|A)P(A)}{P(B)}.$$

Often the relationship is expressed as $P(A|B) \propto P(B|A)P(A)$ where $\propto$ means that the two sides are proportional to each other. Thus the division by $P(B)$ is suppressed.

Applying this to the setting at hand,

$$p(\mathbf{c}|\mathbf{t}) = \frac{p(\mathbf{t}|\mathbf{c})p(\mathbf{c})}{p(\mathbf{t})}.$$

In other words, we answer the question: What is the probability of the coefficients given the observed data?

Since the noise is i.i.d. normal with mean 0 and variance σ^2, we can write for the probability of $\mathbf{t}$ given the design matrix $\mathbf{D}$, coefficients $\mathbf{c}$ and noise variance σ^2

$$p(\mathbf{t}|\mathbf{D}, \mathbf{c}, \sigma^2) = (2\pi)^{-N/2}\sigma^{-N} \exp\left(-\frac{\|\mathbf{t} - \mathbf{Dc}\|}{2\sigma^2} \right).$$

We define a *prior distribution* for $\mathbf{c}$ using all information apart from the data itself, quantifying our belief about the coefficients. For example, a simple assumption is that each coefficient is *a priori* normally distributed with mean zero and variance α^{-1}. The multivariate distribution of $\mathbf{c}$ is given by

$$p(\mathbf{c}|\alpha) = \left(\frac{\alpha}{2\pi}\right)^{M/2} \exp\left(-\frac{\alpha}{2}\mathbf{c}^T\mathbf{c}\right).$$

In the absence of any other knowledge, this is one possible assumption. α is a *hyperparameter* and known as the *precision* of the distribution. If α becomes very large, the distribution becomes peaked at its mean and we have more confidence in the value than if α is small and the width of the distribution large.

After observing N samples of the dependent variable $\mathbf{t}$, the *posterior distribution* of the coefficients $\mathbf{c}$ is given by

$$p(\mathbf{c}|\mathbf{t},\mathbf{D},\alpha,\sigma^2) \propto p(\mathbf{t}|\mathbf{D},\mathbf{c},\sigma^2)p(\mathbf{c}|\alpha).$$

An estimate of $\mathbf{c}$ could be obtained by choosing the $\mathbf{c}$ where the posterior distribution or equivalently its logarithm is maximal,

$$\log p(\mathbf{c}|\mathbf{t},\mathbf{D},\alpha,\sigma^2) \propto \log p(\mathbf{t}|\mathbf{D},\mathbf{c}\,\sigma^2) + \log p(\mathbf{c}|\alpha).$$

The first term is exactly the log likelihood (8.1), which is maximized when $\|\mathbf{t} - \mathbf{Dc}\|^2$ is minimal. The second term is $-\frac{\alpha}{2}\|\mathbf{c}\|^2$ + a constant, and can be regarded as the negative of a penalty term. In fact, taking the negative logarithm of the posterior distribution and then minimizing instead of maximizing has re-created the ridge regression.

More generally, let $\boldsymbol{\alpha}$ contain all parameters governing the joint distribution $p(\mathbf{t},\mathbf{c}|\mathbf{D},\boldsymbol{\alpha},\sigma^2)$. We remove the dependency on the latent variables by integrating over the coefficients $\mathbf{c}$, thus averaging over all possible solutions. This is called *marginalizing* over $\mathbf{c}$ and the result is the *marginal likelihood*

$$\mathcal{L}(\mathbf{t}|\mathbf{D},\boldsymbol{\alpha},\sigma^2) = \int p(\mathbf{t},\mathbf{c}|\boldsymbol{\alpha},\sigma^2)d\mathbf{c},$$

which we aim to maximize. The *Maximum-Likelihood Estimate (MLE)* for $\boldsymbol{\alpha}$ are the values which maximize $\mathcal{L}(\mathbf{t}|\mathbf{D},\boldsymbol{\alpha},\sigma^2)$. In the following section we introduce a technique to find a maximum.

8.12 Expectation–Maximization

In the following, we suppress the dependency of the model on the design matrix $\mathbf{D}$. The *Expectation–Maximization (EM)* algorithm arrives at a maximum iteratively by alternating between two steps. The derivation requires some expansion.

By the *chain rule*, also known as the *product rule*, we have

$$p(\mathbf{t},\mathbf{c}|\alpha,\sigma^2) = p(\mathbf{c}|\mathbf{t},\alpha,\sigma^2)p(\mathbf{t}|\alpha,\sigma^2).$$

Taking the logarithm on both sides, results in

$$\log p(\mathbf{t}, \mathbf{c}|\boldsymbol{\alpha}, \sigma^2) = \log p(\mathbf{c}|\mathbf{t}, \boldsymbol{\alpha}, \sigma^2) + \log p(\mathbf{t}|\boldsymbol{\alpha}, \sigma^2).$$

Both sides can be viewed as functions of the random variable $\mathbf{c}$ and thus we can take the expectation with regards to any distribution $q(\mathbf{c})$,

$$
\begin{aligned}
\int q(\mathbf{c}) \log p(\mathbf{t}, \mathbf{c}|\boldsymbol{\alpha}, \sigma^2)d\mathbf{c} &= \int q(\mathbf{c}) \log p(\mathbf{c}|\mathbf{t}, \boldsymbol{\alpha}, \sigma^2)d\mathbf{c} \\
&+ \int q(\mathbf{c}) \log p(\mathbf{t}|\boldsymbol{\alpha}, \sigma^2)d\mathbf{c} \\
&= \int q(\mathbf{c}) \log p(\mathbf{c}|\mathbf{t}, \boldsymbol{\alpha}, \sigma^2) + \log p(\mathbf{t}|\boldsymbol{\alpha}, \sigma^2),
\end{aligned}
$$

since $\int q(\mathbf{c})d\mathbf{c} = 1$. Rearranging gives

$$
\begin{aligned}
\log p(\mathbf{t}|\boldsymbol{\alpha}, \sigma^2) = \log \mathcal{L}(\mathbf{t}|\boldsymbol{\alpha}, \sigma^2) &= \int q(\mathbf{c}) \log p(\mathbf{t}, \mathbf{c}|\boldsymbol{\alpha}, \sigma^2)d\mathbf{c} \\
&- \int q(\mathbf{c}) \log p(\mathbf{c}|\mathbf{t}, \boldsymbol{\alpha}, \sigma^2)d\mathbf{c}.
\end{aligned}
$$

We can subtract and add the term $\int q(\mathbf{c}) \log q(\mathbf{c})d\mathbf{c}$ on the right hand side and arrive at

$$\log \mathcal{L}(\mathbf{t}|\boldsymbol{\alpha}, \sigma^2) = \int q(\mathbf{c}) \log \frac{p(\mathbf{t}, \mathbf{c}|\boldsymbol{\alpha}, \sigma^2)}{q(\mathbf{c})}d\mathbf{c} - \int q(\mathbf{c}) \log \frac{p(\mathbf{c}|\mathbf{t}, \boldsymbol{\alpha}, \sigma^2)}{q(\mathbf{c})}d\mathbf{c} \quad (8.4)$$

using the properties of the logarithm. $\log \mathcal{L}(\mathbf{t}|\boldsymbol{\alpha}, \sigma^2)$ is also known as the *log evidence*.

The second term (including the minus sign) is the *Kullback–Leibler divergence (KL divergence)* between the distribution $q(\mathbf{c})$ and the posterior distribution $p(\mathbf{c}|\mathbf{t}, \boldsymbol{\alpha}, \sigma^2)$. It is also known as the *discrimination information*, *information divergence*, and *information gain*, since information is obtained from observing $\mathbf{t}$. For two probability density functions p and q, the Kullback–Leibler divergence is defined as

$$D_{KL}(p\|q) = - \int p(x) \log \frac{q(x)}{p(x)}dx = \mathbb{E}_p\left[-\log \frac{q(x)}{p(x)}\right],$$

where $\mathbb{E}_p$ denotes the expectation with respect to the distribution p. Using Jensen's inequality,

$$D_{KL}(p\|q) \geq -\log \mathbb{E}_p\left[\frac{q(x)}{p(x)}\right] = -\log \int p(x) \frac{q(x)}{p(x)}dx = -\log \int q(x)dx = 0,$$

since $\log 1 = 0$ (for more details of the proof see for example [3]). Hence the Kullback–Leibler divergence is always positive, and zero if and only if $p(\mathbf{c}|\mathbf{t}, \boldsymbol{\alpha}, \sigma^2) = q(\mathbf{c})$ almost everywhere.

Therefore, the first term in (8.4),

$$\int q(\mathbf{c}) \log \frac{p(\mathbf{t}, \mathbf{c}|\boldsymbol{\alpha}, \sigma^2)}{q(\mathbf{c})} d\mathbf{c},$$

is a lower bound for $\log \mathcal{L}(\mathbf{t}|\boldsymbol{\alpha}, \sigma^2)$. The maximization is done by maximizing the lower bound by alternating between maximizing with respect to $q(\mathbf{c})$ and with respect to $\boldsymbol{\alpha}$.

Let $\boldsymbol{\alpha}^{(k)}$ be the current estimate of the maximal $\boldsymbol{\alpha}$. The lower bound is maximal with respect to $q(\mathbf{c})$, if the Kullback-Leibler divergence is zero, that is, if $q(\mathbf{c}) = p(\mathbf{c}|\mathbf{t}, \boldsymbol{\alpha}^{(k)}, \sigma^2)$. The next step is to find $\boldsymbol{\alpha}^{(k+1)}$ by maximizing

$$\int p(\mathbf{c}|\mathbf{t}, \boldsymbol{\alpha}^{(k)}, \sigma^2) \log \frac{p(\mathbf{t}, \mathbf{c}|\boldsymbol{\alpha}, \sigma^2)}{p(\mathbf{c}|\mathbf{t}, \boldsymbol{\alpha}^{(k)}, \sigma^2)} d\mathbf{c} =$$
$$\int p(\mathbf{c}|\mathbf{t}, \boldsymbol{\alpha}^{(k)}, \sigma^2) \log p(\mathbf{t}, \mathbf{c}|\boldsymbol{\alpha}, \sigma^2) d\mathbf{c}$$
$$- \int p(\mathbf{c}|\mathbf{t}, \boldsymbol{\alpha}^{(k)}, \sigma^2) \log p(\mathbf{c}|\mathbf{t}, \boldsymbol{\alpha}^{(k)}, \sigma^2) d\mathbf{c}.$$

The second term is independent of $\boldsymbol{\alpha}$. Hence only the first term is relevant for the maximization.

Note: The first term is the *expectation* of the logarithm of the joint probability with respect to the current estimate of the posterior distribution. Hence the name *Expectation–Maximization* algorithm.

8.13   Bayesian Learning

As a specific example, let the prior distribution of c_i be normal with mean zero and variance α_i^{-1}. In other words, each weight has its own hyperparameter. Our starting point is that there is no covariance between different weights. That is the weights do not influence each other. The multivariate prior is given by

$$p(\mathbf{c}|\boldsymbol{\alpha}) = (2\pi)^{-M/2} \prod_{m=1}^{M} \alpha_m^{1/2} \exp\left(-\frac{\alpha_m c_m^2}{2}\right)$$
$$= (2\pi)^{-M/2} |\mathbf{A}|^{1/2} \exp\left(-\frac{1}{2}\mathbf{c}^T \mathbf{A} \mathbf{c}\right),$$

where $\mathbf{A}$ is a diagonal matrix with entries $A_{mm} = \alpha_m$, and $|\mathbf{A}|$ denotes the determinant.

The multivariate posterior distribution $p(\mathbf{c}|\mathbf{t}, \mathbf{D}, \boldsymbol{\alpha}, \sigma^2)$ is given by Bayes' rule and is proportional to the product of the prior and the likelihood function:

$$p(\mathbf{c}|\mathbf{t}, \mathbf{D}, \boldsymbol{\alpha}, \sigma^2) = \frac{p(\mathbf{c}|\boldsymbol{\alpha})\mathcal{L}(\mathbf{t}|\mathbf{D}, \mathbf{c}, \sigma^2)}{\int p(\mathbf{c}|\boldsymbol{\alpha})\mathcal{L}(\mathbf{t}|\mathbf{D}, \mathbf{c}, \sigma^2) d\mathbf{c}}. \tag{8.5}$$

Now,

$$
\begin{aligned}
p(\mathbf{c}|\alpha)\mathcal{L}(\mathbf{t}|\mathbf{D},\mathbf{c},\sigma^2) = \; & (2\pi)^{-M/2}|\mathbf{A}|^{1/2}\exp\left(-\frac{1}{2}\mathbf{c}^T\mathbf{A}\mathbf{c}\right) \\
& (2\pi\sigma^2)^{-N/2}\exp\left(-\frac{1}{2\sigma^2}(\mathbf{t}-\mathbf{D}\mathbf{c})^T(\mathbf{t}-\mathbf{D}\mathbf{c})\right).
\end{aligned}
$$

The multiplicative constants cancel in the fraction and we concentrate on the terms in the exponential:

$$
-\frac{1}{2}\left[\mathbf{c}^T\left(\mathbf{A}+\sigma^{-2}\mathbf{D}^T\mathbf{D}\right)\mathbf{c}-2\sigma^{-2}\mathbf{c}^T\mathbf{D}^T\mathbf{t}+\sigma^{-2}\mathbf{t}^T\mathbf{t}\right]
$$

Considering the square term we see that the covariance matrix of the posterior distribution has to be

$$
\boldsymbol{\Sigma} = \left(\mathbf{A}+\sigma^{-2}\mathbf{D}^T\mathbf{D}\right)^{-1}, \tag{8.6}
$$

which is symmetric and positive definite. We complete the square by introducing the mean $\boldsymbol{\mu}$:

$$
-\frac{1}{2}\left[(\mathbf{c}-\boldsymbol{\mu})^T\boldsymbol{\Sigma}^{-1}(\mathbf{c}-\boldsymbol{\mu})+2\mathbf{c}^T\boldsymbol{\Sigma}^{-1}\boldsymbol{\mu}-\boldsymbol{\mu}^T\boldsymbol{\Sigma}^{-1}\boldsymbol{\mu}-2\sigma^{-2}\mathbf{c}^T\mathbf{D}^T\mathbf{t}+\sigma^{-2}\mathbf{t}^T\mathbf{t}\right]
$$

The linear terms in $\mathbf{c}^T$ need to cancel; that is we need $\boldsymbol{\Sigma}^{-1}\boldsymbol{\mu}=\sigma^{-2}\mathbf{D}^T\mathbf{t}$. Hence the mean of the posterior distribution is

$$
\boldsymbol{\mu} = \sigma^{-2}\boldsymbol{\Sigma}\mathbf{D}^T\mathbf{t}. \tag{8.7}
$$

All remaining terms in the exponential are constant with respect to $\mathbf{c}$. Because of the exponential, they are multiplicative constants and cancel in the fraction. It remains to evaluate the denominator, which after cancellations is

$$
\int \exp\left(-\frac{1}{2}(\mathbf{c}-\boldsymbol{\mu})^T\boldsymbol{\Sigma}^{-1}(\mathbf{c}-\boldsymbol{\mu})\right)d\mathbf{c}.
$$

However, it is known that

$$
\int (2\pi)^{-M/2}|\boldsymbol{\Sigma}|^{-1/2}\exp\left(-\frac{1}{2}(\mathbf{c}-\boldsymbol{\mu})^T\boldsymbol{\Sigma}^{-1}(\mathbf{c}-\boldsymbol{\mu})\right)d\mathbf{c}
$$

evaluates to 1, since it is the integral of the normal probability distribution function. Thus the integral of the denominator is $(2\pi)^{M/2}|\boldsymbol{\Sigma}|^{1/2}$.

To summarize, the posterior distribution is normal with

$$
p(\mathbf{c}|\mathbf{t},\mathbf{D},\boldsymbol{\alpha},\sigma^2) \sim \mathcal{N}(\boldsymbol{\mu},\boldsymbol{\Sigma})
$$

with $\boldsymbol{\mu}$ and $\boldsymbol{\Sigma}$ given by (8.7) and (8.6).

Recall that Ordinary Least Squares calculates the coefficients as

$$
\mathbf{c} = \left(\mathbf{D}^T\mathbf{D}\right)^{-1}\mathbf{D}^T\mathbf{t},
$$

while ridge regression determines them as

$$\mathbf{c} = \left(\lambda \mathbf{I} + \mathbf{D}^T \mathbf{D}\right)^{-1} \mathbf{D}^T \mathbf{t},$$

where λ is a parameter which needs to be suitably chosen. Inserting (8.6) into (8.7) gives

$$\boldsymbol{\mu} = \left(\sigma^2 \mathbf{A} + \mathbf{D}^T \mathbf{D}\right)^{-1} \mathbf{D}^T \mathbf{t}.$$

The addition of $\sigma^2 \mathbf{A}$ guards against $\mathbf{D}^T \mathbf{D}$ being possibly singular. The main distinguishing feature of Bayesian learning is that firstly, instead of returning a specific vector of coefficients, a multivariate posterior probability distribution is given for $\mathbf{c}$. Secondly, the entries of $\mathbf{A}$, i.e. the hyperparameters $\alpha_1, \ldots, \alpha_M$ are optimized by maximizing the marginal likelihood as we will see below.

Once the posterior distribution of the coefficients is found, the measurement t_n for data $\mathbf{x}_n$ can be interpreted probabilistically. Recall that we assume t_n is generated by a process,

$$t_n = \mathbf{d}(\mathbf{x}_n)^T \mathbf{c} + \epsilon_n,$$

where $\mathbf{d}(\mathbf{x}_n)^T$ is the nth row of $\mathbf{D}$. Now, $\mathbf{d}(\mathbf{x}_n)^T \mathbf{c}$ is a linear transformation of the normally distributed variable $\mathbf{c}$ and thus also normally distributed. On the other hand, ϵ_n is normally distributed noise. Hence, t_n is the sum of two normally distributed variables. The probabilistic interpretation of t_n is that it is drawn from a univariate normal distribution with

$$\begin{aligned} \text{mean} \quad & m_n = \mathbf{d}(\mathbf{x}_n)^T \boldsymbol{\mu}, \\ \text{variance} \quad & \sigma_n^2 = \sigma^2 + \mathbf{d}(\mathbf{x}_n)^T \boldsymbol{\Sigma} \mathbf{d}(\mathbf{x}_n). \end{aligned}$$

If the variance is small, it indicates that at this point the model explains the data well. If the variance is large, the model is not adequate at this point. This can indicate that the dictionary of basis functions is unsuitable for these data and needs to be amended.

The most suitable values for the hyperparameters are found by maximizing the the marginal likelihood $\mathcal{L}(\mathbf{t}|\boldsymbol{\alpha}, \sigma^2)$ which is precisely the denominator of (8.5). Keeping the multiplicative constants (which were canceled in the previous analysis), the marginal likelihood is

$$\mathcal{L}(\mathbf{t}|\boldsymbol{\alpha}, \sigma^2) = |\mathbf{A}|^{1/2} (2\pi\sigma^2)^{-N/2} |\boldsymbol{\Sigma}|^{1/2} \exp\left(-\frac{1}{2}\sigma^{-2}\mathbf{t}^T\mathbf{t} + \frac{1}{2}\boldsymbol{\mu}^T\boldsymbol{\Sigma}^{-1}\boldsymbol{\mu}\right).$$

We first consider the terms in the exponential. Inserting (8.7), this becomes:

$$\begin{aligned} -\frac{1}{2}\left[\sigma^{-2}\mathbf{t}^T\mathbf{t} - \boldsymbol{\mu}^T\boldsymbol{\Sigma}^{-1}\boldsymbol{\mu}\right] &= -\frac{1}{2}\left[\sigma^{-2}\mathbf{t}^T\mathbf{t} - \sigma^{-4}\mathbf{t}^T\mathbf{D}\boldsymbol{\Sigma}\boldsymbol{\Sigma}^{-1}\boldsymbol{\Sigma}\mathbf{D}^T\mathbf{t}\right] \\ &= -\frac{1}{2}\mathbf{t}^T\left[(\sigma^2\mathbf{I})^{-1} - (\sigma^2\mathbf{I})^{-1}\mathbf{D}\boldsymbol{\Sigma}\mathbf{D}(\sigma^2\mathbf{I})^{-1}\right]\mathbf{t}, \end{aligned}$$

where $\mathbf{I}$ is the $N \times N$ identity matrix. Using $\mathbf{\Sigma} = \left(\mathbf{A} + \mathbf{D}^T(\sigma^2\mathbf{I})^{-1}\mathbf{D}\right)^{-1}$ and the *Woodbury matrix identity* in Appendix A.1.3, this simplifies to

$$-\frac{1}{2}\mathbf{t}^T \left(\sigma^2\mathbf{I} + \mathbf{D}\mathbf{A}^{-1}\mathbf{D}\right)^{-1}\mathbf{t}.$$

Hence the mean of the marginal likelihood is zero and its covariance matrix is

$$\mathbf{C} = \sigma^2\mathbf{I} + \mathbf{D}\mathbf{A}^{-1}\mathbf{D}^T.$$

We need to check whether the multiplicative constants agree with this. To this end, note

$$|\mathbf{\Sigma}| = |\mathbf{\Sigma}^{-1}|^{-1} = |\mathbf{A} + \mathbf{D}^T(\sigma^2\mathbf{I})^{-1}\mathbf{D}|^{-1}.$$

Using the *matrix determinant lemma* from Appendix A.1.5, we get

$$|\mathbf{\Sigma}| = |\sigma^2\mathbf{I} + \mathbf{D}\mathbf{A}^{-1}\mathbf{D}^T|^{-1}|\mathbf{A}|^{-1}|\sigma^2\mathbf{I}| = |\mathbf{C}|^{-1}|\mathbf{A}|^{-1}(\sigma^2)^N.$$

Combining this result with the other multiplicative constants, we see

$$\mathcal{L}(\mathbf{t}|\boldsymbol{\alpha}, \sigma^2) = (2\pi)^{-N/2}|\mathbf{C}|^{-1/2} \exp\left(-\frac{1}{2}\mathbf{t}^T\mathbf{C}^{-1}\mathbf{t}\right).$$

Instead of maximizing this, we maximize the logarithm of the marginal likelihood, the *log evidence*. In the following, we will denote this *objective function* by

$$\mathcal{L}(\boldsymbol{\alpha}) = -\frac{1}{2}\left(N\log 2\pi + \log|\mathbf{C}| + \mathbf{t}^T\mathbf{C}^{-1}\mathbf{t}\right).$$

We want to maximize this function with respect to the hyperparameters. To this end, we separate the contribution of a single hyperparameter out. First note that since $\mathbf{A}$ is a diagonal matrix, $\mathbf{C}$ can be re-written as

$$\mathbf{C} = \sigma^2\mathbf{I} + \sum_{m=1}^{M} \frac{1}{\alpha_m}\mathbf{d}_m\mathbf{d}_m^T.$$

Let

$$\mathbf{C}_{-i} = \sigma^2\mathbf{I} + \sum_{\substack{m=1 \\ m \neq i}}^{M} \frac{1}{\alpha_m}\mathbf{d}_m\mathbf{d}_m^T = \mathbf{C} - \frac{1}{\alpha_i}\mathbf{d}_i\mathbf{d}_i^T$$

and $\boldsymbol{\alpha}_{-i} = (\alpha_1, \ldots, \alpha_{i-1}, \alpha_{i+1}, \ldots, \alpha_M)^T$. Using the *Sherman–Morrison formula* in Appendix A.1.4, we have

$$\mathbf{C}^{-1} = \mathbf{C}_{-i}^{-1} - \frac{\mathbf{C}_{-i}^{-1}\mathbf{d}_i\mathbf{d}_i^T\mathbf{C}_{-i}^{-1}}{\alpha_i + \mathbf{d}_i^T\mathbf{C}_{-i}^{-1}\mathbf{d}_i}.$$

The *matrix determinant lemma* in Appendix A.1.5 on the other hand gives

$$|\mathbf{C}| = |\mathbf{C}_{-i}| \left(\alpha_i + \mathbf{d}_i^T\mathbf{C}_{-i}^{-1}\mathbf{d}_i\right)\frac{1}{\alpha_i}.$$

Combining these two, we can write

$$
\begin{aligned}
\mathcal{L}(\boldsymbol{\alpha}) &= -\frac{1}{2}\left(N\log(2\pi) + \log|\mathbf{C}_{-i}| + \mathbf{t}^T\mathbf{C}_{-i}^{-1}\mathbf{t} - \log\alpha_i\right.\\
&\quad \left. + \log(\alpha_i + \mathbf{d}_i^T\mathbf{C}_{-i}^{-1}\mathbf{d}_i) - \frac{\mathbf{t}^T\mathbf{C}_{-i}^{-1}\mathbf{d}_i\mathbf{d}_i^T\mathbf{C}_{-i}^{-1}\mathbf{t}}{\alpha_i + \mathbf{d}_i^T\mathbf{C}_{-i}^{-1}\mathbf{d}_i}\right)\\
&= \mathcal{L}(\boldsymbol{\alpha}_{-i}) + \frac{1}{2}\left(\log\alpha_i - \log(\alpha_i + \mathbf{d}_i^T\mathbf{C}_{-i}^{-1}\mathbf{d}_i) + \frac{(\mathbf{d}_i^T\mathbf{C}_{-i}^{-1}\mathbf{t})^2}{\alpha_i + \mathbf{d}_i^T\mathbf{C}_{-i}^{-1}\mathbf{d}_i}\right)\\
&= \mathcal{L}(\boldsymbol{\alpha}_{-i}) + \ell(\alpha_i).
\end{aligned}
$$

Hence we have separated the dependencies on α_i into $\ell(\alpha_i)$, while $\mathcal{L}(\boldsymbol{\alpha}_{-i})$ contains the portion of the log evidence independent of α_i.

We define the quantities

$$
s_i = \mathbf{d}_i^T\mathbf{C}_{-i}^{-1}\mathbf{d}_i \quad \text{and} \quad q_i = \mathbf{d}_i^T\mathbf{C}_{-i}^{-1}\mathbf{t},
$$

which are independent of α_i. With this definition

$$
\ell(\alpha_i) = \frac{1}{2}\left(\log\alpha_i - \log(\alpha_i + s_i) + \frac{q_i^2}{\alpha_i + s_i}\right).
$$

The derivative of the log evidence with respect to α_i is given by

$$
\begin{aligned}
\frac{\partial\mathcal{L}(\boldsymbol{\alpha})}{\partial\alpha_i} = \frac{\partial\ell(\alpha_i)}{\partial\alpha_i} &= \frac{1}{2}\left(\frac{1}{\alpha_i} - \frac{1}{\alpha_i + s_i} - \frac{q_i^2}{(\alpha_i + s_i)^2}\right)\\
&= \frac{1}{2}(\alpha_i + s_i)^{-2}\left(s_i - q_i^2 + \frac{s_i^2}{\alpha_i}\right),
\end{aligned}
\tag{8.8}
$$

since $\mathcal{L}(\boldsymbol{\alpha}_{-i})$ is independent of α_i.

If $s_i - q_i^2$ is non-negative, then the derivative is always positive, since α_i is positive, and $\ell(\alpha_i)$ is monotonically increasing. If on the other hand $s_i - q_i^2$ is negative, then equation (8.8) vanishes for

$$
\alpha_i = \frac{s_i^2}{q_i^2 - s_i},
\tag{8.9}
$$

where $\ell(\alpha_i)$ has a maximum. Figure 8.10 illustrates these two cases. Setting $\alpha_i = \infty$ in the first case and adjusting α_i to the value given in (8.9) in the second case, increases the logarithm of the marginal likelihood. For $\alpha_i = \infty$, the probability density function for $\mathbf{c}_i$ becomes infinitely peaked at zero. In practice, this means that $d_i(\mathbf{x})$ is regarded as not contributing to explaining the data, and therefore can be excluded from the model. This introduces *sparsity* in the model.

To avoid many different matrices $\mathbf{C}_{-i}$, it is convenient to define

$$
\begin{aligned}
Q_i &= \mathbf{d}_i^T\mathbf{C}^{-1}\mathbf{t},\\
S_i &= \mathbf{d}_i^T\mathbf{C}^{-1}\mathbf{d}_i.
\end{aligned}
$$

For infinite α_i, we have $q_i = Q_i$ and $s_i = S_i$, since in this case $\mathbf{C} = \mathbf{C}_{-i}$. While for finite α_i,

$$
\begin{aligned}
s_i &= \mathbf{d}_i^T \mathbf{C}_{-i}^{-1} \mathbf{d}_i = \mathbf{d}_i^T \left(\mathbf{C} + \mathbf{d}_i (-\alpha_i)^{-1} \mathbf{d}_i^T \right)^{-1} \mathbf{d}_i \\
&= \mathbf{d}_i^T \left[\mathbf{C}^{-1} - \mathbf{C}^{-1} \mathbf{d}_i \left(-\alpha_i + \mathbf{d}_i^T \mathbf{C}^{-1} \mathbf{d}_i \right)^{-1} \mathbf{d}_i^T \mathbf{C}^{-1} \right] \mathbf{d}_i \\
&= S_i + \frac{S_i^2}{\alpha_i - S_i} = \frac{\alpha_i S_i}{\alpha_i - S_i}.
\end{aligned}
$$

Similarly,

$$
q_i = \frac{\alpha_i Q_i}{\alpha_i - S_i}.
$$

Let's examine Q_i more closely. The matrices $\boldsymbol{\Sigma}$ and $\mathbf{C}$ are related by the Woodbury matrix identity in Appendix A.1.3,

$$
\begin{aligned}
\mathbf{C}^{-1} &= \left(\sigma^2 \mathbf{I} + \mathbf{D} A^{-1} \mathbf{D}^T \right)^{-1} \\
&= \sigma^{-2} \mathbf{I} - \sigma^{-2} \mathbf{D} \left(\mathbf{A} + \mathbf{D}^T \sigma^{-2} \mathbf{D} \right)^{-1} \mathbf{D}^T \sigma^{-2} \\
&= \sigma^{-2} \left(\mathbf{I} - \sigma^{-2} \mathbf{D} \boldsymbol{\Sigma} \mathbf{D}^T \right).
\end{aligned}
$$

Using this and (8.7),

$$
\mathbf{C}^{-1} \mathbf{t} = \sigma^{-2} \left(\mathbf{t} - \mathbf{D} \boldsymbol{\mu} \right).
$$

However, $\mathbf{D}\boldsymbol{\mu}$ is exactly the mean of our predictive probability distribution for $\mathbf{t}$. Thus, $\mathbf{C}^{-1}\mathbf{t}$ gives the difference between the target and the predictive mean for each sample scaled by the noise variance. If this is small, our current model is adequate to explain the data. If not, we consider $\mathbf{C}^{-1}\mathbf{t}$ an error which needs to be removed. Calculating $Q_i = \mathbf{d}_i^T \mathbf{C}^{-1} \mathbf{t}$ checks how well aligned $\mathbf{d}_i$ is with this error. If it is orthogonal, than $\mathbf{d}_i$ will not help in removing this error. If it is well aligned, then the inclusion of $\mathbf{d}_i$ in the model will be beneficial. Therefore Q_i is known as the *quality factor*.

S_i on the other hand is known as the sparsity factor, since, as we will see below, it helps to evaluate how collinear $\mathbf{d}_i$ is to the other basis functions. The quantity s_i is a penalty term associated with the basis function d_i. It quantifies how well d_i could be modeled by the other basis functions. On the other hand, q_i is a quality term, measuring how well d_i would contribute to explaining the data, given the current model. Both s_i and q_i depend on $\mathbf{C}_{-i}$ which in turn depends on $\boldsymbol{\alpha}_{-i}$. Thus the optimal value for α_i changes as the other hyperparameters $\boldsymbol{\alpha}_{-i}$ are adjusted.

The possibility to optimize the hyperparameters individually in turn gives rise to the following algorithm. It initializes the model with a single basis function and its optimal hyperparameter value, and sets the hyper-parameters of the others notionally to infinity. At the start of the algorithm when only one basis function is in the model, $\mathbf{D}$ has only one column, and $\boldsymbol{\Sigma}$ is a 1×1 matrix. $\mathbf{C}$ is $\sigma^2 \mathbf{I}$ plus the outer product of that single basis function evaluated at the sample points with itself. Its inverse is easily calculated using the Shermann–Morrisson formula from Appendix A.1.4. Then the basis function $d_i(\mathbf{x})$, where

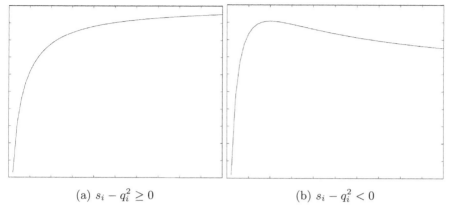

(a) $s_i - q_i^2 \geq 0$ (b) $s_i - q_i^2 < 0$

Figure 8.10: $\ell(\alpha_i)$.

setting its hyper-parameter α_i to its optimal value (given the current model) gives the largest increase in the marginal likelihood, is found and the model updated accordingly.

The algorithm converges, if no significant increase in the logarithm of the marginal likelihood can be achieved anymore. Since the optimal value of α_i can be finite or infinite, there are three different possible updates. If $d_i(\mathbf{x})$ is not in the model and the optimal α_i is finite, it gets added to the model. If $d_i(\mathbf{x})$ is in the model and the optimal α_i is infinite, it gets deleted from the model. The third option is that $d_i(\mathbf{x})$ is in the model and the optimal α_i is finite, in which case α_i is updated to this value. Note that in the case when the optimal α_i is infinite and $d_i(\mathbf{x})$ is already excluded from the model, no action is necessary. In all three cases, $\boldsymbol{\Sigma}$ and $\boldsymbol{\mu}$ have to be updated, since A has changed.

The matrices $\mathbf{D}$ and $\boldsymbol{\Sigma}$ change in size during the run of the algorithm. However, they stay comparatively small compared to the number M of basis functions in the dictionary and the sample number N. The matrix $\mathbf{C}$ is subject to rank 1 updates as basis functions are added, deleted or their hyperparameters re-estimated. These rank 1 updates are the addition or subtraction of the outer product of the basis function evaluated at the sample points with itself with suitable coefficients derived from the hyperparameter updates. However, in practice $\mathbf{C}$ and its inverse are not calculated.

It should be noted here that the dictionary of candidate basis functions does not need to be static. A new candidate basis function can be created, evaluated, and possibly added to the model at any point.

In the update formulae below, the change in the logarithm of the marginal likelihood is denoted by $\Delta\mathcal{L}$ and updated quantities are denoted by a tilde (e.g. $\tilde{\boldsymbol{\mu}}$):

Addition: In this case,

$$
2\Delta\mathcal{L} = \frac{Q_i^2 - S_i}{S_i} + \log\frac{S_i}{Q_i^2},
$$

$$
\tilde{\mathbf{\Sigma}} = \begin{pmatrix} \mathbf{\Sigma} + \sigma^{-4}\Sigma_{ii}\mathbf{\Sigma}\mathbf{D}^T\mathbf{d}_i\mathbf{d}_i^T\mathbf{D}\mathbf{\Sigma} & -\sigma^{-2}\Sigma_{ii}\mathbf{\Sigma}\mathbf{D}^T\mathbf{d}_i \\ -\sigma^{-2}\Sigma_{ii}\mathbf{d}_i^T\mathbf{D}\mathbf{\Sigma} & \Sigma_{ii} \end{pmatrix},
$$

$$
\tilde{\boldsymbol{\mu}} = \begin{pmatrix} \boldsymbol{\mu} - \sigma^{-2}\mu_i\mathbf{\Sigma}\mathbf{D}^T\mathbf{d}_i \\ \mu_i \end{pmatrix},
$$

where $\Sigma_{ii} = (\alpha_i + S_i)^{-1}$ and $\mu_i = \Sigma_{ii}Q_i$.

Deletion: The index j denotes the the column of $\mathbf{D}$ which needs to be removed, when d_i is removed from the current model. $\mathbf{\Sigma}_j$ is the jth column of $\mathbf{\Sigma}$ and Σ_{jj} is the jth diagonal element. The jth element of $\boldsymbol{\mu}$ is μ_j.

$$
2\Delta\mathcal{L} = \frac{Q_i^2}{S_i - \alpha_i} - \log\left(1 - \frac{S_i}{\alpha_i}\right),
$$

$$
\tilde{\mathbf{\Sigma}} = \mathbf{\Sigma} - \frac{1}{\Sigma_{jj}}\mathbf{\Sigma}_j\mathbf{\Sigma}_j^T,
$$

$$
\tilde{\boldsymbol{\mu}} = \boldsymbol{\mu} - \frac{\mu_j}{\Sigma_{jj}}\mathbf{\Sigma}_j.
$$

After these updates the jth row and column needs to be removed from $\tilde{\mathbf{\Sigma}}$ and the jth element from $\tilde{\boldsymbol{\mu}}$.

Re-estimation: Defining

$$
\kappa_j = (\Sigma_{jj} + (\tilde{\alpha}_i - \alpha_i)^{-1})^{-1}
$$

with Σ_{jj} and $\mathbf{\Sigma}_j$ as in the case of deletion, we have

$$
2\Delta\mathcal{L} = \frac{Q_i^2}{S_i + (\tilde{\alpha}_i^{-1} - \alpha_i^{-1})^{-1}} - \log(1 + S_i(\tilde{\alpha}_i^{-1} - \alpha_i^{-1})),
$$

$$
\tilde{\mathbf{\Sigma}} = \mathbf{\Sigma} - \kappa_j\mathbf{\Sigma}_j\mathbf{\Sigma}_j^T,
$$

$$
\tilde{\boldsymbol{\mu}} = \boldsymbol{\mu} - \kappa_j\mu_j\mathbf{\Sigma}_j.
$$

We already noted that a new candidate basis function can be created and evaluated for inclusion in the model at any point in time. The model can also adapt, when a new data sample arrives. Thus, this technique is also suitable for online learning. We can calculate the change in the logarithm of the marginal likelihood for the current model, when a data sample $(t_*, \mathbf{x}_*)$ is added. This means adding a row to the design matrix $\mathbf{D}$ yielding

$$
\mathbf{D}_* = \begin{pmatrix} \mathbf{D} \\ \mathbf{d}(\mathbf{x}_*)^T \end{pmatrix}.
$$

We then have

$$\mathbf{C}_* = \sigma^2 \mathbf{I} + \begin{pmatrix} \mathbf{D} \\ \mathbf{d}(\mathbf{x}_*)^T \end{pmatrix} \mathbf{A}^{-1} \begin{pmatrix} \mathbf{D}^T & \mathbf{d}(\mathbf{x}_*) \end{pmatrix} = \begin{pmatrix} \mathbf{C} & \mathbf{v} \\ \mathbf{v}^T & v \end{pmatrix}$$

where $\mathbf{v} = \mathbf{D}\mathbf{A}^{-1}\mathbf{d}(\mathbf{x}_*)$ and $v = \mathbf{d}(\mathbf{x}_*)^T \mathbf{A}^{-1}\mathbf{d}(\mathbf{x}_*) + \sigma^2$. Note that $\mathbf{C}_*$ is symmetric.

Using the formulae for block matrices we have

$$|\mathbf{C}_*| = |\mathbf{C}||v - \mathbf{v}^T \mathbf{C}^{-1}\mathbf{v}|$$

and

$$\mathbf{C}_*^{-1} = \begin{pmatrix} \mathbf{C}^{-1} + \dfrac{\mathbf{C}^{-1}\mathbf{v}\mathbf{v}^T\mathbf{C}^{-1}}{v - \mathbf{v}^T\mathbf{C}^{-1}\mathbf{v}} & -\mathbf{C}^{-1}\mathbf{v}\dfrac{1}{v - \mathbf{v}^T\mathbf{C}^{-1}\mathbf{v}} \\ -\mathbf{v}^T\mathbf{C}^{-1}\dfrac{1}{v - \mathbf{v}^T\mathbf{C}^{-1}\mathbf{v}} & \dfrac{1}{v - \mathbf{v}^T\mathbf{C}^{-1}\mathbf{v}} \end{pmatrix}.$$

Letting $\mathbf{t}_*^T = (t_1, \ldots, t_N, t_*)$, we can calculate

$$\mathbf{t}_*^T \mathbf{C}_*^{-1}\mathbf{t}_* = \mathbf{t}^T\mathbf{C}^{-1}\mathbf{t} + \frac{1}{v - \mathbf{v}^T\mathbf{C}^{-1}\mathbf{v}}(\mathbf{v}^T\mathbf{C}^{-1}\mathbf{t} - t_*)^2$$

Thus the logarithm of the marginal likelihood $\log \mathcal{L}(\mathbf{t}_*|\boldsymbol{\alpha}, \sigma^2)$ is $\log \mathcal{L}(\mathbf{t}|\boldsymbol{\alpha}, \sigma^2) + \Delta\mathcal{L}$, where

$$\Delta\mathcal{L} = -\frac{1}{2}\left[\log 2\pi + \log|v - \mathbf{v}^T\mathbf{C}^{-1}\mathbf{v}| + \frac{1}{v - \mathbf{v}^T\mathbf{C}^{-1}\mathbf{v}}(\mathbf{v}^T\mathbf{C}^{-1}\mathbf{t} - t_*)^2\right].$$

This change can be interpreted probabilistically. Using the Woodbury matrix identity from Appendix A.1.3 again, $\boldsymbol{\Sigma}$ is related to $\mathbf{C}^{-1}$ by

$$\boldsymbol{\Sigma} = \mathbf{A}^{-1} - \mathbf{A}^{-1}\mathbf{D}^T\mathbf{C}^{-1}\mathbf{D}\mathbf{A}^{-1},$$

The predictive distribution for t_* has variance and mean

$$\begin{aligned} \sigma_*^2 &= \sigma^2 + \mathbf{d}(\mathbf{x}_*)^T\boldsymbol{\Sigma}\mathbf{d}(\mathbf{x}_*) \\ &= \sigma^2 + \mathbf{d}(\mathbf{x}_*)^T(\mathbf{A}^{-1} - \mathbf{A}^{-1}\mathbf{D}^T\mathbf{C}^{-1}\mathbf{D}\mathbf{A}^{-1})\mathbf{d}(\mathbf{x}_*) \\ &= \sigma^2 + \mathbf{d}(\mathbf{x}_*)^T\mathbf{A}^{-1}\mathbf{d}(\mathbf{x}_*)^T - \mathbf{v}^T C^{-1}\mathbf{v} = v - \mathbf{v}^T C^{-1}\mathbf{v}, \\ m_* &= \mathbf{d}(\mathbf{x}_*)^T\boldsymbol{\mu} = \sigma^{-2}\mathbf{d}(\mathbf{x}_*)^T\boldsymbol{\Sigma}\mathbf{D}^T\mathbf{t} = \mathbf{v}^T\mathbf{C}^{-1}\mathbf{t}, \end{aligned} \tag{8.10}$$

where we used the fact that $\mathbf{D}\mathbf{A}^{-1}\mathbf{D}^T = \mathbf{C} - \sigma^2\mathbf{I}$. Thus $(\mathbf{v}^T\mathbf{C}^{-1}\mathbf{t} - t_*)^2$ is the square of the difference of the sample measurement and its mean predicted by the current model.

Thus, the change in the log evidence is

$$\begin{aligned} \Delta\mathcal{L} &= -\frac{1}{2}\left[\log 2\pi + \log \sigma_*^2 + \left(\frac{m_* - t_*}{\sigma_*}\right)^2\right] \\ &= \log \frac{1}{\sqrt{2\pi\sigma_*^2}}\exp\left(-\frac{(m_* - t_*)^2}{2\sigma_*^2}\right). \end{aligned}$$

Hence the change is the logarithm of the likelihood of the new data value t_* at $\mathbf{x}_*$ given the predictive probability distribution $\mathcal{N}(m_*, \sigma_*^2)$.

Since $\sigma_* \geq \sigma$, the change lies between $-\infty$ and $\log(2\pi\sigma^2)^{-1/2}$. It can be positive. In this case, the new sample affirms the model. If the likelihood of the data is small, the marginal likelihood is reduced, indicating that the model should be improved. To do so, all quantities need to be updated first with the new data included.

New data sample: In this case,

$$\Delta\mathcal{L} = \log \frac{1}{\sqrt{2\pi}\sigma_*} \exp\left(-\frac{(m_* - t_*)^2}{2\sigma_*^2}\right),$$

$$\tilde{\mathbf{\Sigma}} = \mathbf{\Sigma} - \frac{1}{\sigma_*^2}\mathbf{\Sigma}\mathbf{d}(\mathbf{x}_*)\mathbf{d}(\mathbf{x}_*)^T\mathbf{\Sigma},$$

$$\tilde{\boldsymbol{\mu}} = \boldsymbol{\mu} - \frac{m_* - t_*}{\sigma_*^2}\mathbf{\Sigma}\mathbf{d}(\mathbf{x}_*).$$

Then the algorithm can continue improving the model by considering the basis functions in turn.

Sparse Bayesian learning infers a predictive distribution for t_* which is $\mathcal{N}(m_*, \sigma_*^2)$ with mean and variance as given in (8.10). This predictive distribution is heavily dependent on the model, since it depends on $\mathbf{d}(\mathbf{x}_*)$ which are the basis functions included in the model evaluated at $\mathbf{x}_*$. It is customary to choose basis functions for the dictionary which decay quickly when moving away from their centre, or basis functions with finite, compact support. Therefore, the degenerate case is possible where $\mathbf{d}(\mathbf{x}_*)$ is close to, or even equal to zero, and in this case the predictive probability distribution becomes $\mathcal{N}(0, \sigma^2)$ which is meaningless. The confidence we place in the predictions should only be informed by the data, and be independent of the dictionary of basis functions.

Let $\mathcal{S}$ be a subset of the samples. This could be all samples or a suitable set of neighbours of $\mathbf{x}_*$. We estimate the probability distribution of t_* to be normal with mean and variance

$$\bar{m} = \underset{\mathbf{x}_i \in \mathcal{S}}{\mathrm{mean}}\{t_i\},$$

$$\bar{\sigma} = \underset{\mathbf{x}_i \in \mathcal{S}}{\mathrm{var}}\{t_i\}.$$

With this estimate, the expected change, when considering $\mathbf{x}_*$, in the log evidence is

$$\mathbb{E}[\Delta\mathcal{L}] = \int_{-\infty}^{\infty}\left[\log\frac{1}{\sqrt{2\pi\sigma_*^2}} - \frac{(t_* - m_*)^2}{2\sigma_*^2}\right]$$
$$\frac{1}{\sqrt{2\pi\bar{\sigma}^2}}\exp\left(-\frac{(t_* - \bar{m})^2}{2\bar{\sigma}^2}\right)dt_*$$
$$= \log\frac{1}{\sqrt{2\pi\sigma_*^2}} - \frac{\bar{\sigma}^2 + (\bar{m} - m_*)^2}{2\sigma_*^2}.$$

The second term is the important one. If the predictive probability distribution does not match well the probability distribution estimated from the data in the neighbourhood, the expected change in the logarithm of the marginal likelihood is negative. This expected change creates an uncertainty map with the largest negative values being the most uncertain regions. The uncertainty map can guide the data gathering, informing us about where additional samples are necessary.

Next we consider the problem of collinearity. When basis functions are added to the model, collinearity can be introduced. In this case the columns of **D** become linearly dependent or nearly linearly dependent. In practical terms, this means that then there are infinitely many models which explain the data equally well, since linear combinations of basis functions can be exchanged for each other. For example, assume that the basis functions are derived from a kernel $k : \mathbb{R}^d \times \mathbb{R}^d \to \mathbb{R}$. That is

$$d_i(\mathbf{x}) = k(\mathbf{x}, \mathbf{x}_i).$$

Thus the model space is dependent on the data. If two sampling points lie close together, then the corresponding basis functions are very similar and the corresponding columns of **D** are nearly linearly dependent, since the basis functions are evaluated at the same sampling points. In the following we derive a technique which checks for collinearity while the algorithm is executed.

Since at initialization only one basis function is employed, the columns of **D** are at initialization linearly independent. Assume that the columns of **D** have remained linearly independent and that $d_i(\mathbf{x})$ is a candidate basis function to be added to the model. Since the columns of **D** are linearly independent, $\mathbf{d}_i$ can be decomposed uniquely into two vectors, one lying in the space spanned by the columns of **D** and one lying in the orthogonal complement of this space:

$$\mathbf{d}_i = \mathbf{D}\mathbf{a} + \mathbf{b},$$

where $\mathbf{D}^T\mathbf{b} = 0$.

If $\|\mathbf{b}\| > 0$, then $\mathbf{d}_i$ is linearly independent of the columns of **D** and can be added to the model. We also want to avoid near linear dependency, or in other words small values of $\|\mathbf{b}\|$.

Since the decomposition was orthogonal, we have

$$\|\mathbf{d}_i\|^2 = \|\mathbf{D}\mathbf{a}\|^2 + \|\mathbf{b}\|^2.$$

While the length of $\mathbf{d}_i$ is known, the length of $\mathbf{D}\mathbf{a}$ and $\mathbf{b}$, however, are unknown.

Consider

$$\begin{aligned}
\sigma^2 S_i &= \sigma^2 \mathbf{d}_i^T \mathbf{C}^{-1} \mathbf{d}_i = \mathbf{d}_i^T \left(\mathbf{I} - \sigma^{-2} \mathbf{D}\mathbf{\Sigma}\mathbf{D}^T \right) \mathbf{d}_i \\
&= \left(\mathbf{a}^T \mathbf{D}^T + \mathbf{b}^T \right) \left(\mathbf{I} - \sigma^{-2} \mathbf{D}\mathbf{\Sigma}\mathbf{D}^T \right) (\mathbf{D}\mathbf{a} + \mathbf{b}) \\
&= \mathbf{a}^T \mathbf{D}^T \mathbf{D} \left(\mathbf{I} - \sigma^{-2} \mathbf{\Sigma}\mathbf{D}^T \mathbf{D} \right) \mathbf{a} + \mathbf{b}^T \mathbf{b},
\end{aligned}$$

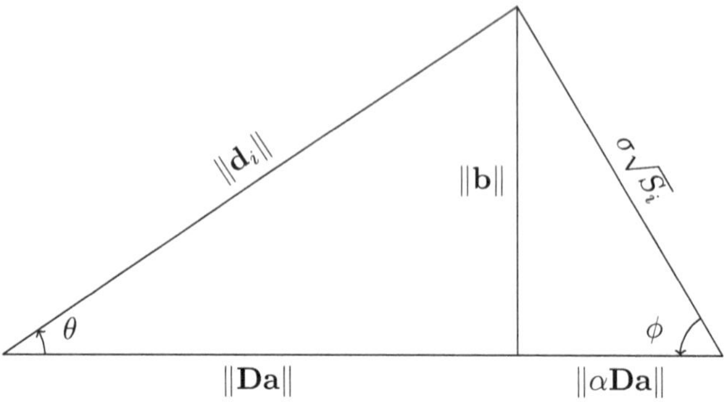

Figure 8.11: Geometric illustration.

due to orthogonality. Using $\sigma^{-2}\boldsymbol{\Sigma} = \left(\sigma^2\mathbf{A} + \mathbf{D}^T\mathbf{D}\right)^{-1}$, we can simplify part of the first term

$$\mathbf{I} - \sigma^{-2}\boldsymbol{\Sigma}\mathbf{D}^T\mathbf{D}$$
$$= \mathbf{I} - \left(\sigma^2\mathbf{A} + \mathbf{D}^T\mathbf{D}\right)^{-1}\left(\sigma^2\mathbf{A} + \mathbf{D}^T\mathbf{D} - \sigma^2\mathbf{A}\right)$$
$$= \boldsymbol{\Sigma}\mathbf{A}.$$

Combining the last two results, we get

$$\sigma^2 S_i = \mathbf{a}^T\mathbf{D}^T\mathbf{D}\boldsymbol{\Sigma}\mathbf{A}\mathbf{a} + \|\mathbf{b}\|^2.$$

However, the first quantity on the right hand side is not known.

To overcome the unknowns, note that we can scale the vector $\mathbf{Da}$ by a factor α such that

$$\|\alpha\mathbf{Da}\|^2 = \mathbf{a}^T\mathbf{D}^T\mathbf{D}\boldsymbol{\Sigma}\mathbf{A}\mathbf{a}.$$

Since $\alpha\mathbf{Da}$ and $\mathbf{b}$ are orthogonal, the square of the length of the vector $\alpha\mathbf{Da} + \mathbf{b}$ is

$$\|\alpha\mathbf{Da} + \mathbf{b}\|^2 = \|\alpha\mathbf{Da}\|^2 + \|\mathbf{b}\|^2$$
$$= \mathbf{a}^T\mathbf{D}^T\mathbf{D}\boldsymbol{\Sigma}\mathbf{A}\mathbf{a} + \|\mathbf{b}\|^2 = \sigma^2 S_i.$$

Figure 8.11 illustrates the relationship between the various vectors and their lengths geometrically. By the law of sines

$$\frac{\sin\theta}{\sin\phi} = \frac{\sigma\sqrt{S_i}}{\|\mathbf{d}_i\|} = \sigma\sqrt{\frac{\mathbf{d}_i^T}{\|\mathbf{d}_i\|}\mathbf{C}^{-1}\frac{\mathbf{d}_i}{\|\mathbf{d}_i\|}}.$$

Since $\sin\phi \le 1$, $\sigma\sqrt{S_i}/\|\mathbf{d}_i\|$ is an upper bound for $\sin\theta$. If this value becomes too small, $\mathbf{d}_i$ is deemed to be nearly linearly dependent to the columns of $\mathbf{D}$, and $d_i(\mathbf{x})$ is not added to the model even if $Q_i^2 > S_i$ indicates that augmenting the model with this basis function increases the logarithm of the marginal likelihood. Since $\mathbf{d}_i$ can be closely approximated by a linear combination of the columns of $\mathbf{D}$ a similar increase can be achieved by updating the coefficients of these.

Note that not including $d_i(\mathbf{x})$ at this stage does not mean that it is excluded from the model altogether. Since the algorithm also deletes basis functions from the model the linear dependency might not be present at a later stage.

A similar collinearity check can be derived when $d_i(\mathbf{x})$ is already in the model and α_i is finite. In this case, if $\sigma\sqrt{s_i}/\|\mathbf{d}_i\|$ is too small, the columns of $\mathbf{D}$ are deemed nearly linearly dependent. This collinearity check is used to postpone the re-estimation of a basis function in favour of other updates, since the near linear dependency might be resolved in the following steps of the algorithm. This increases stability. Otherwise, the algorithm can enter a cycle of re-estimations of linearly dependent basis functions. The hyperparameters of these basis functions are typically small indicating a wide probability distribution of the coefficients of these basis functions and thus uncertainty. Other strategies could be updating a group of basis functions within one step or creating a new basis function from a group of functions.

To summarize, Bayesian learning provides a framework where instead of the model being a vector of coefficients, a posterior probability distribution for $\mathbf{c}$ is obtained. This in turn leads to a predictive probability distribution for the targets and unseen data samples. In the case of unseen data samples, the change in the log-evidence this new data sample causes can be calculated. If this change is positive, it affirms the model; otherwise it indicates that the model should be improved. By using the empirical distribution of the data samples the expectation of the change in likelihood at new points can be calculated, creating an uncertainty map of the predictions which is independent of the current model. This framework copes with new data samples arriving through the pipeline. it also gives the possibility to add new basis functions to the model on the fly, while incorporating a mechanism to avoid collinearity.

8.14   Gaussian Process

In the preceding sections, we have chosen the function space in which our model lies by specifying a finite dictionary of basis functions, and the model $f(\mathbf{x})$ being a linear combination of these basis functions. This is restrictive. The aim of this section is to look for the model in the uncountably infinite space of functions. More specifically the model is a distribution over this space.

In the Dirichlet process, where we were confronted by infinitely many possibilities for the cluster parameters, the infinity is dealt with, because only a finite set of data is seen. The data size gives a maximum number of the necessary parameters. All other infinitely many possibilities are given one joint probability. Similarly here, the infinitely many possibilities are manageable, since only a finite set of data is seen.

A *stochastic process* is a probability distribution over functions $f(\mathbf{x})$, where $\mathbf{x} \in \mathbb{R}^D$, such that, when evaluated at any arbitrary set of points $\mathbf{x}_1, \ldots, \mathbf{x}_N$, the values $f(\mathbf{x}_1), \ldots, f(\mathbf{x}_N)$ follow a specific joint, multivariate probability distribution. If this distribution is a normal distribution, then it is a *Gaussian process*. If $D = 2$, it is also known as a *Gaussian random field*.

Recall that so far the model was specified as

$$f(\mathbf{x}) = \sum_{m=1}^{M} c_m d_m(\mathbf{x}).$$

Let $\mathbf{x}_1, \ldots, \mathbf{x}_N$ be any set of points, and denote $(f(\mathbf{x}_1), \ldots, f(\mathbf{x}_N))^T$ by $\mathbf{f}$. Then

$$\mathbf{f} = \mathbf{Dc},$$

where $\mathbf{D}$ is the design matrix and $\mathbf{c}$ is the vector of coefficients as before. The assumption is that $\mathbf{c}$ is normally distributed. Since $\mathbf{f}$ is a linear transformation of $\mathbf{c}$, it also has a normal distribution. Therefore, our models so far are Gaussian processes.

The prior assumption is that $\mathbf{c}$ has zero mean, and hence $\mathbf{f}$ has also zero mean. We considered the two cases, where the prior covariance matrix of $\mathbf{c}$ is either $\alpha^{-1}\mathbf{I}$, that is all coefficients have the same precision α, or the diagonal matrix $\mathbf{A}^{-1}$, where each coefficient has its own precision. The covariance matrix of $\mathbf{f}$ is then

$$\text{var}[\mathbf{f}] = \mathbb{E}[\mathbf{ff}^T] = \mathbb{E}[\mathbf{Dcc}^T\mathbf{D}^T] = \mathbf{D}\mathbb{E}[\mathbf{cc}^T]\mathbf{D}^T = \begin{cases} \alpha^{-1}\mathbf{DD}^T \\ \mathbf{DA}^{-1}\mathbf{D}^T \end{cases}.$$

Note that the covariance matrix of $\mathbf{f}$ depends on the set $\mathbf{x}_1, \ldots, \mathbf{x}_N$ via the design matrix. Therefore different sets of points will give rise to different multivariate distributions of $\mathbf{f}$. Defining $\mathbf{d}(\mathbf{x}_i)^T = (d_1(\mathbf{x}_i), \ldots, d_M(\mathbf{x}_i))$ as the vector of all basis functions evaluated at a specific point $\mathbf{x}_i$, the (i, j) entry of the covariance matrix is

$$\alpha^{-1}\mathbf{d}(\mathbf{x}_i)^T\mathbf{d}(\mathbf{x}_j) \qquad \text{or} \qquad \mathbf{d}(\mathbf{x}_i)^T\mathbf{A}^{-1}\mathbf{d}(\mathbf{x}_j).$$

The covariance matrices arising in a Gaussian process are completely specified, if the (i, j) entry is given by evaluating a kernel function at $\mathbf{x}_i$ and $\mathbf{x}_j$, i.e. $k(\mathbf{x}_i, \mathbf{x}_j)$. The covariance matrix is then the Gram matrix and positive definite. The above are two specific choices of kernel. The choice of kernel function depends on the application. For possible choices of kernels see Section 5.2 in Chapter 5.

Listing 8.3 implements six kernel functions and illustrates how to generate four draws from a Gaussian process with a given kernel. While a draw from a Gaussian process is a function, the listing generates densely spaced point values. These are a draw from the distribution with the required covariance matrix, which is the Gram matrix $\mathbf{K}$ for the given kernel and points $\mathbf{x}_1, \ldots, \mathbf{x}_N$. Its Cholesky factorization $\mathbf{K} = \mathbf{LL}^T$ is calculated. A vector $\mathbf{v}$ of length N is drawn from the standard, multivariate normal distribution. We set $\mathbf{f} = \mathbf{Lv}$ which is then a random vector from the multivariate normal distribution with

```
x= (−5:0.1:5)'; % Each row is a sample.

k = kGaussian(x,x, [1 1]);
% Note that sometimes it is necessary to add 1e−15*eye(size(x,1))
% to k to ensure positive definiteness due to rounding errors.
L = chol(k,'lower');
figure;
f1 = L*normrnd(0,1, size(x,1),1);
plot(x,f1,'k−');
hold on;
f2 = L*normrnd(0,1, size(x,1),1);
plot(x,f2,'k−−');
f3 = L*normrnd(0,1, size(x,1),1);
plot(x,f3,'k:');
f4 = L*normrnd(0,1, size(x,1),1);
plot(x,f4,'k−.');

% Each row of x and y is one data sample. The functions below
% calculate the covariance matrix for a specific kernel.

% Constant kernel.
function k = kConst(x,y, param)
k =param^2* ones(size(x,1), size(y,1));
end

% Linear kernel.
function k = kLinear(x,y, params)
% The matrix of the inner products of each row of x and each row of y
% is given by x*y'.
k = params(1)^2 + params(2)^2*x*y';
end

% Quadratic kernel.
function k = kQuadratic(x,y, params)
k = (params(1)^2 + params(2)^2*x*y').^2;
end

% Gaussian kernel.
function k = kGaussian(x,y, params)
% Calculate squared distance as the sum of the inner product of one
% row of x with itself and one row of y with itself minus twice the
% inner product of these two rows.
sd = repmat(dot(x,x,2),1,size(y,1)) + ...
    repmat(dot(y,y,2)',size(x,1),1) − 2*x*y';
k = params(1)^2* exp(−sd/(params(2)^2*2));
end

% Exponential kernel.
function k = kExponential(x,y,params)
sd = repmat(dot(x,x,2),1,size(y,1)) + ...
    repmat(dot(y,y,2)',size(x,1),1) − 2*x*y';
k = params(1)^2 * exp(−sqrt(sd)/params(2));
end

% Inverse multiquadric kernel.
function k = kInverseMQ(x,y, params)
sd = repmat(dot(x,x,2),1,size(y,1)) + ...
    repmat(dot(y,y,2)',size(x,1),1) − 2*x*y';
k = params(1)^2./sqrt(1+sd/(params(2)*2));
end
```

Listing 8.3: Draws from a Gaussian process with different kernels.

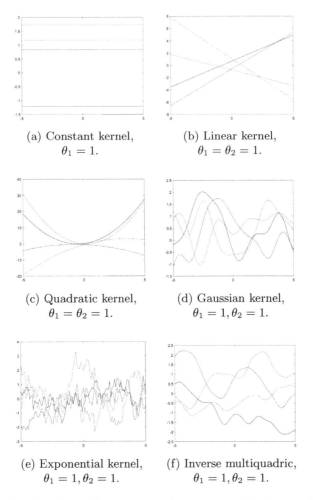

(a) Constant kernel,
$\theta_1 = 1$.

(b) Linear kernel,
$\theta_1 = \theta_2 = 1$.

(c) Quadratic kernel,
$\theta_1 = \theta_2 = 1$.

(d) Gaussian kernel,
$\theta_1 = 1, \theta_2 = 1$.

(e) Exponential kernel,
$\theta_1 = 1, \theta_2 = 1$.

(f) Inverse multiquadric,
$\theta_1 = 1, \theta_2 = 1$.

Figure 8.12: Four draws from a Gaussian process with different kernels.

covariance matrix $\mathbf{K}$. The point values of $\mathbf{f}$ are plotted as functions. Figure 8.12 shows the results of four draws for each of the implemented kernels.

Consider the case, when in our model $d_1(\mathbf{x}) = 1$ and $d_{d+1}(\mathbf{x}) = x_d$, the dth component of $\mathbf{x}$ for $d = 1, \ldots, D$. This means, the model is a linear polynomial. In this case,

$$\alpha^{-1}\mathbf{d}(\mathbf{x}_i)^T\mathbf{d}(\mathbf{x}_j) = \alpha^{-1}(1 + \mathbf{x}_i^T\mathbf{x}_j).$$

Thus this is the linear kernel with $\theta_1 = \theta_2 = \alpha^{-1}$.

On the other hand, if the kernel is the square of the inner product, then

$$
\begin{aligned}
k(\mathbf{x}_i, \mathbf{x}_j) &= (\mathbf{x}_i^T \mathbf{x}_j)^2 = \left(\sum_{d=1}^{D} x_{i,d} x_{j,d} \right) \left(\sum_{e=1}^{D} x_{i,e} x_{j,e} \right) \\
&= \sum_{d,e=1}^{D} (x_{i,d} x_{i,e})(x_{j,d} x_{j,e}) \\
&= \sum_{d,e=1}^{D} d_{d,e}(\mathbf{x}_i) d_{d,e}(\mathbf{x}_j) = \mathbf{d}(\mathbf{x}_i)^T \mathbf{d}(\mathbf{x}_j),
\end{aligned}
$$

if we define $M = D^2$ basis functions as the product of any two components, $d_{d,e}(\mathbf{x}) = x_d x_e$. Or in other words, the basis functions are all D-dimensional *monomials* of degree 2.

More generally, including a term of the inner product to the power of a means that monomials of degree a are part of the model space. As the number of monomials increases quickly with their degree, the kernel is a compact way to write these. A correspondence between kernels and basis functions is not that straightforward in general.

If the kernel contains a positive, additive constant, then the mean of the Gaussian process can be assumed to be zero, because a positive, additive constant corresponds to the constant basis function, also known as the *bias* being included in the model space.

This concludes the description of drawing a function from a Gaussian process. We now turn our attention to drawing a function from the posterior of a Gaussian process, once some data pairs $(\mathbf{x}_1, t_1), \ldots, (\mathbf{x}_N, t_N)$ have been seen. Let $\mathbf{x}_*$ be arbitrary and distinct from $\mathbf{x}_n$, $n = 1, \ldots, N$. Recall that the measurements we see have additive normally distributed noise with mean zero and variance σ^2, that is

$$
t_n = f(\mathbf{x}_n) + \epsilon_n \qquad \text{as well as} \qquad t_* = f(\mathbf{x}_*) + \epsilon_*.
$$

We assume that f is drawn from a Gaussian process with kernel k and mean zero. Therefore, the vector $(t_1, \ldots, t_N, t_*)$ follows a normal distribution with mean zero and covariance matrix

$$
\begin{pmatrix} \mathbf{C} & \mathbf{k}(\mathbf{x}_*) \\ \mathbf{k}(\mathbf{x}_*)^T & c(\mathbf{x}_*) \end{pmatrix},
$$

where the vector $\mathbf{k}(\mathbf{x}_*)$ has elements $k(\mathbf{x}_n, \mathbf{x}_*)$ for $n = 1, \ldots, N$, $c(\mathbf{x}_*) = k(\mathbf{x}_*, \mathbf{x}_*) + \sigma^2$, and the (i, j) entry of $\mathbf{C}$ is

$$
C_{ij} = k(\mathbf{x}_i, \mathbf{x}_j) + \sigma^2 \delta_{ij},
$$

where δ_{ij} is zero for $i \neq j$ and one for $i = j$ and is known as the *Kronecker delta*.

Using the inversion formula for a block matrix in Appendix A.1.1, the precision matrix of the joint distribution is

$$
\begin{pmatrix}
\mathbf{C}^{-1} + \dfrac{\mathbf{C}^{-1}\mathbf{k}(\mathbf{x}_*)\mathbf{k}(\mathbf{x}_*)^T\mathbf{C}^{-1}}{c(\mathbf{x}_*) - \mathbf{k}(\mathbf{x}_*)^T\mathbf{C}^{-1}\mathbf{k}(\mathbf{x}_*)} & \dfrac{-\mathbf{C}^{-1}\mathbf{k}(\mathbf{x}_*)}{c(\mathbf{x}_*) - \mathbf{k}(\mathbf{x}_*)^T\mathbf{C}^{-1}\mathbf{k}(\mathbf{x}_*)} \\[2ex]
\dfrac{-\mathbf{k}(\mathbf{x}_*)^T\mathbf{C}^{-1}}{c(\mathbf{x}_*) - \mathbf{k}(\mathbf{x}_*)^T\mathbf{C}^{-1}\mathbf{k}(\mathbf{x}_*)} & \dfrac{1}{c(\mathbf{x}_*) - \mathbf{k}(\mathbf{x}_*)^T\mathbf{C}^{-1}\mathbf{k}(\mathbf{x}_*)}
\end{pmatrix}.
$$

We can deduce that the conditional probability of t_* given $\mathbf{t}$ is normal and has variance and mean

$$
\begin{aligned}
\sigma^2(\mathbf{x}_*) &= c(\mathbf{x}_*) - \mathbf{k}(\mathbf{x}_*)^T\mathbf{C}^{-1}\mathbf{k}(\mathbf{x}_*) = \sigma^2 + k(\mathbf{x}_*, \mathbf{x}_*) - \mathbf{k}(\mathbf{x}_*)^T\mathbf{C}^{-1}\mathbf{k}(\mathbf{x}_*), \\
m(\mathbf{x}_*) &= \mathbf{k}(\mathbf{x}_*)^T\mathbf{C}^{-1}\mathbf{t}.
\end{aligned}
$$

(8.11)

Since $\mathbf{x}_*$ is chosen arbitrarily (excluding $\mathbf{x}_1, \ldots, \mathbf{x}_N$), this specifies a distribution over functions.

Let $\mathbf{x}_1^*, \ldots, \mathbf{x}_{N^*}^*$ be N^* points where predictions $\mathbf{t}^* = (t_1^*, \ldots, t_{N^*}^*)^T$ are to be made. The joint distribution of $(\mathbf{t}, \mathbf{t}^*)$ has covariance matrix

$$
\begin{pmatrix}
\mathbf{C} & \mathbf{K} \\
\mathbf{K}^T & \mathbf{C}^*
\end{pmatrix},
$$

where $\mathbf{C}$ is as before, $\mathbf{K}$ is a $N \times N^*$ matrix with (n, m) entry given by $k(\mathbf{x}_n, \mathbf{x}_m^*)$ and the (i, j) entry of $\mathbf{C}^*$ is $k(\mathbf{x}_i^*, \mathbf{x}_j^*) + \sigma^2\delta_{ij}$. The conditional probability distribution of $\mathbf{t}^*$ given $\mathbf{t}$ has mean and variance

$$
\begin{aligned}
\mathbf{m}^* &= \mathbf{K}^T\mathbf{C}^{-1}\mathbf{t}, \\
\mathbf{\Sigma}^* &= \mathbf{C}^* - \mathbf{K}^T\mathbf{C}^{-1}\mathbf{K}.
\end{aligned}
$$

As before, we can draw point values from this distribution and plot as functions to illustrate draws from the posterior Gaussian process. Figure 8.13 illustrates four draws using various kernels along with four training data points from the function $1 + x + \sin x$, which is drawn in red. The mean of the conditional distribution is drawn in blue along with a band of two standard deviations in gray. The parameters were kept at $\theta_1 = \theta_2 = 1$. The noise variance was set to 0.00001. The distribution is tightened near the training data, while the uncertainty increases away from the training data points. In the range between two training points the uncertainty is largest in the middle. As can be seen to the right, away from the training data the distribution becomes the prior distribution of the Gaussian process (as for example in Figure 8.12d for the Gaussian kernel) with added noise, since the entries of $\mathbf{K}$ are getting close to zero, because the Gaussian kernel decreases as the distance between points increases.

Comparing (8.11) to (8.10), we see that the results agree when the kernel is chosen to be $k(\mathbf{x}, \mathbf{y}) = \mathbf{d}(\mathbf{x})^T\mathbf{A}^{-1}\mathbf{d}(\mathbf{y})$. In the previous section, however, the entries of $\mathbf{A}$ were determined by maximizing the log evidence. Similarly,

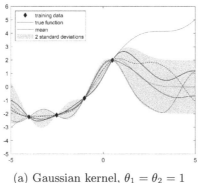

(a) Gaussian kernel, $\theta_1 = \theta_2 = 1$

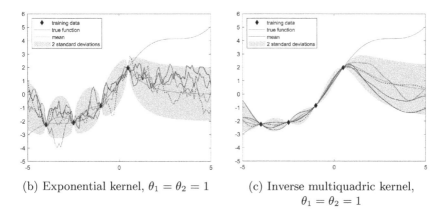

(b) Exponential kernel, $\theta_1 = \theta_2 = 1$ (c) Inverse multiquadric kernel, $\theta_1 = \theta_2 = 1$

Figure 8.13: Draws from the posterior Gaussian process for different kernels.

θ_1 and θ_2 can be chosen by maximizing the log likelihood of the training data, which is given by

$$\mathcal{L} = -\frac{1}{2} \left[\log |\mathbf{C}| + \mathbf{t}^T \mathbf{C}^{-1} \mathbf{t} + N \log(2\pi) \right].$$

Using the formulae in Appendices A.2.4 and A.2.7 for the derivatives of the determinant and inverse of a matrix, when the entries of the matrix depend on a variable, we arrive at

$$\frac{\partial}{\partial \theta_i} \mathcal{L} = -\frac{1}{2} \left[\text{tr} \left(\mathbf{C}^{-1} \frac{\partial \mathbf{C}}{\partial \theta_i} \right) - \mathbf{t}^T \mathbf{C}^{-1} \frac{\partial \mathbf{C}}{\partial \theta_i} \mathbf{C}^{-1} \mathbf{t} \right].$$

The (m, n) entry of the derivative of $\mathbf{C}$ with respect to θ_i is

$$\frac{\partial}{\partial \theta_i} (k(\mathbf{x}_m, \mathbf{x}_n) + \sigma^2 \delta_{mn}) = \frac{\partial}{\partial \theta_i} k(\mathbf{x}_m, \mathbf{x}_n).$$

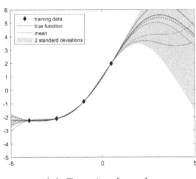

(a) Gaussian kernel,
$\theta_1 \approx 2.81, \theta_2 \approx 2.85$

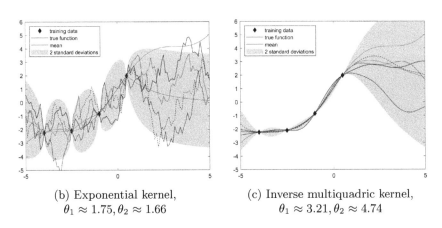

(b) Exponential kernel,
$\theta_1 \approx 1.75, \theta_2 \approx 1.66$

(c) Inverse multiquadric kernel,
$\theta_1 \approx 3.21, \theta_2 \approx 4.74$

Figure 8.14: Draws from the posterior Gaussian process for different kernels.

Thus the derivative of the kernel with regards to its parameters is needed. For the Gaussian kernel

$$k(\mathbf{x}, \mathbf{y}) = \theta_1^2 \exp\left(-\frac{\|\mathbf{x} - \mathbf{y}\|^2}{2\theta_2^2}\right),$$

this is for example

$$\frac{\partial}{\partial \theta_1} k(\mathbf{x}, \mathbf{y}) = 2\theta_1 \exp\left(-\frac{\|\mathbf{x} - \mathbf{y}\|^2}{2\theta_2^2}\right) = \frac{2}{\theta_1} k(\mathbf{x}, \mathbf{y}),$$

$$\frac{\partial}{\partial \theta_2} k(\mathbf{x}, \mathbf{y}) = \frac{\|\mathbf{x} - \mathbf{y}\|^2}{\theta_2^3} \theta_1^2 \exp\left(-\frac{\|\mathbf{x} - \mathbf{y}\|^2}{2\theta_2^2}\right) = \frac{\|\mathbf{x} - \mathbf{y}\|^2}{\theta_2^3} k(\mathbf{x}, \mathbf{y}).$$

Setting the gradient of $\mathcal{L}$ to zero and solving for the parameters is generally not straight forward. However, since the gradient is known, it can be used to approximate the maximum iteratively (see for example [2]).

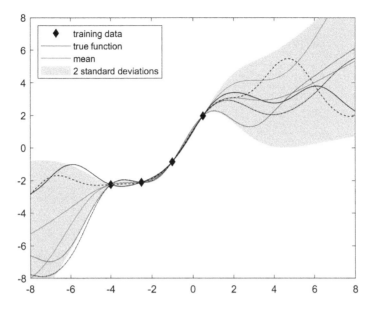

Figure 8.15: Sum of linear and Gaussian kernel with optimized parameters.

Figure 8.14 shows the mean and two standard deviations of the posterior Gaussian process for various kernels, where the parameters have been optimized, as well as four draws. The fit near the data points is excellent, when the Gaussian kernel is used. The inverse multiquadric kernel also performs well, while the exponential kernel seems unsuitable.

The mean of the posterior Gaussian process returns to zero away from the data points due to the choice of kernel, though it does so slower than with the original choice of parameters. Therefore, we combine the Gaussian kernel with a linear kernel. That is, the kernel is given by

$$k(\mathbf{x}, \mathbf{y}) = \theta_1^2 + \theta_2^2 \mathbf{x}^T \mathbf{y} + \theta_3^2 \exp\left(-\frac{\|\mathbf{x} - \mathbf{y}\|^2}{2\theta_4^2}\right).$$

After optimization the parameters are approximately $\theta_1 \approx 10^{-5}, \theta_2 \approx 0.90, \theta_3 \approx 1.13$ and $\theta_4 \approx 1.65$. The goodness of approximation is illustrated in Figure 8.15 over a wider range. While the mean drifts away from the true function, the general trend is kept. A better approximation can be achieved with more data. Which combination of kernels should be used depends on the application and is a field of active research with new kernels being developed for specific usages.

CHAPTER 9

Feature Learning

This chapter concentrates on inferring the building blocks which are essential for an adequate model of the data. Neural networks are revisited and their many variations explained, especially how design choices such as activation (propagation functions) and error functions depend on the task at hand. Error backpropagation is derived in generality as the application of the chain rule when differentiating the different error functions. Autoencoders are introduced as very simple network architecture and illustrated with an extensive example showing how neural networks create a model space for the data. The relationships to other techniques are explored to gain an intuitive understanding of the mechanisms at work. The chapter concludes with a Bayesian treatment of inferring the model space, illustrated by the Indian Buffet Process.

In the previous chapter, the choice of model space was still largely governed by the user, making a choice on basis functions or kernels to be used. The data can inform these choices or optimize parameters of the basis functions or kernels. If the basis functions are kernels centred at the data, the data has influence also this way. However, the shape is still very much up to user choice.

In this chapter, we explore the possibility of the data deciding the model space. To do so, many examples of data generated from the same process are necessary. For example, the MNIST data set of [26], where the generating process was people writing digits, contains $60,000$ images of size 28×28 pixels of handwritten digits. The objective is a technique which generates a model space where all samples are equally well represented, if reconstruction is the objective, or separated for classification.

One drawback is that this is only possible where enough reliable data is seen. While it is possible to have some samples, where some features are not given or given with non-negligible error, there need to be sufficiently many samples where these features have reliable values. This is known as *veracity* and is one of the *four V's of Big Data*. The other three are *volume*, *velocity* and *variety*.

On the one hand, large volumes of data are necessary to cope with missing or corrupted parts of the data. On the other hand, processing such large volumes of data is a challenge in itself. The Sentinel satellites of the European Space Agency are expected to eventually gather four Terabytes of data per day. These are only twelve satellites of the 4000 plus satellites orbiting earth.

The Large Hadron Collider at CERN creates 20,000,000 collisions per second. Thus new data arrives at extraordinary velocity. Only 400 per second are recorded as interesting. This results in 3,000 Terabytes of data per year.

The Electronic Control Units (ECUs) in a car record several thousands of signals approximately every quarter of a second. These relate to many different quantities and states, each of which has its own meaning and interpretation. The essential information needs to be extracted from the *variety of data*.

Sometimes this takes an unexpected form as in Figures 7.3 and 7.4 of the principal component representation of hand-written digits from the MNIST data set. However, there the goal was to make the digits machine identifiable, not recognizable for a human. We will again use this data set in this chapter.

9.1 Neural Networks

As before we are given data pairs $(\mathbf{x}_1, \mathbf{t}_1), \ldots, (\mathbf{x}_N, \mathbf{t}_N)$, where $\mathbf{x}_n \in \mathbb{R}^D$ and each component of $\mathbf{x}_n$ is a feature. Also $\mathbf{t}_n$ can be multidimensional, that is $\mathbf{t}_n \in \mathbb{R}^K$.

Recall that a neural network consists of a set of input neurons, a set of output neurons and possibly several sets of hidden neurons. Each component of $\mathbf{x}_n$ is passed to one input neuron. Therefore, the number of input neurons is D. During training, the results of the output neurons are compared to $\mathbf{t}_n$. Hence, the number of output neurons is K. The synapses connecting one set of neurons to another are a layer.

More specifically, let L be the number of layers. This means that there are $L-1$ sets of hidden neurons. The size of the lth set of hidden neurons is M_l. The latent variables of the hidden neurons are denoted by $z_m^{(l)}, l = 1, \ldots, L-1, m = 1, \ldots, M_l$. In the lth layer, the weight of the synapse connecting the ith neuron on the left to the jth neuron on the right is denoted by $w_{ji}^{(l)}$. This neural

network is illustrated as

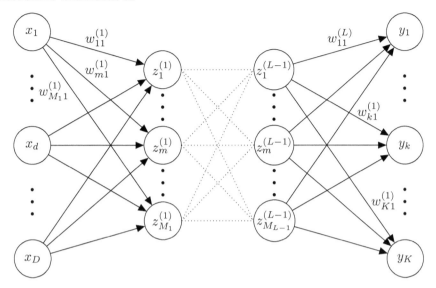

The dotted connections indicate that any combination of hidden neurons and layers is possible.

For $l \neq 1, L$, the result of the *propagation function* is the *activation*

$$a_j^{(l)} = \sum_{i=1}^{M_{l-1}} w_{ji}^{(l)} z_i^{(l-1)} = \mathbf{w}_j^{(l)^T} \mathbf{z}^{(l-1)}.$$

This is passed to the *activation function*, also called *transfer function*, of this layer h_l to generate the latent variable

$$z_j^{(l)} = h_l(a_j^{(l)}) = h_l \left(\mathbf{w}_j^{(l)^T} \mathbf{z}^{(l-1)} \right).$$

For $l = 1$, we have

$$a_j^{(1)} = \mathbf{w}_j^{(1)^T} \mathbf{x} \qquad \text{and} \qquad z_j^{(1)} = h_1 \left(\mathbf{w}_j^{(1)^T} \mathbf{x} \right).$$

For $l = L$, the output variables $y_1, \ldots, y_K$ are calculated,

$$y_k = h_L(a_k^{(L)}) = h_L \left(\mathbf{w}_k^{(L)^T} \mathbf{z}^{(L-1)} \right).$$

A list of possible activation functions is given in Section 5.5. The choice depends on the problem at hand:

- For **regression** problems, the identity, i.e. the *linear* activation function $h_L(x) = x$ is used in the output layer.

- For **binary classification**, $K = 1$ and t_n takes either the value 0 or 1 depending on whether the sample with features $\mathbf{x}_n$ belongs to the negative or positive class respectively. There is only one output neuron and the activation function of the final layer is the *logistic sigmoid* function $h_L(x) = (1 + \exp(-x))^{-1}$. The output is interpreted as the probability of the input with features $x_1, \ldots, x_D$ belonging to the positive class.

- If there are K **separate, binary classification** tasks for the same data, then $\mathbf{t}_n$ is a vector of zeros and ones. The kth component indicates the class membership of the kth classification task. There are K output neurons. Each output neuron can be likened to a binary classifier using the logistic sigmoid function.

- For **multiple classes**, $\mathbf{t}_n$ is a *1-of-K representation*. This means only one of the K components of $\mathbf{t}_n$ is one and all others are zero. A one in the kth position indicates the sample with features $\mathbf{x}_n$ belongs to the kth class. The output activation function is the *softmax* function. It maps the K-dimensional vector of activations $\mathbf{a} = (a_1, \ldots, a_K)^T$ to a K-dimensional vector $\boldsymbol{\sigma}(\mathbf{a})$ with the jth entry of $\boldsymbol{\sigma}(\mathbf{a})$ being

$$\boldsymbol{\sigma}(\mathbf{a})_j = \frac{\exp(a_j)}{\sum_{k=1}^{K} \exp(a_k)}.$$

  The kth output is interpreted as the probability of the sample with features $\mathbf{x} = (x_1, \ldots, x_D)^T$ belonging to the kth class.

If the activation function in all neurons is the logistic sigmoid, then the neural network is commonly known as a *multilayer perceptron*. However, this is a misnomer, since the perceptron uses the discontinuous step function sgn illustrated in Figure 4.7 as the activation function. In the section on backpropagation, we will see the advantages continuity brings. Therefore the logistic sigmoid function is a better choice. To add to the confusion, the term multilayer perceptron is now being used, when arbitrary activation functions are employed, and for both classification or regression tasks. Because of this dilution of the definition, the form of the neural network cannot be deduced, if the term multilayer perceptron is used.

The neural network is a function mapping the input $\mathbf{x} = (x_1, \ldots, x_D)^T$ to the output $\mathbf{y} = (y_1, \ldots, y_K)^T$. The input passing through the network is known as *forward propagation*. Cycles, where information could be passed backwards, are not allowed. The architecture is strictly *feed-forward*.

As the latent variables can be relabeled, which means that neurons are interchanged with each other, there are $\prod_{l=1}^{L-1} M_l!$ equivalent neural networks, which give the same mapping from input to output. Another equivalence arises from the activation functions of the hidden neurons being even functions, i.e. $h(-x) = h(x)$, or odd functions, i.e. $h(-x) = -h(x)$. If the signs of the weights of all synapses going into a particular hidden neuron are flipped, then, if the activation function is even, the mapping from input to output remains the same, while, if the activation function is odd, flipping also the sign for all synapses going out from this hidden neuron also keeps the mapping the same. If such sign flips are possible for all hidden neurons, then this means there are $\prod_{l=1}^{L-1} 2^{M_l}$ equivalent neural networks. These are known as *weight space symmetries*. Altogether there are

$$\prod_{l=1}^{L-1} 2^{M_l} M_l!$$

symmetries.

Not all possible synapses need to be present. If the number of synapses is small, the network is known as *sparse*. Excluding certain synapses is a design decision. An example is a convolutional neural network. If the weight of a synapse becomes zero during the learning process, that is effectively it does not contribute to the final output, this means that the information carried by this synapse is deemed unimportant. While the resulting mapping is the same as excluding the synapse, one is by design, the other is determined from the data by learning. It is also possible for a synapse to stretch over two layers. This is known as a *skip-layer* connection.

The larger the number of synapses, the more training data is necessary to train the network, since we need enough information to decide the values of the weights.

The training of the neural network consists of adjusting the weights until the output for the training data is deemed good enough. Again, the measure of goodness depends on the task at hand.

- Recall that for **regression**, we assume that the data are the result of an underlying process with additive normally distributed noise with mean zero and variance σ^2. In this context, the process is modeled by the neural network. That is, $\mathbf{t}$ is normally distributed with the mean being the output $\mathbf{y}$ for input $\mathbf{x}$ and variance $\sigma^2 \mathbf{I}$, where $\mathbf{I}$ is the $K \times K$ identity matrix. The mean $\mathbf{y}$ depends also on the weights of all the synapses in the network, which we denote by the vector $\mathbf{w}$, in short $\mathbf{y}(\mathbf{x}, \mathbf{w})$. The likelihood of the data given the network can be calculated as

$$\mathcal{L} = \prod_{n=1}^{N} (2\pi\sigma^2)^{-K/2} \exp\left(-\frac{1}{2\sigma^2} \|\mathbf{t}_n - \mathbf{y}(\mathbf{x}_n, \mathbf{w})\|^2\right).$$

The weights are determined by maximizing the likelihood. This is equivalent to minimizing the negative logarithm of the likelihood

$$-\log \mathcal{L} = \frac{NK}{2} \log \sigma^2 + \frac{NK}{2} \log(2\pi) + \frac{1}{2\sigma^2} \sum_{n=1}^{N} \|\mathbf{t}_n - \mathbf{y}(\mathbf{x}_n, \mathbf{w})\|^2.$$

It is convention to state this as minimizing the error function

$$E(\mathbf{w}) = \frac{1}{2} \sum_{n=1}^{N} \|\mathbf{y}(\mathbf{x}_n, \mathbf{w}) - \mathbf{t}_n\|^2$$

known as *error sum of squares (ESS)*.

- For **binary classification**, t_n is zero if the sample with features $\mathbf{x}_n$ belongs to the negative class, and one otherwise. Since the single output of the neural network is interpreted as the probability of belonging to the positive class, the likelihood function in this case is given by

$$\mathcal{L} = \prod_{n=1}^{N} y(\mathbf{x}_n, \mathbf{w})^{t_n} \left(1 - y(\mathbf{x}_n, \mathbf{w})\right)^{1-t_n}.$$

Taking the negative logarithm gives the *cross-entropy* error function

$$E(\mathbf{w}) = - \sum_{n=1}^{N} t_n \log y(\mathbf{x}_n, \mathbf{w}) + (1 - t_n) \log \left(1 - y(\mathbf{x}_n, \mathbf{w})\right).$$

- If the network tackles K **separate, binary classification** tasks on the same data, then the error function is

$$E(\mathbf{w}) = - \sum_{k=1}^{K} \sum_{n=1}^{N} t_{n,k} \log y_k(\mathbf{x}_n, \mathbf{w}) + (1 - t_{n,k}) \log \left(1 - y_k(\mathbf{x}_n, \mathbf{w})\right),$$

where $t_{n,k}$ is the kth component of $\mathbf{t}_n$ and $y_k(\mathbf{x}_n, \mathbf{w})$ is the output of the kth neuron, when $\mathbf{x}_n$ is the input. It is the sum of the errors of the K individual binary classification tasks.

- For K **multiple classes**, $y_k(\mathbf{x}_n, \mathbf{w})$ is taken as the probability of the sample with features $\mathbf{x}_k$ belonging to the kth class and $\mathbf{t}_n$ is a 1-of-K representation. The likelihood of the data is

$$\mathcal{L} = \prod_{n=1}^{N} \prod_{k=1}^{K} y_k(\mathbf{x}_n, \mathbf{w})^{t_{n,k}}.$$

As before, taking the negative logarithm leads to the error function

$$E(\mathbf{w}) = - \sum_{k=1}^{K} \sum_{n=1}^{N} t_{n,k} \log y_k(\mathbf{x}_n, \mathbf{w}).$$

This looks very similar to the one above, but now only one of $t_{n,k}$, $k = 1, \ldots, K$ is non zero.

We illustrate the capability of neural networks by training a two layer neural network with three hidden neurons to represent the function $1+x+\sin x$. The biases are modeled explicitly with dummy neurons which always have the value one. The network diagram is

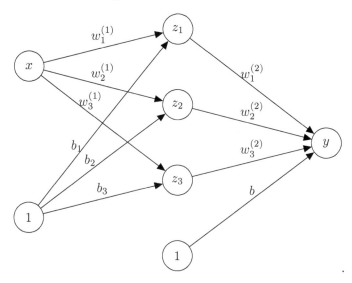

The activation function in the first layer is the hyperbolic tangent in the form

$$h_1(x) = \frac{2}{1 + \exp(-2x)} - 1,$$

also known as the *tan-sigmoid* activation function, since this implementation is faster. In the second layer, it is the linear activation function.

Twenty-one equally spaced training points are used. Once the neural network is trained, it can be evaluated for any input x. Figure 9.1a shows the results, the true function in red, the prediction in blue and the bias b in black. It also displays $z_1 w_1^{(2)}$, $z_2 w_2^{(2)}$ and $z_3 w_3^{(2)}$, which are all functions of the input x. The latent variables z_1, z_2 and z_3 can be interpreted as basis functions the neural network constructed from the data. The prediction is the weighted sum of these and the bias. The error at the training points is shown in 9.1b.

Since the weights are initialized randomly and the optimization procedure might converge in a local and not a global optimum, the training outcome can be very different for different runs. Several runs of the script in Listing 9.1 illustrate this. Modifying the script by for example increasing the number of hidden neurons or the number and selection of training points gives further insight. Such a two layer neural network can approximate any continuous function on a finite interval to arbitrary accuracy as long as there are sufficiently many hidden neurons and training data to calculate the weights.

```
clear;
x = -5:0.5:5;
% True function.
t = 1 + x + sin(x);
% Contstruct two layer network with 3 hidden neurons.
net = fitnet(3);
% Remove normalization and de-normalization.
net.input.processFcns = { };
net.output.processFcns= { };
% Train the neural network.
[net,¬,¬,e] = train(net,x,t);
% Weights in the first layer.
IW = net.IW{1,1};
% Bias weights in the first layer.
b1 = net.b{1};
% Bias weight in the second layer.
b2 = net.b{2};
% Weights in the second layer.
LW = net.LW{2,1};

X = -5:0.1:5;
T = 1 + X + sin(X);
% Predictions.
Y = b2 + LW * tansig( b1 * ones(1,length(X)) + IW * X );
% Constructed basis functions.
Z = tansig( b1 * ones(1,length(X)) + IW * X );
figure;
% Plot true function.
plot(X,T,'r-')
hold on
% Plot predictions.
plot(X,Y,'b-')
% Plot bias.
plot(X,b2*ones(1,length(X)),'k-')
% Plot basis functions with coefficients.
plot(X,LW(1)*Z(1,:),'k:')
plot(X,LW(2)*Z(2,:),'k-.')
plot(X,LW(3)*Z(3,:),'k--')
legend('true function', 'prediction','bias','Location','northwest')
% Plot error at training data.
figure
plot(x,e,'ko', 'MarkerFaceColor','k','MarkerSize',3)
hold on
plot(x,zeros(1,length(x)),'k-')
```

Listing 9.1: Two layer neural network regression.

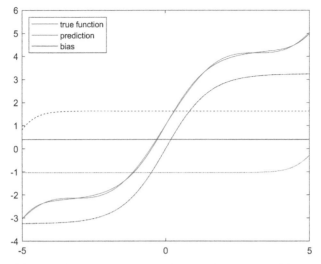

(a) True function, prediction, bias and constructed basis functions.

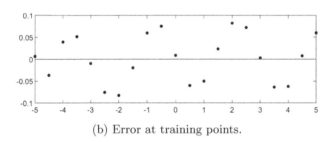

(b) Error at training points.

Figure 9.1: Two layer neural network regression with a set of three hidden neurons.

9.2   Error Backpropagation

For all neural networks, an error function $E(\mathbf{w})$ needs to be minimized. A necessary condition for a minimum is that the gradient $\nabla E(\mathbf{w})$ vanishes, where the gradient is the vector formed from the derivatives of $E(\mathbf{w})$ with respect to each component of $\mathbf{w}$, that is each $w_{ji}^{(l)}$. Hence, one possible technique is to calculate the gradient, set it to zero and solve for the weights. In most cases this is not feasible. Iterative techniques on the other hand use the fact that the gradient evaluated for a particular $\mathbf{w}$ points in the direction of the steepest ascent. Thus, its negative gives the direction in which the function decreases most rapidly. The simplest iterative technique is *gradient descent* and updates the current weight vector $\mathbf{w}$ by a step along negative gradient,

$$\mathbf{w}^{\text{new}} = \mathbf{w} - \eta \nabla E(\mathbf{w}),$$

where the step size $\eta > 0$ is called the *learning rate*. It is also called *steepest descent*. There are more sophisticated methods such as *conjugate gradients* and *quasi-Newton methods*. Optimization is, however, not the purpose of this text. More information on this topic can be found in [14].

The error functions all involve a sum over the training data, which can be written as

$$E(\mathbf{w}) = \sum_{n=1}^{N} E_n(\mathbf{w}).$$

Since the derivative of a sum is the sum of derivatives, it follows

$$\nabla E(\mathbf{w}) = \sum_{n=1}^{N} \nabla E_n(\mathbf{w}).$$

There are now several possibilities to update the weight vector $\mathbf{w}$.

- Calculating $\nabla E_n(\mathbf{w})$ for all n and thus using $\nabla E(\mathbf{w})$ in the chosen optimization method are known as *batch* methods.

- When $\mathbf{w}$ is updated using only a subset $\mathcal{S}$ of training data, that is $\sum_{\mathcal{S}} \nabla E(\mathbf{w})$ is used in the optimization method, then these are *mini-batch methods*. The batch size and how the mini-batches are chosen are design choices. Often the training data is partitioned randomly several times. If the optimization method is gradient descent, this is known as *mini-batch gradient descent*.

- If only one $\nabla E_n(\mathbf{w})$ is used, when updating $\mathbf{w}$, the method is *online*, *sequential* or *stochastic*. The algorithm either cycles through the training data or selects training samples randomly with replacement. It is called *online gradient descent*, *sequential gradient descent* or *stochastic gradient descent*, if the optimization method is gradient descent.

Mini-batch and stochastic methods have the advantage that a local minimum of an individual $E_n(\mathbf{w})$ or a set $\sum_{\mathcal{S}} E(\mathbf{w})$ will generally not be a local minimum for a different choice of n or $\mathcal{S}$. Thus, it escapes from local minima. They also are more efficient with duplicates in the training set, since only a selection of training data is seen in each update anyway, while when all training data are used, the duplicates add computational overhead.

To calculate $\nabla E_n(\mathbf{w})$, the chain rule is used multiple times, since the output is the result of applying a chain of linear propagation functions alternating with activation functions.

We first consider the task of regression. In this case,

$$
\begin{aligned}
E_n(\mathbf{w}) &= \frac{1}{2}\|\mathbf{y}(\mathbf{x}_n, \mathbf{w}) - \mathbf{t}_n\|^2 = \frac{1}{2}\sum_{k=1}^{K}(y_k(\mathbf{x}_n, \mathbf{w}) - t_{nk})^2 \\
&= \frac{1}{2}\sum_{k=1}^{K}\left(a_k^{(L)}(\mathbf{x}_n, \mathbf{w}) - t_{nk}\right)^2,
\end{aligned}
$$

since the linear activation function is used. The dependence of the activation $a_k^{(L)}$ on $\mathbf{x}_n$ and $\mathbf{w}$ is made explicit by writing $a_k^{(L)}(\mathbf{x}_n, \mathbf{w})$.

Taking the derivative with respect to $w_{ji}^{(l)}$, we arrive at

$$
\begin{aligned}
\frac{\partial}{\partial w_{ji}^{(l)}} E_n(\mathbf{w}) &= \sum_{k=1}^{K} \left(a_k^{(L)}(\mathbf{x}_n, \mathbf{w}) - t_{nk} \right) \frac{\partial}{\partial w_{ji}^{(l)}} a_k^{(L)}(\mathbf{x}_n, \mathbf{w}) \\
&= \sum_{k=1}^{K} \left(y_k(\mathbf{x}_n, \mathbf{w}) - t_{nk} \right) \frac{\partial}{\partial w_{ji}^{(l)}} a_k^{(L)}(\mathbf{x}_n, \mathbf{w}).
\end{aligned}
\tag{9.1}
$$

It is the weighted sum of the errors in the output neurons, where the weights are the derivative of the corresponding activation with respect to $w_{ji}^{(l)}$. Note that also in this derivative the chain rule needs to be employed, unless $l = L$, in which case $a_k^{(L)}$ depends directly on $w_{ji}^{(L)}$. We look at this in more detail later. First, we take a look at the other possible machine learning tasks.

If the neural network performs binary classification, there is only one output neuron. We resolve the last step of the forward propagation, that is the application of the final activation function, explicitly in the nth error function,

$$
\begin{aligned}
E_n(\mathbf{w}) &= -t_n \log \frac{1}{1 + \exp(-a^{(L)}(\mathbf{x}_n, \mathbf{w}))} \\
&\quad -(1 - t_n) \log \left(1 - \frac{1}{1 + \exp(-a^{(L)}(\mathbf{x}_n, \mathbf{w}))} \right) \\
&= t_n \log \left(1 + \exp(-a^{(L)}(\mathbf{x}_n, \mathbf{w})) \right) \\
&\quad -(1 - t_n) \log \frac{\exp(-a^{(L)}(\mathbf{x}_n, \mathbf{w}))}{1 + \exp(-a^{(L)}(\mathbf{x}_n, \mathbf{w}))} \\
&= \log \left(1 + \exp(-a^{(L)}(\mathbf{x}_n, \mathbf{w})) \right) + (1 - t_n) a^{(L)}(\mathbf{x}_n, \mathbf{w}).
\end{aligned}
$$

Differentiating with respect to $w_{ji}^{(l)}$ gives

$$
\begin{aligned}
\frac{\partial}{\partial w_{ji}^{(l)}} E_n(\mathbf{w}) &= \frac{\exp(-a^{(L)}(\mathbf{x}_n, \mathbf{w}))}{1 + \exp(-a^{(L)}(\mathbf{x}_n, \mathbf{w}))} \frac{\partial}{\partial w_{ji}^{(l)}} (-a^{(L)}(\mathbf{x}_n, \mathbf{w})) \\
&\quad +(1 - t_n) \frac{\partial}{\partial w_{ji}^{(l)}} a^{(L)}(\mathbf{x}_n, \mathbf{w}) \\
&= \left(\frac{1}{1 + \exp(-a^{(L)}(\mathbf{x}_n, \mathbf{w}))} - 1 \right) \frac{\partial}{\partial w_{ji}^{(l)}} a^{(L)}(\mathbf{x}_n, \mathbf{w}) \\
&\quad +(1 - t_n) \frac{\partial}{\partial w_{ji}^{(l)}} a^{(L)}(\mathbf{x}_n, \mathbf{w}) \\
&= (y(\mathbf{x}_n, \mathbf{w}) - t_n) \frac{\partial}{\partial w_{ji}^{(l)}} a^{(L)}(\mathbf{x}_n, \mathbf{w})
\end{aligned}
$$

It has the same functional form as (9.1) with $K = 1$.

Using the fact that the error function of K simultaneous binary classification tasks is the sum over the error functions of the individual binary classification problems, also for this error function, the derivative of the nth contribution with respect to $w_{ji}^{(l)}$ takes the form of (9.1).

Lastly, we look at classifying K multiple classes, when $\mathbf{t}_n$ has a 1-of-K representation. Let k_n be the index of the component of $\mathbf{t}_n$ which equals one. Since all other components are zero, the nth contribution to the error function is

$$E_n(\mathbf{w}) = -\log \frac{\exp(a_{k_n}^{(L)})}{\sum_{s=1}^{K} \exp(a_s^{(L)})} = \log\left(\sum_{s=1}^{K} \exp(a_s^{(L)})\right) - a_{k_n}^{(L)}.$$

The derivative with respect to $w_{ji}^{(l)}$ is

$$
\begin{aligned}
\frac{\partial}{\partial w_{ji}^{(l)}} E_n(\mathbf{w}) &= \frac{\sum_{r=1}^{K} \exp(a_r^{(L)}) \frac{\partial}{\partial w_{ji}^{(l)}} a_r^{(L)}}{\sum_{s=1}^{K} \exp(a_s^{(L)})} - \frac{\partial}{\partial w_{ji}^{(l)}} a_{k_n}^{(L)} \\
&= \sum_{r=1}^{K} y_r(\mathbf{x}_n, \mathbf{w}) \frac{\partial}{\partial w_{ji}^{(l)}} a_r^{(L)} - t_{nr} \frac{\partial}{\partial w_{ji}^{(l)}} a_r^{(L)} \\
&= \sum_{r=1}^{K} (y_r(\mathbf{x}_n, \mathbf{w}) - t_{nr}) \frac{\partial}{\partial w_{ji}^{(l)}} a_r^{(L)},
\end{aligned}
$$

where we used the 1-of-K representation of $\mathbf{t}_n$. This has the same functional form as (9.1).

The fact that the derivatives of the nth contribution of all error functions result in (9.1) is known as the *canonical link* between the output activation functions and the error function.

The chain rule needs to be employed, until we reach the layer, where there is a direct dependency on $w_{ji}^{(l)}$. Let s be any layer index greater than or equal to l. We will show by backwards induction that

$$\frac{\partial}{\partial w_{ji}^{(l)}} E_n(\mathbf{w}) = \sum_{m=1}^{M_s} \delta_m^{(s)} \frac{\partial}{\partial w_{ji}^{(l)}} a_m^{(s)}(\mathbf{x}_n, \mathbf{w}) \tag{9.2}$$

for suitably chosen $\delta_m^{(s)}$.

This is already the case for $s = L$, due to Equation (9.1), if we set $M_L = K$ and define

$$\delta_k^{(L)} = y_k(\mathbf{x}_n, \mathbf{w}) - t_{nk},$$

which is the error in the kth output.

We assume that (9.2) is true for some $s > l$ and show that it is true for $s - 1$. Since

$$a_m^{(s)}(\mathbf{x}_n, \mathbf{w}) = \sum_{r=1}^{M_{s-1}} w_{mr}^{(s)} z_r^{(s-1)} = \sum_{r=1}^{M_{s-1}} w_{mr}^{(s)} h_{s-1}\left(a_r^{(s-1)}(\mathbf{x}_n, \mathbf{w})\right),$$

the derivative with respect to $w_{ji}^{(l)}$ is

$$\frac{\partial}{\partial w_{ji}^{(l)}} a_m^{(s)}(\mathbf{x}_n, \mathbf{w}) = \sum_{r=1}^{M_{s-1}} w_{mr}^{(s)} h_{s-1}'(a_r^{(s-1)}) \frac{\partial}{\partial w_{ji}^{(l)}} a_r^{(s-1)}(\mathbf{x}_n, \mathbf{w}),$$

where h_{s-1}' is the derivative of h_{s-1}.

Inserting this into (9.2), we obtain

$$\begin{aligned}
\frac{\partial}{\partial w_{ji}^{(l)}} E_n(\mathbf{w}) &= \sum_{m=1}^{M_s} \delta_m^{(s)} \sum_{r=1}^{M_{s-1}} w_{mr}^{(s)} h_{s-1}'(a_r^{(s-1)}) \frac{\partial}{\partial w_{ji}^{(l)}} a_r^{(s-1)}(\mathbf{x}_n, \mathbf{w}) \\
&= \sum_{r=1}^{M_{s-1}} \underbrace{h_{s-1}'(a_r^{(s-1)}) \sum_{m=1}^{M_s} \delta_m^{(s)} w_{mr}^{(s)}}_{\delta_r^{(s-1)}} \frac{\partial}{\partial w_{ji}^{(l)}} a_r^{(s-1)}(\mathbf{x}_n, \mathbf{w}),
\end{aligned}$$

which is of the required form.

The errors $\delta_r^{(s-1)}$ are defined in terms of $\delta_m^{(s)}$, the weights in the sth layer and the derivative of the activation function in layer $s - 1$ evaluated at the activations of this layer. They can be calculated by propagating backwards through the network from one set of neurons to the next. This is known as *error backpropagation*

Once we reach the lth layer, we have

$$\frac{\partial}{\partial w_{ji}^{(l)}} E_n(\mathbf{w}) = \sum_{m=1}^{M_l} \delta_m^{(l)} \frac{\partial}{\partial w_{ji}^{(l)}} a_m^{(l)}(\mathbf{x}_n, \mathbf{w}),$$

and $a_m^{(l)}$ is directly dependent on $w_{ji}^{(l)}$ via

$$a_m^{(l)} = \sum_{s=1}^{M_{l-1}} w_{ms}^{(L)} z_s^{(l-1)}.$$

Differentiation is straightforward and we arrive at

$$\frac{\partial}{\partial w_{ji}^{(l)}} a_m^{(l)}(\mathbf{x}_n, \mathbf{w}) = \left\{ \begin{array}{ll} 0 & \text{for} \quad m \neq j \\ z_i^{(l-1)} & \text{for} \quad m = j \end{array} \right..$$

The derivative of the nth contribution to the error becomes

$$\frac{\partial}{\partial w_{ji}^{(l)}} E_n(\mathbf{w}) = \delta_j^{(l)} z_i^{(l-1)}.$$

It is the error in the jth neuron of the lth set of neurons (letting the output neurons be the Lth set) multiplied by the ith variable in the previous set of neurons, if the synapse connects neuron i to neuron j. Note that, if $l = 1$, then those variables are the input variables, while, when $l > 1$, they are the latent variables.

Schematically, this is illustrated as

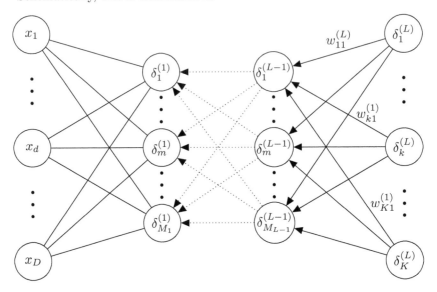

In the process of training a neural network, the weights are first initialized, most often randomly. In forward propagation, the training samples are passed through the neural network, calculating all intermediate activations and latent variables, as well as the output for each. Then the errors of each training point are calculated in a backward pass through the network. After this, the derivative with respect to every weight in the network of the contribution to the error function of each training sample is calculated. These form the gradient, either batch, mini-batch or stochastic. The weights are updated by the chosen optimization method.

To get some intuition of what is happening, consider binary classification. We denote the weights in the last layer by $w_i^{(L)}$, since there is only one output neuron. Because of $a^{(L)} = \sum_{s=1}^{M_L} w_s^{(L)} z_s^{(L-1)}$, the derivatives with respect to the weights in the last layer of the nth contribution to the error sum are

$$\frac{\partial}{\partial w_i^{(L)}} E_n(\mathbf{w}) = (y(\mathbf{x}_n, \mathbf{w}) - t_n) z_i^{(L-1)}$$

Since $y(\mathbf{x}_n, w)$ is interpreted as the probability of $\mathbf{x}_n$ belonging to the positive class, the difference lies in the interval $(-0.5, 0]$ for $t_n = 1$, if the network classifies x_n correctly as belonging to the positive class, and between $[-1, -0.5)$ otherwise. If on the other hand, $t_n = 0$, the difference lies in $(0.5, 1]$, if the network incorrectly puts this sample in the positive class, and in the interval $[0, 0.5)$ otherwise. This means that, when the weights get updated

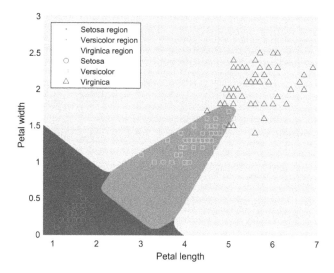

Figure 9.2: Two layer neural network classification with a set of three hidden neurons using the logistic sigmoid activation function.

by this component of the gradient vector, they are adjusted in a positive direction, if the sample is in the negative class, and in a negative direction, if it is in the positive class. The adjustments are weighted by $z_i^{(L-1)} \geq 0$ and the learning rate $\eta > 0$. If the network currently classifies the sample incorrectly, the adjustments are larger. The closer $y(\mathbf{x}_n, \mathbf{w})$ is to t_n, the smaller the adjustment.

This is comparable to the perceptron updates. There a line, plane or hyperplane is pulled, when a sample is misclassified. The direction is such that the sample is more likely to be classified correctly, if it is seen again. If the hidden set of neurons employ continuous, non-linear activation functions, non-linearity is introduced, and the adjustment of weights causes the adjustment of a curve or manifold.

We illustrate such curves by training a two layer neural network with three hidden neurons using the logistic sigmoid as activation function to classify the three classes of the Fisher iris data set. The resulting boundary curves can be seen in Figure 9.2. It is the result of running Listing 5.5 replacing the the heaviside activation function with the logistic sigmoid and reducing the number of hidden neurons to three. All data was used for training.

9.3 Autoencoders

In this section, we examine a group of neural networks known as *autoencoders* (*AE*). The main feature is that the number of output neurons K equals the number of input neurons D, i.e. $K = D$, and the purpose is the reconstruction of the input data, i.e. $\mathbf{y} \approx \mathbf{x}$. The number of layers L is even, meaning the number of sets of hidden neurons is odd. The first $L/2$ layers encode the data. The latent variables of the set of hidden neurons numbered $L/2$ represent the encoded data. The last $L/2$ layers decode these latent variables to arrive at the input. The encoder function ϕ maps the input $\mathbf{x} = (x_1, \ldots, x_D)^T$ to the latent variables $\mathbf{z}^{(L/2)} = (z_1^{(L/2)}, \ldots, z_{M_{L/2}}^{(L/2)})^T$, while the decoder function ψ maps the latent variables $\mathbf{z}^{(L/2)}$ back to $\mathbf{x}$.

While it is possible to implement the bias in each layer with a dummy neuron, which always has the value 1, it is common to denote it explicitly as $\mathbf{b}^{(l)} = (b_1, \ldots, b_{M_l})^T$. It signifies a shift after the linear transformation given by the weights $w_{ji}^{(l)}$, where $i = 1, \ldots, M_{l-1}$ and $j = 1, \ldots, M_l$. Here $M_0 = M_L = D$. The weights of each layer are stored in the weight matrix $\mathbf{W}^{(l)}$, the (j, i) entry being $w_{ji}^{(l)}$. The vector of activations $\mathbf{a}^{(l)} = (a_1, \ldots, a_{M_l})^T$ is then

$$\mathbf{a}^{(l)} = \mathbf{W}^{(l)} \mathbf{z}^{(l-1)} + \mathbf{b}^{(l)},$$

where $\mathbf{z}^{(0)} = \mathbf{x}$. The notation $h^{(l)}(\mathbf{a}^{(l)})$ is used to denote the element-wise application of the activation function $h^{(l)}$ to the entries of $\mathbf{a}^{(l)}$.

The error function for the autoencoder is typically the *mean squared error*,

$$E(\mathbf{w}) = \frac{1}{N} \sum_{n=1}^{N} \|\mathbf{x}_n - \psi(\phi(\mathbf{x}_n))\|^2.$$

There are several variants of autoencoders, which adjust the error function with a particular goal in mind.

Firstly, there is the *denoising autoencoder (DAE)*, where the goal is that a good representation of the sample is still obtained, even if it is corrupted by noise. To this end, the training samples $\mathbf{x}_1, \ldots, \mathbf{x}_n$ first undergo a probabilistic corruption process giving corrupted training samples $\tilde{\mathbf{x}}_1, \ldots, \tilde{\mathbf{x}}_n$. The autoencoder is then trained using the error function

$$E(\mathbf{w}) = \frac{1}{N} \sum_{n=1}^{N} \|\mathbf{x}_n - \psi(\phi(\tilde{\mathbf{x}}_n))\|^2.$$

The corrupted data is encoded and decoded and the results compared to the uncorrupted data.

Secondly, we have *sparse autoencoders (SAE)*. Here, the aim is that only a certain number of latent variables is non-zero. One possibility is to only keep the k largest latent variables in a particular set of hidden neurons and set the others to zero during forward propagation. Backpropagation only passes through the hidden neurons, whose latent variables are non-zero. This is known as a *k-sparse autoencoder*.

The *average activation* of the mth neuron in the lth set of hidden neurons is defined as

$$p_m^{(l)} = \frac{1}{N} \sum_{n=1}^{N} z_m^{(l)}(\mathbf{x}_n).$$

In other words, it averages over all the values the latent variable $z_m^{(l)}$ takes, when the samples $\mathbf{x}_1, \ldots, \mathbf{x}_N$ pass through the network. The aim is to keep this value low so that $z_m^{(l)}$ is zero or close to zero for most inputs and has non-negligible values for only a subset of inputs. This means the neuron only reacts, or "fires", to a small number of inputs, and this way identifying a subset with common features. The error function includes a penalty term for all hidden neurons, which is large, if $p_m^{(l)}$ differs from the predefined *sparsity proportion* p, small, if $p_m^{(l)}$ is close to p, and zero, if they are equal. The *Kullback–Leibler divergence* between p and $p_m^{(l)}$ has the necessary properties.

More specifically, the *sparsity regularization* penalty term

$$\Omega_s = \sum_{l=1}^{L-1} \sum_{m=1}^{M_l} D_{KL}(p \| p_m^{(l)}) = \sum_{l=1}^{L-1} \sum_{m=1}^{M_l} \left(p \log \frac{p}{p_m^{(l)}} + (1-p) \log \frac{1-p}{1-p_m^{(l)}} \right)$$

is added to the error function with a suitable weight β. Suitable values for both β and p can be chosen by monitoring the convergence and reconstruction behaviour or via a validation set.

However, when adding this penalty, the latent variables might become small, while the weights in the subsequent layer become large counteracting the effect. In order to avoid this, another penalty term, the L_2 *regularization* term acting on the weights is added as well,

$$\Omega_w = \frac{1}{2} \sum_{l=1}^{L} \sum_{j=1}^{M_{l-1}} \sum_{i=1}^{M_l} w_{ji}^{(l)},$$

where $M_0 = M_L = D$. As before, this is weighted by a suitable parameter λ. This term penalizes growth in the weights.

Another variation is the *contractive autoencoder*. Let the encoder function be $\phi(\mathbf{x}) = (\phi_1(\mathbf{x}), \ldots, \phi_{M_{L/2}}(\mathbf{x}))^T$. Its *Jacobian matrix* is

$$J_\phi(\mathbf{x}) = \begin{pmatrix} \frac{\partial \phi_1}{\partial x_1}(\mathbf{x}) & \cdots & \frac{\partial \phi_1}{\partial x_D}(\mathbf{x}) \\ \vdots & \ddots & \vdots \\ \frac{\partial \phi_{M_{L/2}}}{\partial x_1}(\mathbf{x}) & \cdots & \frac{\partial \phi_{M_{L/2}}}{\partial x_D}(\mathbf{x}) \end{pmatrix}.$$

If we would have $M_{L/2} = D$, that is the Jacobian matrix is square, then the modulus of its determinant is the factor by which volumes in $\mathbb{R}^D$ under the transformation by ϕ shrink or expand. The aim is to control this volume change favouring reductions in volume, therefore the name contractive autoencoder. However, generally $M_{L/2} \neq D$. In fact, a much smaller number than D is desired. The *Frobenius norm* is a measure of change under these circumstances.

For a general, real $a \times b$ matrix $\mathbf{A}$ with entries A_{ij}, the Frobenius norm is defined as

$$\|\mathbf{A}\|_F = \sqrt{\operatorname{tr}(\mathbf{A}^T \mathbf{A})} = \sqrt{\sum_{i=1}^{a} \sum_{j=1}^{b} A_{ij}^2}.$$

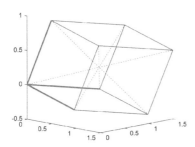
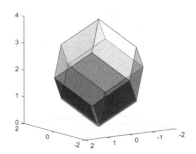

(a) Image of the three dimensional unit cube and its four space diagonals under the action of a 3×3 matrix $\mathbf{A}$.

(b) Image of the four dimensional unit hypercube under the action of a 3×4 matrix $\mathbf{A}$.

Figure 9.3: Images of the unit hypercube under mappings from $\mathbb{R}^3$ (left) and $\mathbb{R}^4$ (right) to $\mathbb{R}^3$. The red edges are the images of the standard basis vectors.

To motivate that this is a measure of change, consider $a = b = 3$. The squared Frobenius norm is then the average of the squared lengths of the space diagonals of the parallelepiped spanned by the columns of $\mathbf{A}$. Figure 9.3a gives an example for

$$\mathbf{A} = \begin{pmatrix} 1 & 0 & 1/2 \\ 1/2 & 1 & 0 \\ 0 & -1/2 & 1 \end{pmatrix}.$$

This parallelepiped is the image of the unit cube under the action of $\mathbf{A}$. A space diagonal is the the line segment connecting two vertices which are not on the same face. They are given by the vectors

$$\begin{aligned} \mathbf{d}_1 &= \mathbf{a}_1 + \mathbf{a}_2 + \mathbf{a}_3 \\ \mathbf{d}_2 &= \mathbf{a}_1 + \mathbf{a}_2 - \mathbf{a}_3 \\ \mathbf{d}_3 &= \mathbf{a}_2 + \mathbf{a}_3 - \mathbf{a}_1 \\ \mathbf{d}_4 &= \mathbf{a}_3 + \mathbf{a}_1 - \mathbf{a}_2. \end{aligned}$$

Calculating $(\mathbf{d}_1^T\mathbf{d}_1 + \mathbf{d}_2^T\mathbf{d}_2 + \mathbf{d}_3^T\mathbf{d}_3 + \mathbf{d}_4^T\mathbf{d}_4)/4$ results in $\mathbf{a}_1^T\mathbf{a}_1 + \mathbf{a}_2^T\mathbf{a}_2 + \mathbf{a}_3^T\mathbf{a}_3$ which is exactly the square of the Frobenius norm of $\mathbf{A}$.

Figure 9.3b illustrates the action of a 3×4 matrix on the four dimensional hypercube. The result is a polyhedron. In particular, the matrix is

$$\mathbf{A} = \begin{pmatrix} 1 & -1 & -1 & 1 \\ 1 & 1 & -1 & -1 \\ 1 & 1 & 1 & 1 \end{pmatrix}.$$

The red edges are the columns of $\mathbf{A}$ and span the polyhedron. All other vertices are sums of these vectors.

For the contractive autoencoder, the penalty term

$$\Omega_F = \frac{1}{2}\sum_{n=1}^{N}\|J_\phi(\mathbf{x}_n)\|_F^2 = \sum_{n=1}^{N}\operatorname{tr}(J_\phi(\mathbf{x}_n)^T J_\phi(\mathbf{x}_n)) = \sum_{n=1}^{N}\sum_{m=1}^{M_{L/2}}\sum_{d=1}^{D}[\frac{\partial\phi_m}{\partial x_d}(\mathbf{x}_n)]^2$$

is added to the error function, again multiplied by a weight λ.

The autoencoders so far produce, once trained, latent variables $\mathbf{z}^{(L/2)}$ deterministically from which the output is calculated in a deterministic way. The *variational autoencoder* takes a different approach. The assumption is that the data are sampled from some distribution $p(\mathbf{x})$, where

$$p(\mathbf{x}) = \int p\left(\mathbf{x}|\mathbf{z}^{(L/2)}\right)p\left(\mathbf{z}^{(L/2)}\right)d\mathbf{z}^{(L/2)}.$$

Further, it is assumed that $p(\mathbf{x}|\mathbf{z}^{(L/2)})$ follows a normal distribution with mean $\psi(\mathbf{z}^{(L/2)})$ and variance $\sigma^2\mathbf{I}$, where $\mathbf{I}$ is the $D\times D$ identity matrix. The prior probability distribution of $\mathbf{z}$ is the standard multivariate normal distribution. The overall probability $p(\mathbf{x})$ can be estimated by by sampling many values $\mathbf{z}_1^{(L/2)},\ldots,\mathbf{z}_K^{(L/2)}$ and then calculating

$$p(\mathbf{x}) = \frac{1}{K}\sum_{k=1}^{K}p(\mathbf{x}|\mathbf{z}_K^{(L/2)}) = \frac{1}{K}\sum_{k=1}^{K}(2\pi\sigma^2)^{-D/2}\exp\left(-\frac{1}{2\sigma^2}\|\mathbf{x}-\psi(\mathbf{z}^{(L/2)})\|^2\right).$$

The complete data likelihood is estimated as

$$\prod_{n=1}^{N}p(\mathbf{x}_n) = \prod_{n=1}^{N}\frac{1}{K}\sum_{k=1}^{K}(2\pi\sigma^2)^{-D/2}\exp\left(-\frac{1}{2\sigma^2}\|\mathbf{x}_n-\psi(\mathbf{z}^{(L/2)})\|^2\right).$$

Maximizing this is equivalent to minimizing the negative logarithm. However, this does not result in a convenient sum of squares as before because of the sum resulting from the estimation of $p(\mathbf{x}_n)$ Another computational difficulty is that K might need to be large (especially in high dimensions) to obtain a good estimate. In addition, sampling is not a continuous operation and thus not differentiable. So error backpropagation cannot be applied.

Variational autoencoders address these problems. The first step is to reduce K, by choosing $p(\mathbf{z}^{(L/2)})$ such that only values of $\mathbf{z}^{(L/2)}$ are sampled which are likely to produce samples of $\mathbf{x}$. In other words, the aim is to find $p(\mathbf{z}^{(L/2)}|\mathbf{x})$. For this, the encoder is used. To make this clear in notation the encoder gives an estimate $p_\phi(\mathbf{z}^{(L/2)}|\mathbf{x})$ as a normal multivariate distribution. The function ϕ does not return $\mathbf{z}^{(L/2)}$, but

$$(\boldsymbol{\mu}(\mathbf{x}),\boldsymbol{\Sigma}(\mathbf{x})) = \phi(\mathbf{x}).$$

To keep the number of variables manageable, $\boldsymbol{\Sigma}(\mathbf{x})$ is restricted to be diagonal, that is $\boldsymbol{\Sigma}(\mathbf{x}) = \operatorname{diag}(\sigma_1^2,\ldots\sigma_{M_{L/2}}^2)$. Hence, $\phi(\mathbf{x})$ calculates $2M_{L/2}$ latent variables.

In order to use back propagation, the sampling is moved to the input neurons, where a vector $\boldsymbol{\eta} = (\eta_1, \dots \eta_{M_{L/2}})^T$ is sampled from the standard normal distribution in $M_{L/2}$ dimensions. With this,

$$\mathbf{z}^{(L/2)} = \boldsymbol{\mu}(\mathbf{x}) + \boldsymbol{\Sigma}^{1/2}(\mathbf{x})\boldsymbol{\eta} = \begin{pmatrix} \mu_1(\mathbf{x}) + \sigma_1(\mathbf{x})\eta_1 \\ \vdots \\ \mu_{M_{L/2}}(\mathbf{x}) + \sigma_{M_{L/2}}(\mathbf{x})\eta_{M_{L/2}} \end{pmatrix}.$$

The schematics are

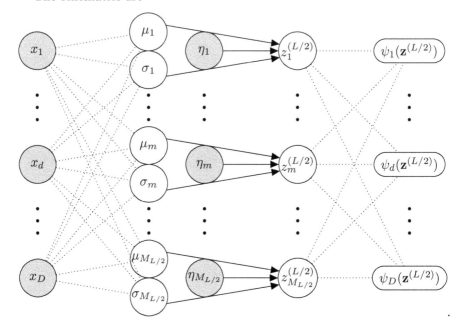

For clarity the input neurons are shaded grey. As before, the dotted lines indicate that any combination of hidden neurons, layers and transfer functions is possible, making use of the capability of neural networks to model any function.

The error function is the sum over all training data $\mathbf{x}_n$ of $-\log p(\mathbf{x}_n)$ and the Kullback-Leibler divergence between $p_\phi(\mathbf{z}^{(L/2)}|\mathbf{x}_n)$ and $p(\mathbf{z}^{(L/2)}|\mathbf{x}_n)$. One term of this sum is given by

$$\begin{aligned} E_n &= -\log p(\mathbf{x}_n) + D_{KL}(p_\phi(\mathbf{z}^{(L/2)}|\mathbf{x}_n) \| p(\mathbf{z}^{(L/2)}|\mathbf{x}_n)) \\ &= \int p_\phi(\mathbf{z}^{(L/2)}|\mathbf{x}_n) \left[-\log p(\mathbf{x}_n) - \log \frac{p(\mathbf{z}^{(L/2)}|\mathbf{x}_n)}{p_\phi(\mathbf{z}^{(L/2)}|\mathbf{x}_n)} \right] d\mathbf{z}^{(L/2)}. \end{aligned}$$

Using Bayes' rule

$$p(\mathbf{z}^{(L/2)}|\mathbf{x}_n) = \frac{p(\mathbf{x}_n|\mathbf{z}^{(L/2)})p(\mathbf{z}^{(L/2)})}{p(\mathbf{x}_n)},$$

which simplifies to

$$
\begin{aligned}
E_n &= \int p_\phi(\mathbf{z}^{(L/2)}|\mathbf{x}_n) \left[-\log \frac{p(\mathbf{z}^{(L/2)})}{p_\phi(\mathbf{z}^{(L/2)}|\mathbf{x}_n)} - \log p(\mathbf{x}_n|\mathbf{z}^{(L/2)}) \right] d\mathbf{z}^{(L/2)} \\
&= D_{KL}(p_\phi(\mathbf{z}^{(L/2)}|\mathbf{x}_n)\|p(\mathbf{z}^{(L/2)})) \\
&\quad - \int p_\phi(\mathbf{z}^{(L/2)}|\mathbf{x}_n) \log p(\mathbf{x}_n|\mathbf{z}^{(L/2)}) d\mathbf{z}^{(L/2)}.
\end{aligned}
$$

The first term is the Kullback–Leibler divergence between the normal distribution with mean $\boldsymbol{\mu}(\mathbf{x}_n)$ and variance $\boldsymbol{\Sigma}(\mathbf{x}_n)$ and the standard normal distribution which is the prior of $\mathbf{z}^{(L/2)}$. This is given by

$$
\begin{aligned}
&D_{KL}(p_\phi(\mathbf{z}^{(L/2)}|\mathbf{x}_n)\|p(\mathbf{z}^{(L/2)})) = \\
&\qquad \frac{1}{2}\left[\operatorname{tr}\boldsymbol{\Sigma}(\mathbf{x}_n) + \boldsymbol{\mu}(\mathbf{x}_n)^T \boldsymbol{\mu}(\mathbf{x}_n) - M_{L/2} - \log|\boldsymbol{\Sigma}(\mathbf{x}_n)|\right] = \\
&\qquad \frac{1}{2}\sum_{m=1}^{M_{L/2}}\left[\sigma_m^2(\mathbf{x}_n) + \mu_m^2(\mathbf{x}_n) - \log \sigma_m^2(\mathbf{x}_n)\right] - \frac{M_{L/2}}{2},
\end{aligned}
$$

where the last equation is due to choosing the variance to have diagonal form.

The second term is the negative expectation of $\log p(\mathbf{x}_n|\mathbf{z}^{(L/2)})$ with respect to $\mathbf{z}^{(L/2)}$. Now $\mathbf{z}^{(L/2)} = \boldsymbol{\mu}(\mathbf{x}_n) + \boldsymbol{\Sigma}^{1/2}(\mathbf{x}_n)\boldsymbol{\eta}$, where $\boldsymbol{\eta}$ follows the standard normal distribution and is sampled in additional input neurons. It is therefore customary to enrich the training samples with samples drawn from the standard normal distribution to arrive at new training samples $\mathbf{x}_{nk} = (\mathbf{x}_n, \boldsymbol{\eta}_k)^T$ for $n = 1, \ldots, N$ and $k = 1, \ldots, K$ for some suitably chosen K. The second term is then estimated as

$$
\begin{aligned}
&-\log p(\mathbf{x}_n|\boldsymbol{\mu}(\mathbf{x}_n) + \boldsymbol{\Sigma}^{1/2}(\mathbf{x}_n)\boldsymbol{\eta}_k) = \\
&\qquad \frac{D}{2}\log(2\pi\sigma^2) + \frac{1}{2\sigma^2}\|\mathbf{x}_n - \psi(\boldsymbol{\mu}(\mathbf{x}_n) + \boldsymbol{\Sigma}^{1/2}(\mathbf{x}_n)\boldsymbol{\eta}_k)\|^2.
\end{aligned}
$$

Each term in the sum over k is treated separately in the error function. Removing constant terms the nk^{th} term of the error function, which needs to be minimized, is given by

$$
\begin{aligned}
E_{nk} &= \frac{1}{2}\sum_{m=1}^{M_{L/2}}\left[\sigma_m^2(\mathbf{x}_n) + \mu_m^2(\mathbf{x}_n) - \log \sigma_m^2(\mathbf{x}_n)\right] \\
&\quad + \frac{1}{2}\|\mathbf{x}_n - \psi(\boldsymbol{\mu}(\mathbf{x}_n) + \boldsymbol{\Sigma}^{1/2}(\mathbf{x}_n)\boldsymbol{\eta}_k)\|^2.
\end{aligned}
$$

This is continuous and therefore error backpropagation can be used to calculate gradients to be used in the minimization as explained in Section 9.2.

9.4 Autoencoder Example

We apply a sparse autoencoder to the MNIST data set of handwritten digits by [26] and compare and link this method to the techniques we have already encountered. For this data, $D = 784$, and the components of $\mathbf{x}$ are the individual pixel values.

The simplest autoencoder has two layers. Schematically, it is represented as

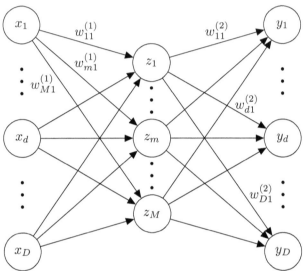

In this simple case, the encoder and decoder functions are

$$\phi(\mathbf{x}) = h^{(1)}(\mathbf{W}^{(1)}\mathbf{x} + \mathbf{b}^{(1)}) \qquad \text{and} \qquad \psi(\mathbf{z}) = h^{(2)}(\mathbf{W}^{(2)}\mathbf{z} + \mathbf{b}^{(2)}).$$

If $\mathbf{W}^{(2)} = \mathbf{W}^{(1)^T}$, then the weights are *tied*. However, in this example we do not enforce this, but let them be independent from each other.

Both $h^{(1)}$ and $h^{(2)}$ are chosen to be the logistic sigmoid applied element-wise. Since the output range of the logistic sigmoid is $[0, 1]$ and autoencoders reconstruct the input, the pixel values of the input images need to be rescaled to lie between zero and one as a pre-processing step.

Valuable insight can be gained from visualizing how the activations and latent variables are separated spatially. This is easily possible for $M = 1, 2$ or 3. In the following, we show one technique of how this can be done for larger M.

The activations in the output layer are given by

$$\mathbf{W}^{(2)}\mathbf{z} + \mathbf{b} = z_1\mathbf{w}_1^{(2)} + \ldots + z_M\mathbf{w}_M^{(2)} + \mathbf{b},$$

where $\mathbf{w}_1^{(2)}, \ldots, \mathbf{w}_M^{(2)}$ are the columns of $\mathbf{W}^{(2)}$. The elements of the mth column are the weights of the synapses connecting the mth hidden neuron to the output neurons. Since the activations are clearly a linear combination of the

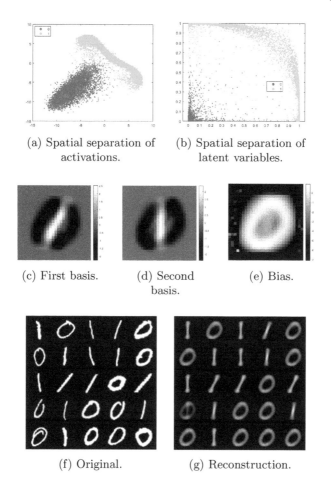

(a) Spatial separation of activations.

(b) Spatial separation of latent variables.

(c) First basis.

(d) Second basis.

(e) Bias.

(f) Original.

(g) Reconstruction.

Figure 9.4: Autoencoder with two hidden neurons applied to two digits.

columns and the bias, the columns can be seen as the basis functions of the model space from which the data is generated. The bias is a shift in the data space, necessary to reconstruct all training data. If the data is *mean centred*, such a shift is unnecessary and the bias is zero. The application of the transfer function ensures that the output lies again in the same range as the input.

The error function to minimize is

$$E(\mathbf{w}) = \frac{1}{N} \sum_{n=1}^{N} \|\mathbf{x}_n - \psi(\phi(\mathbf{x}_n))\|^2 + \beta \Omega_s + \lambda \Omega_w,$$

where the *sparsity regularizer* Ω_s and the L_2 *regularizer* Ω_w are penalty terms to ensure sparsity and control the growth of the weights. After some trials, β was set to 4, while $\lambda = 0.004$. Ω_s also depends on the *sparsity proportion p* which was chosen to be 0.15.

While it is desirable in neural networks to choose the number of hidden neurons smaller than the dimension of the input data, the assumption is that it needs to be chosen large enough so that the distinguishing features in the data are captured. However, what does large enough mean? To understand the inner workings of an algorithm it is helpful to look at smaller problems. In our first experiment, we therefore only consider the digits zero and one. Since only two classes need to be distinguished, we set the number of hidden neurons M to two. Figure 9.4 shows the results.

Firstly, in Figure 9.4a it can be seen that the activations of the training data are separated. It is a two dimensional space, since there are two hidden neurons. This space becomes close to the space spanned by the first two principal components of the training data, because there the data is most easily separated and this is the goal of the encoder. The transfer function emphasizes this separation, which can be seen in Figure 9.4b.

Secondly, the latent space is spanned by the two bases seen in Figures 9.4c and 9.4d. The numeric values at each pixel location are indicated by the colour bar. The first basis resembles a slanted one written over a zero, while the second is similar to a straight one over a zero. The bias resembles a zero. Different runs can arrive at different configurations of the bases and bias, since the neural network is initialized randomly.

Since the autoencoder strives to reconstruct the input, the basis functions it chooses resemble digits or over-laid digits to the human eye, while principal component analysis focuses on the largest eigenvalues and the principal components do not resemble digits and neither do the resulting reconstructions.

In Figure 9.5 another digit was included, but the set-up was kept the same. The reconstructions in Figure 9.5g are discernible, which cannot be said about Figure 7.3b where the reconstruction is done using principal component analysis. One might imagine a two in Figure 9.5c, a one in Figure 9.5d and a zero as the bias in Figure 9.5e. The separations in Figure 9.5b are not clear for digits zero and two, while they are still well separated in Figure 9.5a. This is mostly due to the position of the zero on the horizontal axis, since for zero the logistic sigmoid function evaluates to 0.5 making both classes equally likely. However, samples of one and two are mostly present there and these are separated along the vertical axis. The separation can be improved by shifting the activations right such that the zero point of the horizontal axis lies where the digits zero and two separate.

Alternatively, the number of hidden neurons can be increased to three. The results are shown in Figure 9.5. The reconstructions are good and the basis functions resemble elements of digits. In Figure 9.5b, the latent variables of the digit zero are packed at the $(0,0,0)$ corner of the unit cube, while those of the digit one lie along the edges connecting the $(0,1,0), (0,1,1), (0,0,1)$ and $(1,0,1)$ corners. Hence they are separated.

To examine this further, we consider all ten digits and increase M to ten. To visualize the separation, Figures 9.7a to 9.7j display the stacked histograms (with one hundred bins) of the activations for each of the digits along the ten

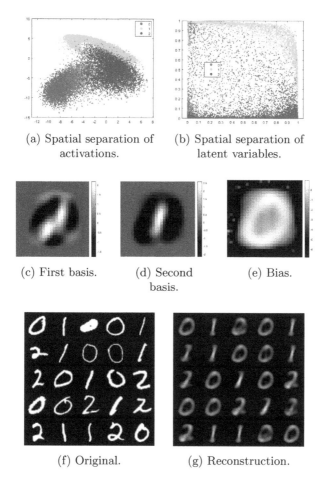

(a) Spatial separation of activations.

(b) Spatial separation of latent variables.

(c) First basis.

(d) Second basis.

(e) Bias.

(f) Original.

(g) Reconstruction.

Figure 9.5: Autoencoder with two hidden neurons applied to three digits.

dimensions of the model space. While the set of digits as a whole seems to be centred around a mean in each of the dimensions, some digits move away from this. For example, in Figure 9.7a digits two, three and eight lie further to the right in this dimension, while in Figure 9.7c it is digits two and six. In Figure 9.7i the digits to the right are mostly one. Applying the logistic sigmoid maps activations to the right close to one, while those to the left are mapped close to zero. Figures 9.7k and 9.7l show the proportions of digits which get mapped close to zero and one in each of the ten dimensions corresponding to the hidden neurons. While nearly all digits are present close to zero, near one some digits are dominant. This means these neurons "fire" for particular digits.

The following table indicates for which digits an individual neuron "fires".

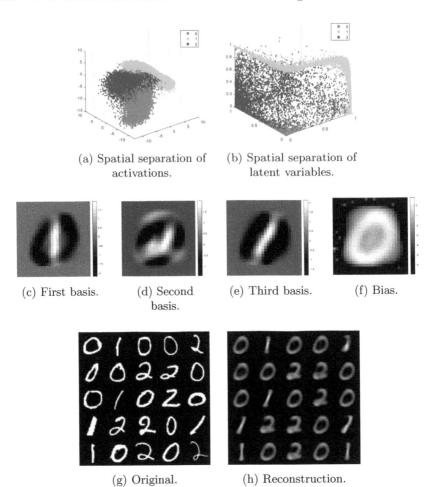

(a) Spatial separation of activations.

(b) Spatial separation of latent variables.

(c) First basis.

(d) Second basis.

(e) Third basis.

(f) Bias.

(g) Original.

(h) Reconstruction.

Figure 9.6: Autoencoder with three hidden neurons applied to three digits.

Neuron \ Digit	0	1	2	3	4	5	6	7	8	9
1			×	×					×	
2	×			×		×			×	
3			×				×			
4		×							×	
5	×							×		
6	×					×			×	
7	×			×		×	×			
8					×					×
9		×	×							
10								×		×

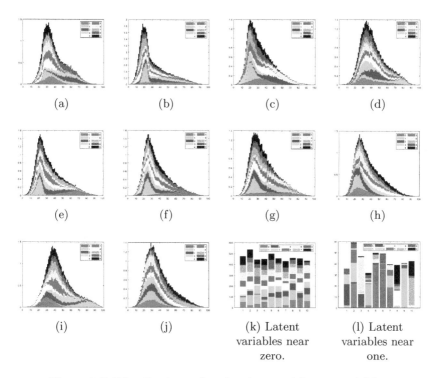

(a)   (b)   (c)   (d)

(e)   (f)   (g)   (h)

(i)   (j)   (k) Latent variables near zero.   (l) Latent variables near one.

Figure 9.7: Distributions of activations and latent variables.

The table can also be read in the other direction telling us the input from which neurons is necessary to construct a particular digit. For example, digit four is mostly constructed from input from the eighth neuron, digit seven mostly from input from the tenth neuron, while digit nine is a mixture of the eighth and tenth neuron. We can now compare this to basis functions corresponding to each neuron given in Figure 9.8. In particular for these examples, the basis functions are shown in Figures 9.8h and 9.8j and indeed digits four, seven and nine can be seen in these. Similarly, digit one can be seen in Figures 9.8d and 9.8i. The other basis functions are not so clearly linked to a particular digit, but it has to be kept in mind that the requirement is that a digit can be reconstructed by a linear combination of the basis functions, not that a single basis function corresponds to a digit.

The bias is displayed in Figure 9.8k. It can be interpreted as the summary of what all samples have in common. Since the digits are centred in the image, the bias is this centre.

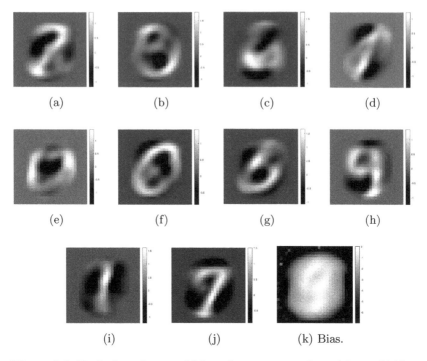

(a)   (b)   (c)   (d)

(e)   (f)   (g)   (h)

(i)   (j)   (k) Bias.

Figure 9.8: Basis functions and bias of an autoencoder with ten hidden neurons.

9.5 Relationship to Other Techniques

In this section, we examine how neural networks relate to some of the techniques we have seen before. We already touched on this in Section 5.5 with regards to classification where a neural network describes different regions as sets of points lying on different sides of a set of lines.

In Chapter 8, the assumption is that all samples are generated by an underlying process of the form

$$t = Dc + \epsilon,$$

where ϵ is additive Gaussian noise with zero mean and variance σ^2, the $D \times M$ matrix D is the design matrix, c is the vector of coefficients, which depends on both D and t.

Recall that the prediction of the ordinary least squares solution, where D is chosen beforehand, is

$$Dc = D(D^T D)^{-1} D^T t.$$

Schematically, this can be visualized as

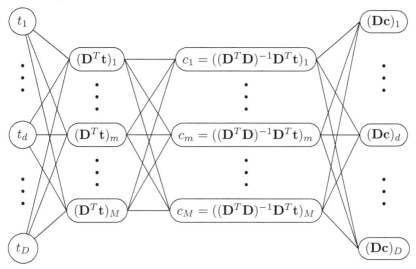

This is reminiscent of a three layer neural network with two sets of hidden neurons and the linear activation function, which is the identity, in each neuron. The weights of the synapses from the input neurons to the first set of hidden neurons are given by the entries of the matrix $\mathbf{D}^T$, while the weights of the synapses from the first to the second set of hidden neurons are given by the entries of the matrix $(\mathbf{D}^T\mathbf{D})^{-1}$. Lastly, the weights from the second set of hidden neurons to the output neurons are given by the matrix $\mathbf{D}$. The variables of the second set of hidden neurons are the coefficients $c_1, \ldots, c_M$.

In a neural network, however, the weights of the synapses are updated iteratively in a learning process, until the output is similar enough to the input. In ordinary least squares, all is entirely determined by the design matrix $\mathbf{D}$, which only depends on the chosen basis functions, but not $t_1, \ldots, t_D$. However, if a set of training data $\mathbf{t}_1, \ldots \mathbf{t}_N$ is available, they can be fed through this neural network. The error sum of squares between the predictions and the input is

$$E(\mathbf{D}) = \sum_{n=1}^{N} \left(\mathbf{t}_n^T \mathbf{t}_n - \mathbf{t}_n^T \mathbf{D}(\mathbf{D}^T\mathbf{D})^{-1}\mathbf{D}^T\mathbf{t}_n \right),$$

since it is the ordinary least squares solution. The derivative with regards to the entry D_{ij} of $\mathbf{D}$ is

$$\frac{\partial}{\partial D_{ij}} E(\mathbf{D}) = \sum_{n=1}^{N} \left((\mathbf{D}^T\mathbf{D})^{-1}\mathbf{D}^T\mathbf{t}_n \right)_j \left(\mathbf{D}(\mathbf{D}^T\mathbf{D})^{-1}\mathbf{D}^T\mathbf{t}_n - \mathbf{t}_n \right)_i,$$

where the subscript of the brackets denotes the component of the vector in the brackets. It is the sum over the training data of the products of the latent variable of the jth neuron in the second set of hidden neurons with the error in the ith output neuron, when the nth training sample passes through the network. The thus obtained gradient can be used to update the entries of $\mathbf{D}$. Note that this effectively means only updating the weights in the final layer giving a new matrix $\mathbf{D}$, since the weights of the other layers are given by $\mathbf{D}^T$ and $(\mathbf{D}^D)^{-1}$. This inverse might not exist, which is an inherent problem of ordinary least squares, which can be addressed by ridge regression as seen in equation 8.3. Thus the design matrix can be inferred from the data, if there are enough data samples $\mathbf{t}_n$.

The matrix $\mathbf{D}\mathbf{D}^T$ is symmetric and positive semi-definite. It can be diagonalized by an orthogonal matrix $\mathbf{Q}$ such that $\mathbf{D}\mathbf{D}^T = \mathbf{Q}\mathbf{P}\mathbf{Q}^T$, where $\mathbf{P}$ is a diagonal matrix with non-negative entries. Thus the predictions are

$$
\begin{aligned}
\mathbf{D}(\mathbf{D}^T\mathbf{D})^{-1}\mathbf{D}^T\mathbf{t} &= \mathbf{D}(\mathbf{Q}\mathbf{P}^{1/2}\mathbf{P}^{1/2}\mathbf{Q}^T)^{-1}\mathbf{D}^T\mathbf{t} \\
&= (\mathbf{D}\mathbf{Q}\mathbf{P}^{-1/2})(\mathbf{P}^{-1/2}\mathbf{Q}^T\mathbf{D}^T)\mathbf{t} \\
&= (\mathbf{D}\mathbf{Q}\mathbf{P}^{-1/2})(\mathbf{D}\mathbf{Q}\mathbf{P}^{-1/2})^T\mathbf{t}.
\end{aligned}
$$

This is a tied two-layer autoencoder with the linear function as activation function. The weights of the first layer are given by the entries of $(\mathbf{D}\mathbf{Q}\mathbf{P}^{-1/2})^T$ and those of the second layer by $\mathbf{D}\mathbf{Q}\mathbf{P}^{-1/2}$. This illustrates again the link of the weights of an autoencoder to the model space which was originally defined as the columns of $\mathbf{D}$ being functions evaluated at the points, where data samples are taken. The application of $\mathbf{P}^{-1/2}$ are different scalings of the hidden variables, while $\mathbf{Q}$ is either a rotation or a reflection, since it is an orthogonal matrix.

Continuing in this vein, the mean of the predictive distribution for $\mathbf{t}$ in Section 8.13 is given by

$$\mathbf{D}\boldsymbol{\mu} = \sigma^{-2}\mathbf{D}\boldsymbol{\Sigma}\mathbf{D}^T\mathbf{t}.$$

Schematically, this can be drawn as

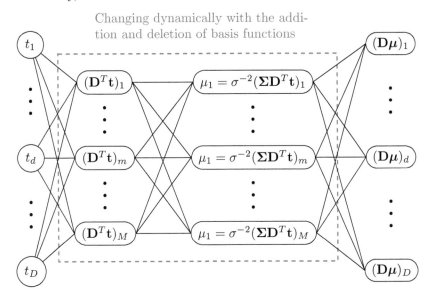

Again, it can be viewed as a neural network with two sets of hidden neurons. As before, the weights from the input neurons to the first set of hidden neurons are given by the entries of $\mathbf{D}^T$, and from the second set of hidden neurons to the output neurons by the entries of $\mathbf{D}$. The weights between the two sets of hidden neurons are given by the entries of $\sigma^{-2}\boldsymbol{\Sigma}$. Since $\boldsymbol{\Sigma} = \left(\mathbf{A} + \sigma^{-2}\mathbf{D}^T\mathbf{D}\right)^{-1}$ and since the entries of $\mathbf{A}$ are determined by maximizing the log evidence of $\mathbf{t}$, the weights between the hidden neurons depend on both the design matrix $\mathbf{D}$ and $t_1, \ldots, t_D$. The number of hidden neurons in each set, M, can change dynamically which is equivalent to the addition and deletion of basis functions. Equally, the number of input and output neurons can increase, if samples are taken at previously unseen locations.

As before with ordinary least squares, a set of training data $\mathbf{t}_1, \ldots, \mathbf{t}_N$ can be used to infer the entries of $\mathbf{D}$. Whenever $\mathbf{D}$ is updated, $\mathbf{A}$ needs to be updated by maximizing the log evidence of $\mathbf{t}$ again.

9.6   Indian Buffet Process

The last section suggests a technique of inferring a model space with variable dimension. In this section, we consider the data generation process similar to the Chinese Restaurant Process in Section 6.6 where the number of clusters is unknown. Now the dimension of the latent feature space is unknown.

As in the previous section, let $\mathbf{t}_1, \ldots, \mathbf{t}_N$ be a set of data samples in a D dimensional space. We assume that for $n = 1, \ldots, N$

$$\mathbf{t}_n = \mathbf{D}\mathbf{c}_n + \epsilon_n,$$

where ϵ_n is additive Gaussian noise with zero mean and variance $\sigma^2\mathbf{I}$ specific to this data sample, the columns of the $D \times M$ matrix $\mathbf{D}$ represent latent features, and $\mathbf{c}_n$ is the sample specific vector of coefficients.

Commonly, this is represented as

$$\mathbf{T} = \mathbf{C}\mathbf{D}^T + \boldsymbol{E},$$

where $\mathbf{T}$ is the $N \times D$ matrix of data samples with each row being one sample, $\mathbf{C}$ is the $N \times M$ matrix of coefficients, and $\boldsymbol{E}$ is a $N \times D$ matrix with each entry being drawn from a normal distribution with zero mean and variance σ^2. In this context, $\mathbf{D}$ is the *factor loading matrix* and its columns are the *factor loadings*, while $\mathbf{C}$ is the *coordinate matrix*.

From the data generation viewpoint, we allow the number of columns in $\mathbf{C}$ and number of rows in $\mathbf{D}^T$, which is M, to be infinite. While infinity is conceptually a difficult concept, it can be dealt with by imposing that only a finite, small number of entries in each row of $\mathbf{C}$ are non-zero. In practical terms, this means that we only see a finite amount of data and each data sample is generated as a finite linear combination of features. Other features from the infinite set are those which we have not yet encountered in our data.

To this end, the matrix $\mathbf{C}$ is written in terms of a *binary indicator matrix* $\mathbf{Z}$ and a *weight matrix* $\mathbf{W}$. That is

$$\mathbf{C} = \mathbf{Z} \odot \mathbf{W},$$

where $\odot$ denotes the element-wise, *Hadamard product*.

The data generation is likened to customers choosing dishes from an Indian Buffet, hence the name *Indian Buffet Process (IBP)*. The customers are indexed by $1, \ldots, n, \ldots, N$. There are infinitely many dishes, but not all have been chosen yet. These are the latent features indexed by m. M is the number of different dishes chosen so far.

The number of dishes i_1 the first customer chooses is drawn from a Poisson distribution with rate λ, which is the expected number of dishes each customer chooses. In matrix $\mathbf{Z}$, the first i_1 entries of the first row are set to one and $M = i_1$. M is the number of distinct dishes chosen so far. The amount he takes from each dish is recorded in the first row of matrix $\mathbf{W}$.

The nth customer chooses from the M dishes any of the previous customers have already chosen with probability j_m/n, if dish m was chosen j_m times. In other words, j_m is the number of non-zero entries in column m above row n in matrix $\mathbf{Z}$. If he chooses a particular dish m, then a one is placed in the mth column of the nth row of $\mathbf{Z}$. He also tries i_n new dishes, where i_n is drawn from a Poisson distribution with rate λ/n. These are indicated by ones in columns with numbers $M + 1, \ldots, M + i_n$ in the nth row. M is updated to $M + i_n$. Again, the amount of each dish is recorded in the nth row of $\mathbf{W}$.

Since λ/n decreases with every new customer, the probability that he will choose even just one new dish decreases, but does not become zero. This means that if our model space does not have enough features to explain the data, there is a non-zero probability to create a feature which will help explain the data.

In each row, the expected number of non-zero entries is λ. This is true for the first row, since the expectation of a Poisson distribution is given by the rate λ. Assume that the expected number of non-zero entries for rows $1, \ldots, n-1$ is λ. The expected number of nonzero entries in the nth row is

$$
\begin{aligned}
\mathbb{E}\left[\sum_{m=1}^{M} \frac{j_m}{n}\right] + \frac{\lambda}{n} &= \frac{1}{n}\mathbb{E}\left[\sum_{m=1}^{M}\sum_{k=1}^{n-1} Z_{km}\right] + \frac{\lambda}{n} \\
&= \frac{1}{n}\sum_{k=1}^{n-1}\mathbb{E}\left[\sum_{m=1}^{M} Z_{km}\right] + \frac{\lambda}{n} \\
&= \frac{(n-1)\lambda}{n} + \frac{\lambda}{n} = \lambda,
\end{aligned}
$$

where Z_{km} denotes the (k, m) entry in matrix $\mathbf{Z}$. By induction, this proves that the expected number of non-zero entries in each row is λ.

Since the probability of choosing a particular dish depends on the number of times it has been the choice of any of the previous customers, we again encounter the *rich-get-richer* phenomenon. Since the number of customers N is finite, so is the final number of sampled dishes M. Those chosen by many customers relate to features which are common in most of the data samples. These are sometimes referred to as *global features*, while those dishes chosen only by a few customers indicate *local features*, which can be used to distinguish between data samples.

This process of generating the data is independent of the order the customers arrive. A different order of customers means multiplying $\mathbf{Z}$ by a permutation matrix from the left which changes the row order. We follow this up with a permutation matrix from the right which swaps columns so that the number of leading zeros in each column is increasing from left to right, because each subsequent customer can choose new dishes. Such a permutation is equivalent to relabeling the dishes. The matrix of weights $\mathbf{W}$ is multiplied by the same permutation matrices from left and right as $\mathbf{Z}$. These permutations are not performed explicitly, but serve to explain that the choice of order in which samples are generated does not influence the final outcome.

The previous paragraph indicates a preferred order of dishes. More specifically, given an order of customers (rows of $\mathbf{Z}$), the columns of $\mathbf{Z}$ are ordered using their binary nature. Each column is interpreted as a binary number with the most significant bit in the first row and the least significant one in the last. The columns are ordered in decreasing magnitude of these binary numbers. Two columns might represent exactly the same binary number. This means two dishes were chosen by exactly the same set of customers. In this case, we have freedom of choice in the order of the columns representing the same binary number. This is known as *left-ordering of binary matrices*.

There are $M!$ possible labelings of the dishes. We need to divide this by the number of possible equivalences. Let K be the number of distinct columns and let m_k be the number of occurrences of the distinct column k for $k = 1, 2, \ldots, K$. That is

$$M = \sum_{k=1}^{K} m_k.$$

The number of possible equivalences in relabeling is

$$\prod_{k=1}^{K} m_k!.$$

Hence the number of different ways the same data can have been generated is

$$\frac{M!}{\prod_{k=1}^{K} m_k!}.$$

The above describes a prior distribution $p(\mathbf{Z})$. Also a prior distribution for $\mathbf{W}$ needs to be specified. This is done by specifying distributions for the elements of $\mathbf{W}$. Possible choices are, for example, the normal distribution as seen in Figure 2.2 or the Laplace distribution shown in Figure 2.18. Lastly, prior distributions for each of the columns of $\mathbf{D}$ need to be defined.

The learning process updates the parameters of these distributions according to the data, and specific instances are sampled. More specifically, $\mathbf{D}$ is kept fixed while the distributions of $\mathbf{Z}$ and $\mathbf{W}$ are updated. After instances of $\mathbf{Z}$ and $\mathbf{W}$ are sampled, they are kept fixed and the distributions of $\mathbf{D}$ are updated. Several passes through the data are necessary.

The process is similar to the Dirichlet process described in Section 6.7. The exact workings are, however, beyond the scope of this book. An introduction and review can be found in [19]. An example in image processing is provided by [11], where the building blocks making up the image are inferred from the image itself in contrast to engineered blocks as shown in Figure 7.1.

Appendix A: Matrix Formulae

A.1 Determinants and Inverses

A.1.1 Block Matrix Inversion

For an $m \times m$ invertible, square matrix $\mathbf{A}$, an $n \times n$ invertible square matrix $\mathbf{D}$, an $m \times n$ matrix $\mathbf{B}$ and an $n \times m$ matrix $\mathbf{C}$, we have

$$
\begin{pmatrix} \mathbf{A} & \mathbf{B} \\ \mathbf{C} & \mathbf{D} \end{pmatrix}^{-1} = \begin{pmatrix} \mathbf{A}^{-1} + \mathbf{A}^{-1}\mathbf{B}\mathbf{E}^{-1}\mathbf{C}\mathbf{A}^{-1} & -\mathbf{A}^{-1}\mathbf{B}\mathbf{E}^{-1} \\ -\mathbf{E}^{-1}\mathbf{C}\mathbf{A}^{-1} & \mathbf{E}^{-1} \end{pmatrix}
$$
$$
= \begin{pmatrix} \mathbf{F}^{-1} & -\mathbf{F}^{-1}\mathbf{B}\mathbf{D}^{-1} \\ -\mathbf{D}^{-1}\mathbf{C}\mathbf{F}^{-1} & \mathbf{D}^{-1} + \mathbf{D}^{-1}\mathbf{C}\mathbf{F}^{-1}\mathbf{B}\mathbf{D}^{-1} \end{pmatrix},
$$

where $\mathbf{E} = \mathbf{D} - \mathbf{C}\mathbf{A}^{-1}\mathbf{B}$ and $\mathbf{F} = \mathbf{A} - \mathbf{B}\mathbf{D}^{-1}\mathbf{C}$ need to be non-singular.

A.1.2 Block Matrix Determinant

For matrices as above,

$$
\begin{vmatrix} \mathbf{A} & \mathbf{B} \\ \mathbf{C} & \mathbf{D} \end{vmatrix} = |\mathbf{A}||\mathbf{D} - \mathbf{C}\mathbf{A}^{-1}\mathbf{B}| = |\mathbf{A} - \mathbf{B}\mathbf{D}^{-1}\mathbf{C}||\mathbf{D}|.
$$

A.1.3 Woodbury Identity

For an $m \times m$ invertible, square matrix $\mathbf{A}$, an $n \times n$ invertible square matrix $\mathbf{C}$, an $m \times n$ matrix $\mathbf{B}$ and an $n \times m$ matrix $\mathbf{D}$, we have

$$
(\mathbf{A} + \mathbf{B}\mathbf{C}\mathbf{D})^{-1} = \mathbf{A}^{-1} - \mathbf{A}^{-1}\mathbf{B}(\mathbf{C}^{-1} - \mathbf{D}\mathbf{A}^{-1}\mathbf{B})^{-1}\mathbf{D}\mathbf{A}^{-1}.
$$

A.1.4 Sherman–Morrison Formula

In particular, if $n = 1$ in the above Woodbury identity, then $\mathbf{C} = c$ is a scalar and $\mathbf{B} = \mathbf{b}$ and $\mathbf{D} = \mathbf{d}^T$ are vectors. The identity becomes

$$
(\mathbf{A} + c\mathbf{b}\mathbf{d}^T)^{-1} = \mathbf{A}^{-1} - \frac{1}{1/c - \mathbf{d}^T\mathbf{A}^{-1}\mathbf{b}}\mathbf{A}^{-1}\mathbf{b}\mathbf{d}^T\mathbf{A}^{-1},
$$

known as the Sherman–Morrison formula. For this to be valid, we need $1/c - \mathbf{d}^T \mathbf{A}^{-1}\mathbf{b} \neq 0$.

A.1.5 Matrix Determinant Lemma

For matrices as in the above Woodbury identity, the determinant can be calculated as

$$|\mathbf{A} + \mathbf{BCD}| = |\mathbf{C}^{-1} - \mathbf{DA}^{-1}\mathbf{B}||\mathbf{C}||\mathbf{A}|.$$

A.2 Derivatives

A.2.1 Derivative of Squared Norm

Let $\mathbf{x}$ be a general vector; then the derivative of the squared Euclidean norm is

$$\frac{\partial}{\partial \mathbf{x}}\|\mathbf{x}\|_2^2 = \frac{\partial}{\partial \mathbf{x}}\mathbf{x}^T\mathbf{x} = 2\mathbf{x}.$$

A.2.2 Derivative of Inner Product

Let $\mathbf{x}$ be a general vector and $\mathbf{a}$ a constant vector of the same length; then the derivative of the inner product between them is

$$\frac{\partial}{\partial \mathbf{x}}\mathbf{x}^T\mathbf{a} = \frac{\partial}{\partial \mathbf{x}}\mathbf{a}^T\mathbf{x} = \mathbf{a}.$$

A.2.3 Derivative of Second Order Vector Product

Let $\mathbf{x}$ be a general vector of length m and $\mathbf{b}$ and $\mathbf{e}$ constant vectors of length n. Further let $\mathbf{A}$ and $\mathbf{D}$ be constant $n \times m$ matrices and $\mathbf{C}$ a constant $n \times n$ matrix; then

$$\frac{\partial}{\partial \mathbf{x}}(\mathbf{Ax} + \mathbf{b})^T\mathbf{C}(\mathbf{Dx} + \mathbf{e}) = \mathbf{A}^T\mathbf{C}(\mathbf{Dx} + \mathbf{e}) + \mathbf{D}^T\mathbf{C}^T(\mathbf{Ax} + \mathbf{b}).$$

A.2.4 Derivative of Determinant

Let $\mathbf{X}$ be a square matrix and let $|\cdot|$ denote the determinant. Then

$$\frac{\partial}{\partial \mathbf{X}}|\mathbf{X}| = |\mathbf{X}|(\mathbf{X}^{-1})^T.$$

Let $\mathbf{X}$ depend on a variable x, then

$$\frac{\partial}{\partial x}|\mathbf{X}| = |\mathbf{X}|\mathrm{tr}\left(\mathbf{X}^{-1}\frac{\partial \mathbf{X}}{\partial x}\right),$$

where $\mathrm{tr}(\cdot)$ denotes the trace.

A.2.5 Derivative of Matrix Times Vectors

Let $\mathbf{X}$ be a $m \times n$ matrix and $\mathbf{a}$ and $\mathbf{b}$ constant vectors of length m and n respectively. Then

$$\frac{\partial}{\partial \mathbf{X}}(\mathbf{a}^T \mathbf{X} \mathbf{b}) = \mathbf{a}\mathbf{b}^T.$$

A.2.6 Derivative of Transpose Matrix Times Vectors

Let $\mathbf{X}$ be a $m \times n$ matrix and $\mathbf{a}$ and $\mathbf{b}$ constant vectors of length n and m respectively. Then

$$\frac{\partial}{\partial \mathbf{X}}(\mathbf{a}^T \mathbf{X}^T \mathbf{b}) = \mathbf{b}\mathbf{a}^T.$$

A.2.7 Derivative of Inverse

Let $\mathbf{X}$ be an invertible matrix, where the entries depend on a variable x. Then

$$\frac{\partial}{\partial x}\mathbf{X}^{-1} = -\mathbf{X}^{-1}\frac{\partial \mathbf{X}}{\partial x}\mathbf{X}^{-1}.$$

A.2.8 Derivative of Inverse Times Vectors

Let $\mathbf{X}$ be an invertible $n \times n$ matrix and $\mathbf{a}$ and $\mathbf{b}$ constant vectors of length n, then

$$\frac{\partial}{\partial \mathbf{X}}(\mathbf{a}^T \mathbf{X}^{-1} \mathbf{b}) = -(\mathbf{X}^{-1})^T \mathbf{b}\mathbf{a}^T (\mathbf{X}^{-1})^T.$$

A.2.9 Derivative of Trace of Second Order Products

Let $\mathbf{X}$ be a general $m \times n$ matrix, $\mathbf{A}, \mathbf{B}$ and $\mathbf{C}$ constant matrices of dimensions $k \times m, n \times n$ and $m \times k$ respectively. Then

$$\frac{\partial}{\partial \mathbf{X}}(\mathbf{A}\mathbf{X}\mathbf{B}\mathbf{X}^T\mathbf{C}) = \mathbf{A}^T\mathbf{C}^T\mathbf{X}\mathbf{B}^T + \mathbf{C}\mathbf{A}\mathbf{X}\mathbf{B}.$$

A.2.10 Derivative of Trace of Product with Diagonal Matrix

Let $\mathbf{A}$ be a constant square matrix and $\mathbf{X}$ a *diagonal* matrix of the same dimension. $\mathrm{tr}(\cdots)$ denotes the trace. Then the derivative of the trace of the product is

$$\frac{\partial}{\partial \mathbf{X}}\mathrm{tr}(\mathbf{A}\mathbf{X}) = \mathrm{diag}(\mathbf{A}),$$

where $\mathrm{diag}(\mathbf{A})$ is a diagonal matrix with the same diagonal as $\mathbf{A}$.

Bibliography

[1] D.R. Appleton, J.M.French, and M.P.J. Vanderpump. Ignoring a Covariate: An Example of Simpson's Paradox. *The American Statistician*, 50(4):340–341, 1996.

[2] D. Bertsimas and J.N. Tsitsiklis. *Introduction to Linear Optimization*. Athena Scientific, 1997.

[3] C.M. Bishop. *Pattern Recognition and Machine Learning*. Springer, 2007.

[4] C. Blakemore and G.F. Cooper. *Development of the Brain Depends on the Visual Environment*, volume 228. Nature, 1970.

[5] L. Breiman, J. Friedman, R. Olshen, and C. Stone. *Classification and Regression Trees*. Wadsworth and Brooks, 1984.

[6] M.D. Buhmann. *Radial Basis Functions*. Cambridge University Press, 2003.

[7] C.J.C. Burges. Geometry and Invariance in Kernel Based Methods. *Advances in Kernel Methods - Support Vector Learning*, MIT Press, pages 89–116, 1999.

[8] L. Campbell. *The Life Of James Clerk Maxwell*. Andesite Press, 2015.

[9] B. Chen, Y. Zhu, J. Hu, and J.C. Principe. *System Parameter Identification*. Elsevier, 2013.

[10] L. Devroye. *Non-Uniform Random Variate Generation*. Springer, 1986.

[11] M. Zhou et al. Nonparametric Bayesian Dictionary Learning for Analysis of Noisy and Incomplete Images. *IEEE Transactions on Image Processing*, 21(1):130–144, 2012.

[12] A.C. Faul. *A Concise Introduction to Numerical Analysis*. CRC Press, 2016.

[13] W. Feller. *An Introduction to Probability Theory and its Applications*, volume 2. Wiley, second edition, 1966.

[14] R. Fletcher. *Practical Methods of Optimization*. John Wiley & Sons, second edition, 2000.

[15] G. Galilei. *Dialogues Concerning Two New Sciences*. Dover Books on Physics, 2003.

[16] A. Gelman, J.B. Carlin, H.S. Stern, D.B. Dunson, A. Vehtari, and D.B. Rubin. *Bayesian Data Analysis*. CRC Press, third edition, 2013.

[17] J.K. Ghosh and R.V. Ramamoorthi. *Bayesian Nonparametrics*. Springer, 2003.

[18] R.C. Gonzales and R.E. Woods. *Digital Image Processing*. Prentice Hall, 2008.

[19] T. L. Griffiths and Z. Ghahramani. The Indian Buffet Process: An Introduction and Review. *Journal of Machine Learning Research*, 12:1185–1224, 2011.

[20] M. Haenlein and A. Kaplan. A Beginner's Guide to Partial Least Squares Analysis. *Understanding Statistics*, 3(4):283–297, 2004.

[21] H. He and Y. Ma. *Imbalanced Learning: Foundations, Algorithms, and Applications*. Wiley, 2013.

[22] F. Hollows and P. Corris. *Fred Hollows : An Autobiography*. Kerr, 1991.

[23] A. Hyvärinen, J. Karhunen, and E. Oja. *Independent Component Analysis*. Wiley-Interscience, first edition, 2001.

[24] M.I. Jordan. An Introduction to Probabilistic Graphical Models. http://people.eecs.berkeley.edu/ jordan/prelims/, 2004. Accessed: 2018-10-22.

[25] J.R. Kane, G. Baglivi, and J.L. Baird. *The Chronicle of Salimbene de Adam*. Binghampton, N.Y., Medieval & Renaissance Texts & Studies, 1986.

[26] Y. LeCun, C. Cortes, and C.J.C. Burges. The MNIST Database of Handwritten Digits. http://yann.lecun.com/exdb/mnist/, 1998. Accessed: 2018-10-22.

[27] A. Lyon. Why are normal distributions normal? *The British Journal for the Philosophy of Science*, 3(4):283–297, 2014.

[28] Sharon Bertsch McGrayne. *The Theory that Would Not Die: How Bayes' Rule Cracked the Enigma Code, Hunted Down Russian Submarines, & Emerged Triumphant from Two Centuries of Controversy*. Yale University Press, 2012.

[29] G. McLachlan and T. Krishnan. *The EM Algorithm and Extensions*. Wiley-Interscience, 2008.

[30] T.P. Minka. Automatic choice of dimensionality for PCA. *Advances in Neural Information Processing Systems*, 13:598–604, 2001.

[31] D.C. Montgomery. *Introduction to Statistical Quality Control*. John Wiley & Sons, seventh edition, 2012.

[32] M. Neuhaus and H. Bunke. *Bridging the Gap Between Graph Edit Distance and Kernel Machines*. Series in machine perception and artificial intelligence. World Scientific Publishing Company Pte Limited, 2007.

[33] J.R. Norris. *Markov Chains*. Cambridge University Press, 1998.

[34] William of Ockham. *Quaestiones et decisiones in quattuor libros Sententiarum*. Johannes Trechsel, 1495.

[35] C.P. Robert. *The Bayesian Choice: From Decision-Theoretic Foundations to Computational Implementation*. Springer, 2007.

[36] S. Ross. *A First Course in Probability*. Pearson, eighth edition, 2008.

[37] K.P. Soman, R. Loganathan, and V. Ajay. *Machine Learning with SVM and Other Kernel Methods*. PHI Learning, 2009.

[38] R. Wolf. *The Higgs Boson Discovery at the Large Hadron Collider*. Springer, 2015.

Index

9 780815 384205